IMPACT MODELS TO ASSESS REGIONAL ACIDIFICATION

edited by

JUHA KÄMÄRI

*International Institute for Applied Systems Analysis,
Laxenburg, Austria*

KLUWER ACADEMIC PUBLISHERS
DORDRECHT / BOSTON / LONDON

INTERNATIONAL INSTITUTE FOR APPLIED SYSTEMS ANALYSIS

ISBN-13: 978-94-010-6744-7   e-ISBN-13: 978-94-009-0571-9
DOI: 10.1007/978-94-009-0571-9

Published by Kluwer Academic Publishers,
P.O. Box 17, 3300 AA Dordrecht, The Netherlands.

Kluwer Academic Publishers incorporates
the publishing programmes of
D. Reidel, Martinus Nijhoff, Dr W. Junk and MTP Press.

Sold and distributed in the U.S.A. and Canada
by Kluwer Academic Publishers,
101 Philip Drive, Norwell, MA 02061, U.S.A.

In all other countries, sold and distributed
by Kluwer Academic Publishers Group,
P.O. Box 322, 3300 AH Dordrecht, The Netherlands.

Printed on acid-free paper

IMPACT MODELS TO ASSESS REGIONAL ACIDIFICATION

THE INTERNATIONAL INSTITUTE FOR APPLIED SYSTEMS ANALYSIS

is a nongovernmental research institution, bringing together scientists from around the world to work on problems of common concern. Situated in Laxenburg, Austria, IIASA was founded in October 1972 by the academies of science and equivalent organizations of twelve countries. Its founders gave IIASA a unique position outside national, disciplinary, and institutional boundaries so that it might take the broadest possible view in pursuing its objectives:

To promote international cooperation in solving problems arising from social, economic, technological, and environmental change
To create a network of institutions in the national member organization countries and elsewhere for joint scientific research
To develop and formalize systems analysis and the sciences contributing to it, and promote the use of analytical techniques needed to evaluate and address complex problems
To inform policy advisors and decision makers about the potential application of the Institute's work to such problems

The Institute now has national member organizations in the following countries:

Austria
The Austrian Academy of Sciences

Bulgaria
The National Committee for Applied Systems Analysis and Management

Canada
The Canadian Committee for IIASA

Czechoslovakia
The Committee for IIASA of the Czechoslovak Socialist Republic

Finland
The Finnish Committee for IIASA

France
The French Association for the Development of Systems Analysis

German Democratic Republic
The Academy of Sciences of the German Democratic Republic

Federal Republic of Germany
Association for the Advancement of IIASA

Hungary
The Hungarian Committee for Applied Systems Analysis

Italy
The National Research Council

Japan
The Japan Committee for IIASA

Netherlands
The Netherlands Organization for Scientific Research, NWO

Poland
The Polish Academy of Sciences

Sweden
The Swedish Council for Planning and Coordination of Research

Union of Soviet Socialist Republics
The Academy of Sciences of the Union of Soviet Socialist Republics

United States of America
The American Academy of Arts and Sciences

Foreword

Sensitive receptors such as soil and fresh bodies of water are at the end of a long chain of events in the process of regional acidification. This chain begins thousands of kilometers upwind at the emitters of acidifying pollutants. The topics covered in this book are important in the study of regional acidification for two reasons. First, it is important to assess the sensitivity of terrestrial and aquatic ecosystems to the deposition of acidifying pollutants. If the sensitivity of an ecosystem is known, then international control strategies can be developed to reduce deposition in the receptor areas of greatest importance. This is an important factor in designing the most effective strategies because of the very high costs of reducing emissions of acidifying pollutants.

Second, it is important to be able to predict changes in ecosystems for decades into the future, whether it be an improvement owing to decreases in acidifying emissions or, alas, a further deterioration because control strategies are nonexistent or inadequate. In either event, it is important to be able to judge the results of our actions.

Decision makers tend to be mistrustful of models unless they can judge their reliability. The application and testing of the models in Part III of this book cover, therefore, an important facet of model building.

This book is an ideal companion to another book that is forthcoming from the Transboundary Air Pollution Project at IIASA: *The RAINS Model of Acidification: Science and Strategies in Europe.* The latter book is a description of the development and use of the Regional Acidification INformation and Simulation (RAINS) model, an integrated assessment model for developing and determining control strategies to reduce regional acidification in Europe. Much of the research described in this book forms part of the foundation of the RAINS model. These two books cover a great deal of the present knowledge about assessing and dealing with a very important environmental problem in Europe – regional acidification.

R.W. Shaw
Leader
Transboundary Air Pollution Project

Introduction

Acidification of the Environment

Acidification of the environment is, like most recent environmental problems, a consequence of change in the natural cycles of elements. Some chemical substances that during millions of years leaked from the biosphere are again being emitted to the atmosphere. Owing to these anthropogenic emissions, fluxes of acidifying compounds to the soils and waters have increased manyfold since preindustrial times.

The whole problem of air pollution is transboundary in nature; its geographic dimensions are difficult to define. Pollution sources are scattered in many countries, the consequences extending far away. Acidifying sulfur and nitrogen oxides can persist in the air up to a few days, long enough to be transported hundreds or even thousands of kilometers from the place they were emitted. During this transit time, gaseous pollutants undergo a complicated chemical conversion into nitric and sulfuric acids, which are deposited as particles or as droplets of water. Thus, practically all industrial countries are responsible for the problem. And, since the sources are international, the resolution of the problem likewise requires cooperative international action.

Air pollutants cause numerous changes in the biotic and abiotic environment. The series of chemical reactions, starting from the emissions of sulfur dioxide and nitrogen compounds to the depletion of the natural acid-neutralizing capacity of soils and waters, has been established in extensive research programs of Scandinavia and North America. This depletion of acid-neutralizing capacity is termed *acidification*. Two factors appear sufficient to explain the magnitude and the geographic distribution of soil and water acidification: the amount of acidic atmospheric deposition and the inherent sensitivity of the recipient.

Acidification has been recognized as a fairly complex process involving many mechanisms both in the terrestrial and the aquatic catchment. Although some of these mechanisms are not understood quantitatively, research is continuously expanding the level of knowledge of various aspects of the problem. Numerous attempts have already been made to identify and quantify lake acidification in order to estimate changes over past decades when historical data are lacking or unreliable.

Scientific knowledge on the whole acidification phenomenon has increased tremendously since Professor Svante Oden of Sweden mapped the changing acidity of precipitation as a regional phenomenon in Europe and postulated this acidity as a probable cause of decline in fish populations. Since that time acidification research has developed from a few single studies to a new field of science. Before any theories were established to account for acidification, there remained competing hypotheses and alternative explanations for the widely observed regional surface water acidification.

This development of acidification science resembles the Kuhnian route to normal science. Kuhn described normal science as research firmly based upon past scientific achievements, paradigms, which some particular scientific community acknowledges as supplying the foundation for its further practice. Before any paradigms, laws, or theories could be established in acidification, competing hypotheses and alternative explanations for the widely observed regional surface water acidification had to be resolved.

Now consensus seems to be near that acidic deposition plays an important role in soil and water acidification and that certain soil chemical processes regulate acidification. The scientific community takes these paradigms for granted, and there is no longer need to start anew from first principles and justify each concept used.

Modeling Acidification

The difficulty of scientific prediction has been well demonstrated in serious attempts to estimate, for example, future oil prices, currencies, or weather. The performance of predictive methods depends on how well the laws governing the deterministic system or the initial conditions of the system are known. Prediction becomes quite reliable when the laws of the system have been clarified; e.g., solar eclipses can be calculated centuries ahead of time with great accuracy. For any such well-established system the factor limiting predictive power is knowledge of initial conditions. When observation techniques allow the initial conditions to be determined only approximately, predictions based on deterministic laws are also approximations.

Natural systems are extremely complex. Nevertheless, to avoid the destruction of sensitive ecosystems, we must understand both these systems themselves and the consequences of human actions on them. Only in recent years, with the availability of efficient computers, has the analysis of ecological systems become a practical possibility. The purpose of such computer-based studies, termed systems analyses, is the better understanding of a given system, usually to permit reliable predictions and better management decisions based on them. This implies a description of the system and its processes, i.e., a development of a model representing the real system, the behavior of which resembles that of the real system as closely as possible.

A model can obviously never be as complex as the real system itself. Models work with aggregated representations of reality. Uncountable interactions among system components are reduced in the model description to a few mathematical equations. Model development is a selective and, hence, a partly subjective procedure that involves several sequential steps. In model conceptualization there are choices regarding the level of spatial and temporal aggregation; the separation between chemical, physical, and ecological elements; as well as the processes to be incorporated into the model structure.

After the decision has been made as to how the system will be represented, the model builder has to choose from several possible model types. There is a choice, for example, between dynamic and steady state models, distributed and lumped models, as well as between internally descriptive (mechanistic) and black-box (input–output) models. The model builder's selections depend upon the goals and objectives of the final model application and on the availability of data. A study investigating the future long-term effects of land-use changes on water quality, for instance, necessitates a different kind of model than a study attempting to predict the effects of a particular sewage discharge.

Predictive models can in a very broad sense be classified into two categories on the basis of their application objectives. First, models can be applied as research tools to indicate further directions of investigation. Especially in early stages of research on a badly defined system, time series data are likely to be scarce, and the only way to progress in this situation is to use some form of simulation model in the hypothesis-generating role. Necessarily, the simulation model produces no immediate practical applications; whereas the second type, management models, assumes that the applications are known and carefully specified. The term management can in this context include objectives such as long-term planning, designing environmental policies, estimating environmental consequences of human actions, and designing treatment facilities for pollutants.

Conceptualization of the acidification phenomenon has naturally been based on recent findings in soil and water chemistry. Yet, since we are dealing with a new problem area and thus with a fairly poorly defined system, the modelers have not always identified the same processes as the key mechanisms that determine system behavior. Additionally, the computational approaches chosen to represent these various processes and observed water quality patterns have varied. The modeling approaches used to date represent practically all existing model types, ranging from simple to complex, from dynamic to steady state, and from black-box to highly deterministic models.

Concerning process-oriented mechanistic approaches, there is a significant convergence of thought about the key processes involved. The models developed so far contain many similar assumptions about a number of soil chemical processes. The models use fluxes of strong acid anions, mainly sulfate, as the most important driving variable. Most of these models more or less explicitly incorporate the anion mobility concept. The importance of sulfate adsorption, cation exchange, and aluminum dissolution, as well as weathering reactions, can be easily gleaned from recent modeling efforts. The convergence of opinion on the importance of these mechanisms would not have been possible without a great number of model applications, performance evaluations, and further model development. It has been not only the successes of various applications that have increased our knowledge on the acidification mechanisms; failures to predict some phenomena have also given us valuable information by allowing us to reject a certain hypothesis and move forward.

To date, model applications have been mainly performed to provide information on the importance of different processes in determining the dynamics of studied catchments. These models have been research tools that have offered one way of quantifying the effect of poorly understood processes. For the early "pre-paradigm" phase of acidification science, these applications have been invaluable. Simulation models, by predicting satisfactorily, for example, some observed patterns of surface water acidity, have given support to the hypothesis that acidic deposition is the most important cause of widely observed regional acidification. Models have confirmed that under some circumstances acidic deposition may dominate soil and freshwater chemistry. Moreover, model applications have generated new ideas and working hypotheses for experimental scientist to test.

Divergence in modeling philosophy is unavoidable and, in fact, desirable. The existence of a wide range of different model types can be explained by the different goals and objectives of the model applications. The decision as to what is going to be predicted deserves a critical evaluation, because that choice, more than anything else, affects the model to be constructed. Research models tend to describe in more detail the processes involved, so that different phenomena and hypotheses can be evaluated and the short-term dynamics can be reproduced. Prediction models designed for management purposes use a more simple "lumped parameter" approach, which is thought to be suitable for reproducing the long-term system behavior that is required from a tool attempting to assist in a decision-making process.

Acidification Models as Management Tools

Descriptions of the quantitative consequences of alternative scenarios can assist in formulating policies for emission control. Unfortunately, to date, there has been only a tenuous link between political decisions and scientific evidence concerning acidification. For example, the most commonly discussed policy in Europe for controlling acidification impacts has been a percentage-based reduction of sulfur emissions by 1993, relative to their 1980 level. Although this kind of policy will be costly for virtually every European country, the actual benefits of the strategy in protecting the natural environment have rarely been investigated. But even augmenting scientific information about the problem will not necessarily lead to identification of suitable policies for controlling acidification of the environment. This information must also be structured in a form usable to decision makers. Recently developed integrated simulation models attempt to provide such a structure.

There are two basic ways of using such an integrated model: (1) scenario analysis and (2) optimization analysis. When creating scenarios, the model user essentially moves from top to bottom through the model, first specifying an energy pattern and an emission control strategy. This information is installed in models that calculate emissions, deposition resulting from these emissions, and the resultant environmental impact. After the model has been run, implications of the policy assumptions are carefully examined both for the performance of the scenario and for the adequacy of the model. Performance refers to the degree to which the scenario produces appealing results. The second criterion, adequacy of the model, refers to the model's validity and the uncertainties associated with the produced scenarios.

A consistent set of energy use patterns, sulfur emissions, sulfur deposition, and environmental impacts is called a scenario; and the type of analysis is termed *scenario analysis*. Scenario analysis gives flexibility to the model user, who can examine the consequences of many different pollution control programs. The whole procedure of constructing scenarios is done in an iterative interactive fashion. In this way the model user can quickly compare the likely impacts of many different policies.

In the other operational mode, *optimization analysis*, the user in a sense inverts the scenario analysis procedure by starting with environmental protection goals and having the model work "backwards" to determine a cost-effective scenario for reducing emissions to accomplish these goals. For example, the user can run the model by setting an environmental or deposition objective and then compute a desirable emissions reduction plan according to specified cost and institutional constraints. These computations are accomplished by mathematical "searching techniques," which draw on linear programming or other similar mathematical algorithms.

Numerous nations have taken the first important step, by signing the Sofia Protocol for controlling NO_x emissions, toward determining cost-efficient ways to reduce emissions. The signatory nations agreed in this protocol to commence negotiations on further steps to reduce emissions, taking into account internationally accepted critical loads. Still to be defined by researchers are loads or levels of air pollutants deemed not to be significantly harmful for specified receptors. Negotiations for further reductions in air pollutant emissions, defined on these bases, will begin in 1989, within the framework of the Geneva Convention on Long-range Transboundary Air Pollution. Therefore, optimization analyses may play an important role in future international negotiations.

Scope of this Volume

There is little hope of capturing the state of the art in acid precipitation modeling in a traditional publication. Modeling, especially in this field, is dynamic and changing very rapidly. Lately, however, modelers have been devoting more and more attention to the possible use of their acidification models in a decision-making context. To provide meaningful information for management purposes, the models should be dynamic in nature; they should analyze the long-term behavior of the environment; they should be applicable on a large regional scale; and, finally, they should produce as output well-defined illustrative information that can easily be related to the effectiveness of the scenario being evaluated. A book treating these decision-related criteria is long overdue.

This volume is largely based on results of the Task Force Meeting on Environmental Impact Models to Assess Regional Acidification, cosponsored by the International Institute for Applied Systems Analysis (IIASA) and the Institute for Meteorology and Water Management of Poland, held in spring 1987 in Warsaw. The meeting had the objective of discussing critical topics regarding modeling of long-term acidification of the environment, with particular emphasis on model applications on a large regional scale. Twenty scientists from eight countries were invited to the meeting to present and discuss results from their "new kinds of models" that are no longer calibrated just for individual sites, but can make use of regional data. In the following 15 chapters, 31 authors carefully describe the best examples of the art of regional acidification modeling.

The first section of the volume is devoted to terrestrial effects of acidification. Chapter 1, by Dr. W. de Vries, describes a model for long-term soil acidification, which is intended to be used in the Netherlands both as a research and management tool. The general lack of input data poses a severe problem with respect to regional applications of process-oriented models. Dr. de Vries indicates how this problem can (partly) be solved by using soil maps and associated soil survey information.

Chapter 2, by Dr. K. Brodersen and co-workers, deals with effects of acid deposition on agricultural soils. The model is applied to Danish poor-quality soils to analyze acidification when the soils are, after a period of intensive farming, left untended. The use of soil models for estimating effects of acid rain on a regional scale is also discussed in general terms.

Chapter 3, by Professor P. Hari and Dr. M. Holmberg, presents a theoretically based model for analyzing regional changes in environmental factors controlling forest growth. Based on analysis of those changes, this study provides scenarios of forest growth in Finland.

The next seven chapters of the volume focus on aquatic effects of acidification. Chapter 4, by Dr. M. Holmberg and co-workers, evaluates qualitatively the sensitivity of European groundwater aquifers to acidification. The geographical distribution of the sensitivity indicators is compiled from existing map information and combined by a stepwise procedure to produce a map of groundwater sensitivity for Europe.

Chapter 5, by Dr. C.K. Minns, examines the response of fisheries to acid deposition by linking a simple empirical steady state chemical model to relationships describing the response of some important fisheries indicators – species richness, presence–absence, and production. Impact scenarios are presented for lake resources in Ontario, Canada.

Chapter 6, prepared by Dr. C. Bernes and Dr. E. Thörnelöf, is based on an empirical steady state acidification model similar to the one described in Chapter 5. The model is applied to synoptic lake survey data to yield quantitative estimates of the present and future acidification status of Swedish surface waters.

Chapter 7, by Professor B.J. Cosby and colleagues, presents a "regionalization" methodology that uses a conceptual model of long-term responses to acidic deposition (MAGIC) in a Monte Carlo simulation framework. Observed distributions of water quality variables, derived from regional water surveys in 1974 in southern Norway, are reproduced, and model projections are shown for the 12 years following the survey.

Chapter 8, by Dr. T.J. Musgrove and co-workers, applies the regionalized conceptual model (MAGIC) presented in Chapter 7 to simulate regional-scale surface water characteristics of the Galloway area of southwest Scotland. Monte Carlo techniques are employed with a large data set from the study region to produce a wide-ranging model response encompassing the observed chemical features of the region.

Chapter 9, by Dr. M. Posch and the volume editor, introduces a "regional calibration" methodology for a simple process-oriented lake acidification model developed for the integrated RAINS model system of IIASA. This RAINS Lake Module is regionalized by selecting random parameter vectors from *a priori* frequency distributions, and by using Monte Carlo techniques to filter out those parameter vectors that produce the surveyed joint lake pH-alkalinity distribution for regions in Finland. Thus, the model is driven by historical deposition patterns.

Chapter 10, by Professor M. Small and co-workers, identifies a methodology that allows a simple "Direct Distribution model" to serve as a summary representation of a more complex mechanistic model presented in Chapter 9. The parameters of the Direct Distribution model are derived from the observed behavior of the RAINS Lake Module; and model results for the southernmost region of Finland produced by the two models are compared.

The last five substantive chapters of the volume explore issues of model reliability and utility. Chapter 11, by Dr. R.H. Gardner and co-workers, evaluates the reliability of regional predictions by comparing the RAINS Lake Module results, discussed in Chapters 9 and 10, with data from Fennoscandia. Chapter 11 identifies and quantifies errors associated with extrapolation of site-specific information, tests our ability to perform useful extrapolations, and establishes guidelines for the role of models and data in regional assessments.

Chapter 12, written by Dr. M.C. Sutton, presents an approach for evaluating the performance of stochastic calibration procedures that rely on use of synthetic, user-generated systems. The performance of the Monte Carlo calibration procedure, described in Chapter 9, is assessed by applying it to a synthetic region. Dr. Sutton demonstrates that synthetic systems can be used to estimate the maximum achievable model accuracy, given perfect information. This has bearing on the value of obtaining additional data from the system to be modeled.

Chapter 13, by Professor G.M. Hornberger and colleagues, addresses the uncertainty inherent in both regional hindcasts and forecasts produced by the regionalized conceptual model (MAGIC) presented in Chapter 7. Estimates of parameter uncertainty are obtained using a bootstrap resampling scheme. Sensitivity of the parameters of the regionally calibrated model, as well as the uncertainty in the hindcast and forecast variables, are evaluated.

Chapter 14, written by Professor E.S. Rubin and four collaborators, serves as an example of integrated assessment models. Their analysis focuses on the effect of future emissions on regional aquatic acidification in eastern North America. The extent to which harmful effects of sulfur emissions can be reduced and reversed as a result of policy measures is estimated. Determining the magnitude and sources of uncertainties in predicted impacts is a key element of the integrated analysis.

In the final chapter, the editor summarizes the various approaches to regionalizing and evaluating acidification predictions, as presented in earlier chapters. The major accomplishments of the modelers deserve to be, and are, highlighted in this summary. An attempt is made to identify some important directions for future research in the broad field of modeling regional acidification of the environment.

J. Kämäri
National Board of Waters and the Environment
Helsinki, Finland

Contents

PART I

Terrestrial Effects

CHAPTER 1

Philosophy, Structure, and Application Methodology of a Soil Acidification Model for the Netherlands

W. de Vries

1.1. Introduction

Governments in Europe and North America are currently under increasing pressure to take remedial action against the acidification of the environment. However, policy decisions regarding emissions require insight into the effectiveness of abatement strategies in reducing long-term environmental effects. In this connection, models provide an important management tool to assist decision makers in their evaluation of strategies to control sulfur and nitrogen emissions. Consequently, the Dutch Priority Program on Acidification has initiated the development of a model system (DAS – Dutch Acidification Model) that quantitatively predicts environmental impacts, in given geographical areas over a given time, for alternative emission scenarios.

DAS has a modular construction: each aspect of the problem is represented by a separate compartment. The ultimate aim is to evaluate protection strategies, based on a quantitative description of the linkages between energy-related or other types of emission, atmospheric deposition, soil acidification and effects on terrestrial and aquatic ecosystems, agricultural production, materials, and, possibly, public health. Furthermore, the model will also be used as a research tool to direct further investigation. A schematic diagram of the model compartments and submodels illustrating the procedure of model use is given in *Figure 1.1*. Similar model systems, although less comprehensive, have been developed for Denmark (Petersen, 1984; Christensen *et al.*, 1985) and for Europe as a whole (Alcamo *et al.*, 1985).

J. Kämäri (ed.), Impact Models to Assess Regional Acidification, 3–21.
© 1990 *International Institute for Applied Systems Analysis.*

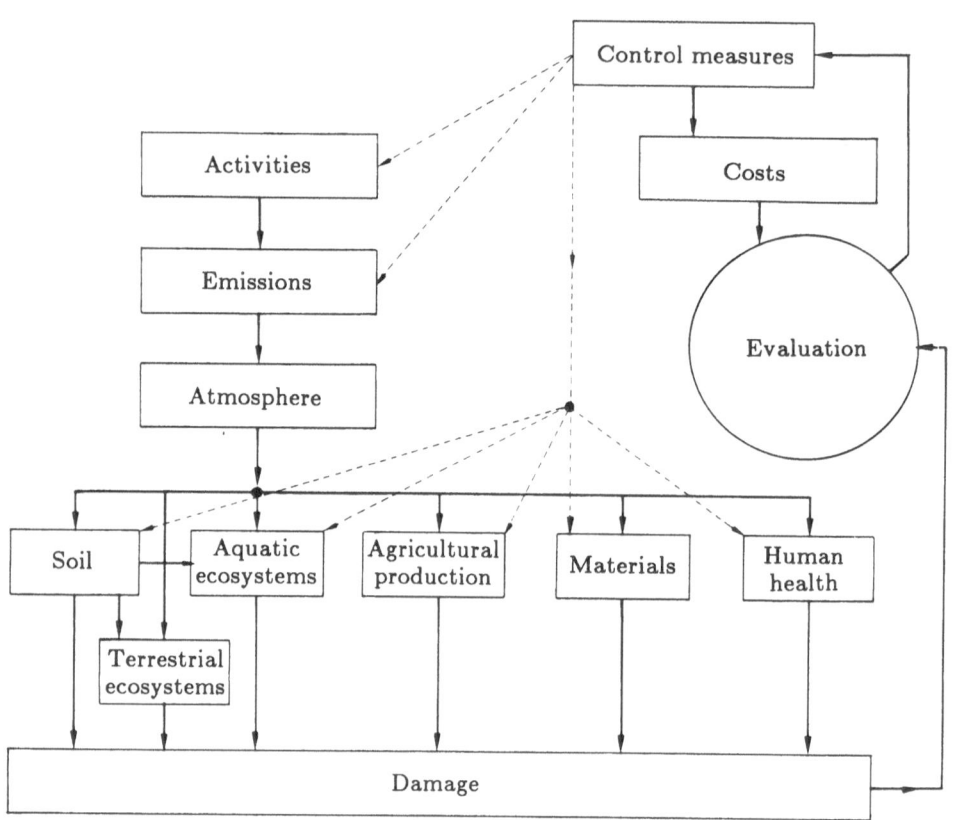

Figure 1.1. Structure of the Dutch acidification system model (DAS).

Within the overall framework of DAS, the soil (and groundwater) acidification submodel forms a very important linkage between atmospheric deposition and effects on terrestrial and aquatic ecosystems. This chapter describes the structure, approach, and application methodology of this regional soil acidification model (RESAM). Special attention is given to the philosophy underlying the modeling approach, in relation to the data needed to run the model.

1.2. The Relation between Modeling Approach and Model Application

The philosophy behind a modeling approach is largely determined by the goals and objectives of the model application. In this respect, a distinction can be made between research models and management models. Research models are used as a research tool to gain insight into, or generate hypotheses about, system behavior and to direct further investigations. Management models aim to help decision makers in designing environmental policies (Kämäri, 1987). Contrary to research models, management models have an immediate practical application.

Examples of different modeling approaches or types are (Kämäri, 1987):

- Stochastic versus deterministic models.
- Empirical (black-box) versus mechanistic (process-oriented) models.
- Steady state versus dynamic models.
- Lumped versus distributed parameter models.
- Simple process-oriented versus complex process-oriented models.

The models relating water quality response to atmospheric inputs are based on practically all existing approaches. These vary from a simple static lumped empirical model (Wright and Henriksen, 1983) to a complex dynamic distributed process-oriented model (Chen *et al.*, 1983). At present, most water quality models are process-oriented. Surface water acidification is predicted by describing processes in the soil system that have a major influence on water quality response. Well-known examples are the models by Reuss and Johnson (1985), the Birkenes model (Christophersen *et al.*, 1982), ILWAS (Chen *et al.*, 1983; Gherini *et al.*, 1985), and MAGIC (Cosby *et al.*, 1985a,b). These so-called soil-oriented balance models (Reuss *et al.*, 1986) are based on the concept of anion mobility through the charge balance and incorporate common geochemical processes, such as cation exchange, sulfate adsorption, and dissolution of aluminum from a hydroxide.

Apart from ILWAS, these models do not simulate the cycling of nutrients in the vegetation through biochemical processes such as uptake, litter fall, and mineralization. The same holds true for models that are specifically developed to predict soil acidification on a regional scale (Bloom and Grigal, 1985; Kauppi *et al.*, 1986). However, incorporation of the nutrient cycle is of major importance in predicting the concentration of nitrogen (NH_4 and NO_3) and base cations (Ca, Mg, K) in the upper soil layers. Since RESAM is linked with a model describing the effect of acidification on terrestrial ecosystems, these predictions are very important. Consequently, incorporating biochemical processes, as has been done in ILWAS, has been considered necessary. The reason for not adopting ILWAS itself is the detailed data input required.

In general, the disadvantage of relatively complex, multi-parameter management models is the general lack of input data necessary for application on a regional scale. Consequently, assuming that the model structure is correct (or at least adequately represents current knowledge), the uncertainty in the output of these models may still be large because of the uncertainty of input data such as forcing functions (source/sink terms), initial conditions of soil variables, and parameter values (Hornberger *et al.*, 1986a). This dichotomy between detail and reliability of information obtained in research models and regional applicability in management models is illustrated in *Figure 1.2*.

The desired degree of spatial resolution in model output is of crucial importance when selecting the level of detail that is appropriate for both the model and its parameters. In this connection, there are two important aspects that must be considered when assessing the adequacy of processor-oriented simulation models for application on a regional scale. The first aspect concerns the degree to which the multitude of hydrological, biochemical, and geochemical processes

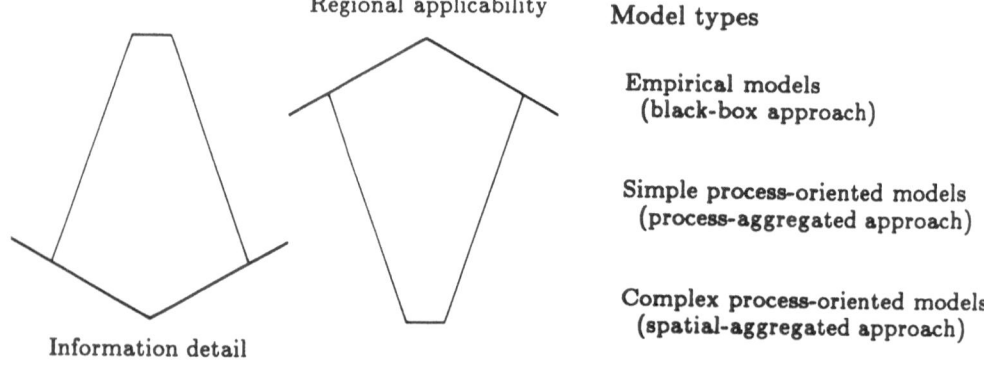

Figure 1.2. Schematic diagram illustrating the contrast between model complexity and regional applicability.

that occur in soils can be represented by simplifications of those processes (process aggregation). The second aspect concerns the extent to which spatial heterogeneity of soil properties can be lumped (spatial aggregation) (Cosby *et al.*, 1985a).

For example, insight into the hydrological and chemical response of surface water to inputs (e.g., precipitation, acid deposition) is generally obtained from physically based lumped parameter models, which "average" or lump spatially distributed physical and chemical processes in a catchment (spatial aggregation) (Cosby *et al.*, 1985a). The use of "distributed" models allowing for spatial variations within a catchment or region is a new development in the field of soil science (Chen *et al.*, 1983; Bathurst 1986). However, the response of soil water chemistry to input of pollutants has to be ascertained using a physically based distributed parameter model, because soil types and soil horizons influence the hydrological and chemical behavior of the soil system.

To render RESAM useful both as a research tool and a management tool, the model is characterized by:

(1)   *A deterministic approach*: Variability in parameters and variables is implicitly accounted for by using spatially distributed values [see (5)].
(2)   *A process-oriented approach*: Description of the processes influencing soil acidification gives insight into system behavior and helps to direct further research.
(3)   *Simple process descriptions*: Process description is simplified to a fair degree to minimize data input requirements.
(4)   *Dynamic simulation*: Dynamic simulation allows predictions to be made under changing conditions, which is necessary to evaluate the effect of alternative emissions scenarios.
(5)   *Spatially distributed parameter input*: The spatial distribution of data related to atmospheric input, soil use, and soil type is taken into account by using geographic information available in databases, remote sensing data, topographic maps, and soil maps. The model is described in terms of

parameters that can be related to these data with so-called transfer functions. This is especially true with respect to soil survey information. Consequently, the model can be used to identify areas vulnerable to acidification within a receptor area.

In conclusion, RESAM may be classified as a simple process-oriented model (indicated in *Figure 1.2*) with spatially distributed and soil survey-oriented parameters. At the Netherlands Soil Survey Institute, we consider this intermediate level of generalization to be most appropriate. In this case, the complexity of the model and the associated data requirements are in balance with the availability of data from soil research and soil surveys. Furthermore, the model allows quantitative and reproducible predictions of water quality based on current knowledge of important processes.

1.3. The Model

1.3.1. Characteristics

The ultimate aim of the soil acidification model is the prediction of the effect of acid deposition on soil water chemistry for characteristic combinations of forests and soil types.

The model output is confined to major ions in soil water relevant to the growth and vitality of forests and to groundwater quality. Ions included are all macronutrients except P (NH_4, NO_3, SO_4, Ca, Mg, K) plus H, Al, Na, Cl, and RCOO. The concentration of Al and the ratio of Al to base cations (Ca, Mg, K) is important for forest vitality, because of the effects of aluminum toxicity and base cation deficiency. The concentration of organic anions (RCOO) is also important in this respect, because of the complexation and detoxification of Al (Ulrich and Matzner, 1983). Na and Cl are included to get charge balance. In order to make these predictions, the model input includes atmospheric deposition of SO_2, NO_x, NH_3, Ca, Mg, K, Na, and Cl. The deposition of Al and RCOO are ignored.

Forests are represented by five characteristic groups: Douglas fir, Scots pine, other coniferous trees, oak, and other deciduous trees. Forest soils are confined to acid sandy soils, as these cover 80% to 90% of the Dutch forest area. Furthermore, these soils are sensitive to acidification because of the mobilization of aluminum, mainly from relatively soluble aluminum hydroxides. Since the pH of these soils is always below 5, HCO_3 is not included in the model. The soils are characterized by 14 representative profiles based on a recent 1:25,000 soil map of the Netherlands (Steur *et al.*, 1986). They cover a broad range in terms of soil properties, such as the organic matter content, texture, *CEC*, and groundwater level, which influence major hydrological, biochemical, and geochemical processes in the soil.

The soil compartment is confined by the mean lowest groundwater level (*MLW*). The vertical heterogeneity of the soil profile is included by representative soil layers (horizons). The spatial resolution of the receptor (deposition)

Figure 1.3. The receptor (deposition) areas in the Netherlands considered in the model.

areas is delineated by 20 areas with relevant statistical information for the different submodels (*Figure 1.3*).

The temporal resolution of the model input and output is one year, as the model is meant to reveal long-term soil chemical responses (50–100 years) of characteristic forest ecosystems to acid deposition.

1.3.2. Concept and structure

The acidification process is conceptualized as a disturbance in the sulfur and nitrogen cycle in forests, resulting from the deposition of SO_2, NO_x, and NH_3. This induces a significant excess of strong acid anions (SO_4 and NO_3) over basic cations (Ca, Mg, K, and Na) associated with a strong proton production (Reuss and Johnson, 1986). In acid soils, this acid load is mainly neutralized by aluminum mobilization, leading to toxic levels of aluminum in the soil solution.

Figure 1.4 shows the general concept and structure of the model in a relatively simple relational diagram. As stated before, the model is process-oriented. State variables depict the quantity of water or chemical constituent in each compartment at any given time. Rate variables depict the processes influencing

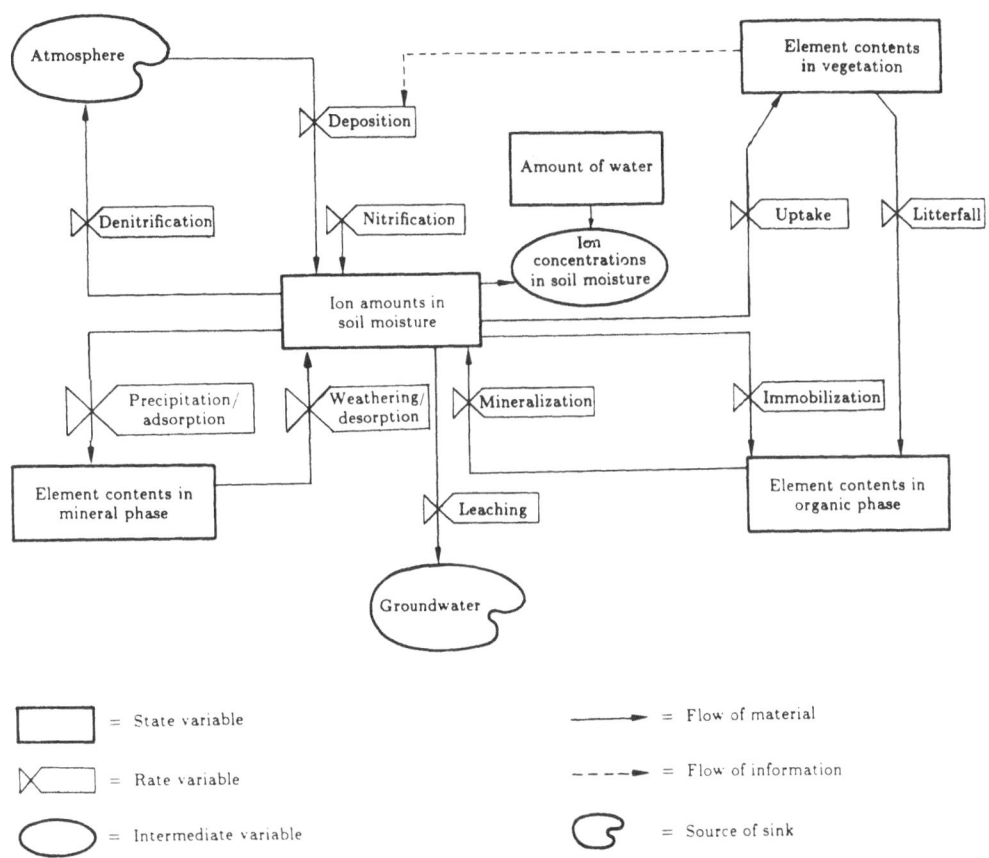

Figure 1.4. A simplified relational diagram of the regional soil acidification model (RESAM).

state variables (including soil water chemistry). Processes in the forest canopy affecting the element and proton input to the soil (throughfall) are filtering and uptake of gaseous air pollutants (SO_2, NO_x, NH_3) and exudation of base cations (Ca, Mg, K, Na) associated with proton consumption. Retention of potential acid sulfur and nitrogen compounds in biomass (vegetation and soil organic matter) is influenced by processes related to element cycling, i.e., litter fall, root decay, net mineralization, and root uptake. The nitrogen cycle is further influenced by nitrification and denitrification. The proton load resulting from atmospheric deposition affected by the mentioned biochemical processes is mainly neutralized by mineral weathering, which is a slow buffer process. Furthermore, protonation of organic anions may be important in the upper soil layer. Finally, fast buffer processes, such as sulfate adsorption and cation exchange, can also play an important role in neutralizing incoming acid, but their capacity in sandy soils is restricted. Both the release of base cations, associated with a decrease in acid-neutralizing capacity (ANC), and the retention of acid anions,

Table 1.1. Summary of the processes in RESAM in relation to the elements considered.

Process	Al	Ca+Mg+K	Na	NH_4[a]	NO_3[a]	SO_4[a]	Cl	RCOO
Atmospheric deposition	−	+	+	+	+	+	+	−
Water flow	+	+	+	+	+	+,b	+	+
Foliar uptake	−	−	−	+	+	(+)[b]	−	−
Foliar exudation	−	+	−	−	−	−	−	−
Litterfall	−	+	−	+	+	(+)	−	−
Root decay	−	+	−	+	+	(+)	−	−
Mineralization	−	+	−	+	+	(+)	−	−
Root uptake	−	+	−	+	+	(+)	−	−
Nitrification	−	−	−	+	+	−	−	−
Denitrification	−	−	−	−	+	−	−	−
Protonation	−	−	−	−	−	−	−	+
Weathering	+	+	+	−	−	−	−	−
Cation exchange	(+)	(+)	(+)	(+)	−	−	−	−
Anion adsorption	−	−	−	−	−	(+)	−	−

[a]NH_4 refers to NH_3, NO_3 to NO_x and SO_4 to SO_2 with respect to foliar uptakes; NH_4 and NO_3 refer to organic N, and SO_4 to organic S, with respect to litterfall, root decay, and mineralization.

[b]Parentheses denote that these processes are not included in the preliminary version of RESAM described in this chapter.

associated with an increase in base-neutralizing capacity (BNC), can be viewed as soil acidification (de Vries and Breeuwsma, 1987).

The various processes included in the model in relation to the elements considered are given in *Table 1.1.* An overview of the process descriptions is given in the following section.

1.3.3. Process formulations

General

Ion concentrations in a soil layer i at time t $[c(i,t)]$ are calculated with a mass balance equation assuming negligible influence of diffusion and dispersion of mass flow, homogeneous isotropic soil layers with a constant density, and perfect mixing within soil layers,

$$c_{i,t} = \frac{\Theta_{i,t} \cdot T_i \cdot c_{i,t-1} + (Fw_{i-1,t} \cdot c_{i-1,t} - \rho_i \cdot T_i \cdot I_{i,t}) \cdot \Delta t}{\Theta_{i,t} \cdot T_i + Fw_{i,t} \cdot \Delta t} \tag{1.1}$$

where c is the ion concentration $(mol_c\ m^{-3})$; ρ is the density $(kg\ m^{-3})$; Θ is the moisture content $(m^3\ m^{-3})$ T = layer thickness (m); Fw is the water flux $(m\ yr^{-1})$ Δt = time step (yr); and I is the mobilization or immobilization by biochemical and geochemical interactions $(mol_c\ kg^{-1}\ yr^{-1})$.

This modeling approach, which is also used in the ILWAS model (Chen *et al.*, 1983), has the following major facets:

(1)   The vertically differentiated soil system is represented by a series of continuously stirred tank reactors (CSTRs).
(2)   The time-dependent behavior of a CSTR (soil layer) is analyzed with an implicit numerical method, by using the concentration at the end of each time step to compute removal of the constituent by outflow from CSTR.
(3)   The equations to simulate multiple constituents and reactions are solved simultaneously.

Once initial conditions of ion concentrations $(cX_{i,t-1})$ for $(t = 1)$ and soil parameters (Θ_i, ρ_i, T_i) have been established, the equation can be solved when:

(1)   The water flux (Fw) in a forest/soil ecosystem is provided to the model.
(2)   The element input at the soil surface $(cX_{i-1,t}$ for $i = 1)$ is given.
(3)   The biochemical and geochemical interactions in all soil layers $(I_{i,t})$ are described.

Hydrological Processes

Since the model is meant to reveal the average long-term soil response to various emission (deposition) scenarios, water flow is assumed to be stationary on a yearly basis – the time step of the model. Furthermore, in the preliminary version of the model, described in this chapter, hydrological processes are assumed to occur at the soil surface. This is done so that the hydrology of a forest/soil ecosystem can be characterized simply by a representative, time-independent precipitation surplus (PS). In the final model, this latter assumption will not be made; instead, the hydrology of the forest ecosystems under consideration will be characterized by an average yearly water flux and moisture content per soil layer.
    Values of PS and Θ are based on calculations made with a separate hydrological simulation model called SWATRE (Belmans *et al.*, 1981), using the 30-year means of precipitation and evaporation at 15 meteorological stations in the Netherlands. The derivation of crucial physical properties of the soil that affect water transport – *viz.*, the moisture retention curve $(\Theta - h)$ and the hydraulic conductivity curve $(K - h)$ – will be discussed later (Section 1.6).

Biochemical Processes

In the preliminary version of the model, biochemical processes are assumed, like hydrological processes, to occur at the soil surface. Consequently, these processes only affect the net element input and the associated proton load to the soil. This mainly affects model predictions in the upper unsaturated zone, but not the leachate concentrations that are important for groundwater quality.
    Biochemical processes are described by two types of equations: a linear "relationship" between two dependent processes or a first-order reaction. For

example, mineralization of fresh organic material is simulated by a linear relationship between litter decay and litter fall

$$FXm, ltf = fXm, ltf \cdot FXlf \qquad (1.2)$$

where FXm, ltf is the flux of nutrient X induced by mineralization of fresh litter $(\mathrm{mol_c\ m^{-2}\ yr^{-1}})$; $FXlf$ is the flux of nutrient X induced by litter fall $(\mathrm{mol_c\ m^{-2}\ yr^{-1}})$; and fXm, ltf is the mineralization fraction of fresh litter.

Mineralization of nutrients from litter already present (older than one year) is described as a first-order reaction (Hagin and Amberger, 1974; van Veen, 1977)

$$FXm, lt = kXm, lt \cdot AXlt \qquad (1.3)$$

where FXm, lt is the flux of nutrient X induced by mineralization of (old) litter $(\mathrm{mol_c\ m^{-2}\ yr^{-1}})$; $AXlt$ is the amount of nutrient X in litter $(\mathrm{mol_c\ m^{-2}})$; and kXm, lt is the mineralization rate constant of (old) litter $(\mathrm{yr^{-1}})$.

Foliar uptake, mineralization of fresh litter and of dead roots, nitrification, denitrification, and root uptake are simulated as a fraction of element fluxes induced by atmospheric deposition, litter fall, root decay, and element input, respectively. Foliar exudation is limited to Ca, Mg, and K; whereas element recycling by litter fall, root decay, mineralization, and root uptake is limited to the same elements plus nitrogen. The net element input is calculated as the sum of deposition, foliar exudation, and mineralization minus the sum of foliar uptake and root uptake. The net proton load is derived from this element input by subtracting the sum of NH_4 and base cations (Ca, Mg, K, Na) from the sum of acid anions (SO_4, NO_3, Cl, RCOO).

The reaction rate of several biochemical processes, such as mineralization, nitrification, and denitrification, is influenced by environmental factors, such as temperature, moisture, and pH. The influence of these factors is accounted for in the model by reducing the maximum values of fractions and constants used. The values of the reduction factors are mainly based on the literature (e.g., Davidson *et al.*, 1978; Bhat *et al.*, 1980). The influence of the moisture content is related to the mean highest and mean lowest groundwater levels (MHW and MLW). The pH regulation of different biochemical processes is an important dynamic feature of the model because pH is calculated during each time step by comparing the acid stress rate (proton load) and the acid neutralization rate (feedback mechanism).

In the final version of the model the processes of nitrification, denitrification, and nutrient uptake by roots will be described by first-order reactions for each specific soil layer. The proton load associated with these biochemical processes will then be calculated for each layer as the sum of cation uptake and nitrification minus the sum of anion uptake and denitrification. At present, these biochemical processes only affect the proton input at the soil surface.

Geochemical Processes

In the preliminary model, geochemical interaction of elements in soil layers is still limited to weathering, because the dissolution of alumino-silicates ultimately controls the consumption of protons in acid soils. Protonation of organic anions is also included, but this process is assumed to occur near the soil surface. The effect of fast buffer processes, such as cation exchange and sulfate adsorption, has not yet been incorporated because of their complexity, whereas the capacity of these buffer mechanisms in the sandy soils under consideration is limited on a long time scale. Monitoring experiments (van Breemen *et al.*, 1986) and column experiments (de Vries *et al.*, forthcoming) suggest that most forest soils in the Netherlands are sulfate-saturated.

In the preliminary version of the model, the rate of proton consumption induced by mineral weathering is described as a first-order reaction:

$$\frac{\Delta ANC}{\Delta t} = kw \cdot ANC \cdot (cH - cH_e) \tag{1.4}$$

where ANC is the acid-neutralizing capacity $(\text{mol}_c \text{ kg}^{-1})$; kw is the weathering constant $(\text{mol}^3 \text{ mol}_c^{-1} \text{ yr}^{-1})$; and cH_e is the equilibrium concentration of H $(\text{mol}_c \text{ m}^{-3})$. The value of cH_e is calculated on the basis of gibbsite solubility. The weathering stoichiometry is assumed to be constant (pH-independent). This is a reasonable assumption for the acid sandy soils considered (Section 1.4.3). The rate of release of an element X ($X = $ Al, Ca, Mg, K, Na) by weathering $(\Delta AX/\Delta t)$ is thus described as:

$$\frac{\Delta AX}{\Delta t} = -fXw \cdot \frac{ANC}{\Delta t} \tag{1.5}$$

where AX is the amount of element X in mineral soil $(\text{mol}_c \text{ kg}^{-1})$ and fXw is the equivalent fraction of element X released by weathering.

Using equations (1.4) and (1.5), weathering of both aluminum and base cations cease when equilibrium with gibbsite is attained. This is not a severe restriction when predicting aluminum concentrations in upper soil layers of the acid sandy soils considered, since weathering is strongly dominated by dissolution of aluminum from hydroxides. However, the aluminum concentration in lower layers of soil with a deep water table will be overpredicted in this case, since base cation weathering increases almost linearly with depth. Consequently, in the final model the weathering of aluminum and base cations will also be described separately. In this situation, only aluminum weathering from hydroxides will cease when equilibrium with gibbsite is attained. Furthermore, cation exchange and sulfate adsorption/desorption will also be included using the Gaines–Thomas equation and the Langmuir adsorption equation, respectively.

1.4. Data Required and Their Derivation

1.4.1. Data requirements

The data required by a physically based model of water quality can be divided
into: (i) model inputs or forcing functions, such as weather data and input of
pollutants; (ii) variables, such as amounts of elements in vegetation, organic
matter, minerals, and soil solution; and (iii) parameters, such as crop factors,
rate constants, and physical and chemical soil properties. The various inputs,
variables, and parameters in RESAM for the different compartments of a forest
ecosystem, (i.e., atmosphere, forest canopy, litter layer, and mineral soil) are
summarized in *Table 1.2*.
  The general form of the variables (including inputs) and parameters in the
model is:

Entity	Constituent	Process	Compartment
F = flux	N = nitrogen	dep = deposition	l = leaves
A = amount	Ca = calcium	lf = litter fall	r = roots
c = concentration	Mg = magnesium	rd = root decay	lt = old litter
k = constant	etc.	e = exudation	ltf = fresh litter
f = fraction		u = uptake	
		m = mineralization	
		n = nitrification	
		den = denitrification	
		p = protonation	
		w = weathering	

Variables refer to combinations of constituents, processes and/or compartments with fluxes (F), amounts (A), or concentrations (c); whereas parameters
refer to combinations with constants (k) or fractions (f). Generally, fluxes, constants, and fractions are related to processes (e.g., $FNlf$, $kCae$, fn, etc.) and
amounts and concentrations to compartments (e.g., $ANlt$, $ACar$, etc.). Sometimes, a combination of process and compartment does occur (e.g., FNm, lt).
  The meaning of the various soil parameters and hydrological parameters
has been given before (Section 1.3.3). The derivation of the variables and
parameters will be discussed below. Special attention will be given to the
specific soil research that is needed to obtain values for constants and fractions
and to the soil survey research that is needed to obtain values for soil parameters
and hydrological parameters.

1.4.2. Derivation of variables

The derivation of variables in relation to the different compartments is summarized in *Table 1.3*. Soil characteristics and hydrological characteristics are also
included to limit the discussion of parameters to constants and fractions only.
  Atmospheric depositions of SO_2, NO_x, and NH_3 are derived from the output of the atmospheric submodel. Deposition values of Ca, Mg, K, Na, and Cl

Table 1.2. Data required by RESAM.

Atmosphere		
Inputs	Fluxes	FSO_2dep, FNO_xdep, FNH_3dep, $FCadep$, $FMgdep$, $FKdep$, $FNadep$, $FCldep$ PS
Canopy		
Variables	Amounts	ANl, $ACal$, $AMgl$, AKl
Parameters	Constants	klf, $kCae$, $kMge$, kKe
	Fractions	fu, l
Litter layer		
Variables	Amounts	ANr, $ACar$, $AMgr$, AKr, $ANlt$, $ACalt$, $AMglt$, $AKlt$
Parameters	Constants	km, lt, krd
	Fractions	fm, ltf, fn, fd, fu, r
Mineral soil		
Variables	Amounts	ANC^a
	Concentrations in solution	cH, cAl, cCa, cMg, cK, cNa, cHN, cNO_3, cSO_4, cCl, $cRCOO$
Parameters	Constants	kw
	Fractions	$fAlw$, $fCaw$, $fMgw$, fKw, $fNaw$
	Soil characteristics	Θ, T, ρ
	Hydrological characteristics	MHW, MLW

[a] *ANC* is the summed amount of the weatherable cations, Al, Fe, Ca, Mg, K, Na.

Table 1.3. Derivation of variables for RESAM.

Compartment	Variables	Derivation
Atmosphere	(SO_2, NO_x, NH_3) *dep*	Output atmospheric submodel
	(Ca, Mg, K, Na, Cl) *dep*	Weather stations
Canopy	A (N, Ca, Mg, K) l	Literature
Litter layer	A (N, Ca, Mg, K) r, lt	Literature
Mineral soil	PS, Θ	Hydrological submodel
	MHW, MLW	Soil map
	ρ, T, ANC	Soil survey information

are gathered from representative stations in each deposition area. Values of *PS* and Θ per soil layer are derived from a hydrological submodel (see hydrological processes). Initial amounts of N, Ca, Mg, and K in leaves and needles, roots, and litter strongly depend on tree species and tree age. At present the model is implemented with preliminary data for Douglas fir and oak, based on literature information.

Information on hydrological characteristics, such as mean highest (*MHW*) and mean lowest water tables (*MLW*), and soil characteristics, such as the thickness of a soil horizon, texture, and organic matter content, are routinely measured during soil surveys. Data on soil properties, such as density and acid-neutralizing capacity, can be derived from these characteristics using so-called transfer functions (Bouma *et al.*, 1986). This will be discussed in more detail in Section 1.6. At present the model is implemented with data of a representative sandy Entic Haplorthod with an *MLW* of 20 m and a sandy Typic Humaquept with an *MLW* of 1.10 m. The soil data used are given in *Table 1.4*.

Table 1.4. Soil data used in the present application of RESAM.

Entic Haplorthod				Typic Humaquept			
Horizon	*T(m)*	$\rho(kg\ m^{-3})$	$ANC(mol_c kg^{-1})$	*Horizon*	*T(m)*	$\rho(kg\ m^{-3})$	$ANC(mol_c kg^{-1})$
A0	0.02	150	–	A0	0.02	150	–
A1	0.15	1340	2.0	Apg	0.25	1360	3.4
B2	0.25	1490	2.5	ACC	0.10	1450	3.0
B3	0.20	1560	2.7	Clg	0.75	1640	2.8
C1	19.40	1610	2.5				

1.4.3. Derivation of parameters

The methodology for deriving parameter values of constants and fractions that control the various element fluxes in a forest ecosystem is summarized in *Table 1.5*, which also gives the values used in the present applications of the model.

Table 1.5. Parameter values used in the present application of the model, including the methodology for deriving them: A = literature information; B = descriptive research; C = process research; D = model calibration.

Compartment	Parameters	Derivation				Values	
		A	*B*	*C*	*D*	*Douglas fir*	*Oak*
Canopy	*fu, l*				X	0.2	0.1
	klf	X				0.3	1.0
	kCae		X			0.02	0.12
	kMge		X			0.05	0.3
	kKe		X			0.19	1.2
Litter layer	*fm, ltf*	X				0.4	0.8
	km, lt	X				0.2	0.25
						Haplorthod	*Humaquept*
Mineral soil	*kw*			X	X	0.003	0.003
	fBw[a]		*X*	*X*		*0.05*	*0.05*
	fAlw		X	X		0.8	0.8

[a] B = Ca, Mg, K, Na.

Parameters Related to Biochemical Processes

The value for the uptake fraction of NO_x and NH_3 by leaves (fu, l) is a rough guess, which needs further refinement. Values of the litter fall constants and the mineralization constants are derived from literature information (Ågren and Kauppi, 1983; Kolenbrander, 1974; Meentemeijer, 1979; van Veen *et al.*, 1981). The values of the mineralization fraction of fresh litter (fm, ltf) and the mineralization rate constant of old litter (km, lt) refer to optimal conditions with respect to moisture and pH.

The exudation constants are calculated from the litter fall constants given in *Table 1.5*, and ratios of the element flux induced by foliar exudation compared with the element flux induced by litter fall, according to

$$kXe = klf \cdot FXe/FXlf \tag{1.6}$$

where kXe is the exudation rate constant of nutrient X (yr^{-1}) and FXe is the flux of nutrient X induced by foliar exudation ($\text{mol}_c \text{ m}^{-2} \text{ yr}^{-1}$).

Equation (1.6) is based on the assumption that both litter fall and foliar exudation can be modeled as a first-order reaction, (see also Section 1.3.3), according to

$$FXlf = klf \cdot AXl \tag{1.7}$$

$$FXe = KXe \cdot AXl \tag{1.8}$$

where AXl is the amount of nutrient X in leaves or needles ($\text{mol}_c \text{ m}^{-2}$).

Ratios of the fluxes of Ca, Mg, and K induced by both processes are derived from monitoring data on nutrient budgets of different forest ecosystems in the Netherlands (van Breemen *et al.*, 1986).

Parameters Related to Geochemical Processes

The rate (kw) and stoichiometry (fXw) of mineral weathering is partly derived from column-type or batch-type titration experiments (process research). In these experiments, the rate of acid neutralization with progressive proton consumption is directly obtained by measuring the rate at which acid must be added to maintain a constant pH. The release of cations with proton consumption is derived from intermittent solution analysis (de Vries *et al.*, forthcoming).

Values of fXw can also be obtained from historical profile analysis (descriptive research). The principle of the historical approach is the analysis of the cation composition of a soil profile. The weathering stoichiometry is ascertained by comparing the distribution of a given element (in $\text{k mol}_c \text{ ha}^{-1}$) between the parent material (C horizon) and the overlying A and B horizons (de Vries, 1987). The results of column and batch experiments (conducted at pH 3) and of a historical profile analysis for several podzolic soils in the Netherlands show that mineral weathering is largely dominated by aluminum release in the acid sandy soils, even under natural conditions (de Vries, 1987). This indicates that the influence of pH on the weathering stoichiometry in strongly acid soils, which are the dominant forest soils in the Netherlands, is relatively small.

The values of $fAlw$ and fBw given in *Table 1.5* are based on mean values of weathering fractions obtained by both methods. The value of kw is mainly based on calibrating the model output against an average pH profile of acid forest soils. The value is less than the values obtained from column experiments, although the order of magnitude is consistent.

1.5. Model Output

Possible model output is related to process-induced element fluxes; amounts of elements in leaves, roots, and litter; and ion concentrations in characteristic soil layers. As an example, *Table 1.6* gives an overview of the cycles of major nutrients including the proton budget of a characteristic forest/soil ecosystem in the Netherlands for representative deposition rates.

Table 1.6. Prediction of major nutrient cycles in a Douglas fir forest on a sandy Entic Haplorthod for yearly averaged deposition rates of SO_2, NO_x, and NH_3 in the Netherlands. Values are equivalent ion fluxes in k mol_c ha^{-1} yr^{-1}.

Compartment/ process	Element load to[a]	H	Ca	Mg	K	NH_4	NO_3	SO_4
Canopy								
Deposition		2.10	0.32	0.28	0.06	2.20	1.30	3.00
Foliar uptake		0.18	–	–	–	−0.44	−0.26	–
Foliar exud.		−0.61	0.10	0.06	0.44	–	–	–
	LL	1.67	0.42	0.34	0.50	1.76	1.04	3.00
Litter layer								
Mineralization		0.00	1.86	0.63	0.79	0.74	3.79	–
Nitrification		2.94	–	–	–	−1.47	1.47	–
Denitrification		0.00	–	–	–	–	0.00	–
Uptake		−0.59	−1.852	−0.70	−1.03	−0.90	−4.96	–
	MS	4.02	0.46	0.17	0.26	0.13	1.34	3.00
Mineral soil								
Weathering		−3.88	0.19	0.19	0.19	–	–	–
	GW	0.20	0.65	0.36	0.45	0.13	1.34	3.00

[a]Abbreviations: LL = litter layer, MS = mineral soil, and GW = groundwater.

The production of H associated with deposition is calculated by assuming complete oxidation of SO_2 and NO_x and negligible nitrification of NH_3 (converted to NH_4) on the forest canopy. Ion concentrations in throughfall are influenced by uptake of NO_x and NH_3 and exudation of Ca, Mg, and K leading to acid consumption. The predicted release of base cations during mineralization slightly exceeds the mineralization of strong acid anions (NO_3). This is compensated for by the production of organic anions (not shown in *Table 1.6*). The ammonium input via throughfall and mineralization is mainly nitrified in the soil, causing a strong acid production, whereas denitrification is predicted to be negligible in this deeply drained soil. Predictions for the Typic Humaquept mainly differ by a lower nitrification and a higher denitrification because of wetter circumstances. Root uptake induces proton consumption, because of a surplus of nitrate over the uptake of basic cations. The resulting proton load to the mineral soil is mainly neutralized by Al mobilization (not shown in *Table 1.6*) and to a small extent by the weathering of Ca, Mg, K, and Na.

Another example is given in *Table 1.7*. It illustrates the influence of tree species and soil type on leachate concentrations of major cations and anions percolating to the groundwater. Except for NO_3, ion concentrations increase in the direction oak–Haplorthod < oak–Humaquept < Douglas fir–Haplorthod < Douglas fir–Humaquept. This is mainly caused by a decreasing precipitation surplus for these ecosystems (0.30 m yr^{-1}, 0.20 m yr^{-1}, 0.15 m yr^{-1}, and 0.05 m yr^{-1}, respectively). The nitrate concentration is strongly influenced by denitrification in the "wet" Humaquept in contrast to the "dry" Haplorthod.

Table 1.7. Prediction of the concentration of major cations and anions (equivalent ion concentrations in mol m^{-3}) for different forest/soil combinations at a yearly average deposition rate of SO_2, NO_x, and NH_3 in the Netherlands.

Soil type	Douglas fir				Oak			
	cAl	cB[a]	cSO$_4$	cNO$_3$	cAl	cB[a]	cSO$_4$	cNO$_3$
Entric Haplorthod	1.38	1.91	1.60	0.90	0.81	1.06	0.90	0.60
Typic Humaquept	3.90	4.72	4.78	1.09	1.64	1.05	1.34	0.20

[a] $B =$ base cations (Ca + Mg + K + Na).

1.6. Model Calibration and Regional Application

Calibration of dynamic simulation models, such as RESAM, requires that extensive time series of at least inputs and outputs of the system are available (Hornberger *et al.*, 1986a). The most valuable data sets that will be used to calibrate the final model are monitoring data on element fluxes in various forest ecosystems in the Netherlands (van Breemen *et al.*, 1986; Mulder *et al.*, 1987; Konsten *et al.*, 1987). However, these records are not extensive enough in time to provide a rigorous test of a long-term model. Additional valuable information that will be used in this respect are data on pH changes during a time interval of 15 to 30 years. Finally, data on ion concentrations in soil water (Kleijn and de Vries, 1987) and shallow groundwater (Hoeks, 1986) in forested areas will be used for a regional model calibration with a method called "Regionalized Sensitivity Analysis" (Cosby *et al.*, 1985a, 1985b; Hornberger *et al.*, 1986a). The uncertainty in long-term model predictions owing to uncertainties in parameter values will also be evaluated with this method.

To be able to forecast soil acidification on a regional scale, information is required on the geographical distribution of soils (including the water regime) and forests within the receptor areas. This can be obtained by superimposing the soil map over the topographic map of the same scale and determining soil units and water table classes under forests at intersection points of coordinates in a grid of a specified spatial resolution (Breeuwsma and Schoumans, 1987). However, this is time-consuming; furthermore, the topographic map does not contain information of all relevant forest types.

Consequently, at the Netherlands Soil Survey Institute, the digitized 1:250,000 soil map is being transferred to a grid-oriented database with a spatial resolution of 250 m × 250 m. This will be coupled to a grid-oriented database containing information on forest type, age, etc., with a spatial resolution of 500 m × 500 m. Furthermore, remote sensing techniques will be used to obtain information on soil use with an even higher spatial resolution.

Another important aspect related to regional application is the spatial variability of input data. With respect to deposition, receptor areas are chosen in such a way that variation in deposition is relatively small – generally within 15%. Variation in soil properties also poses a problem. A very important aid in this respect is the soil map with associated soil survey information. Data on hydrological characteristics (*MHW* and *MLW*) and soil characteristics, such as horizon thickness and organic matter content, can be obtained directly from this soil information system.

Furthermore, physical and chemical soil properties, as described above, can often be related to soil characteristics using transfer functions. This is illustrated in *Figure 1.5*.

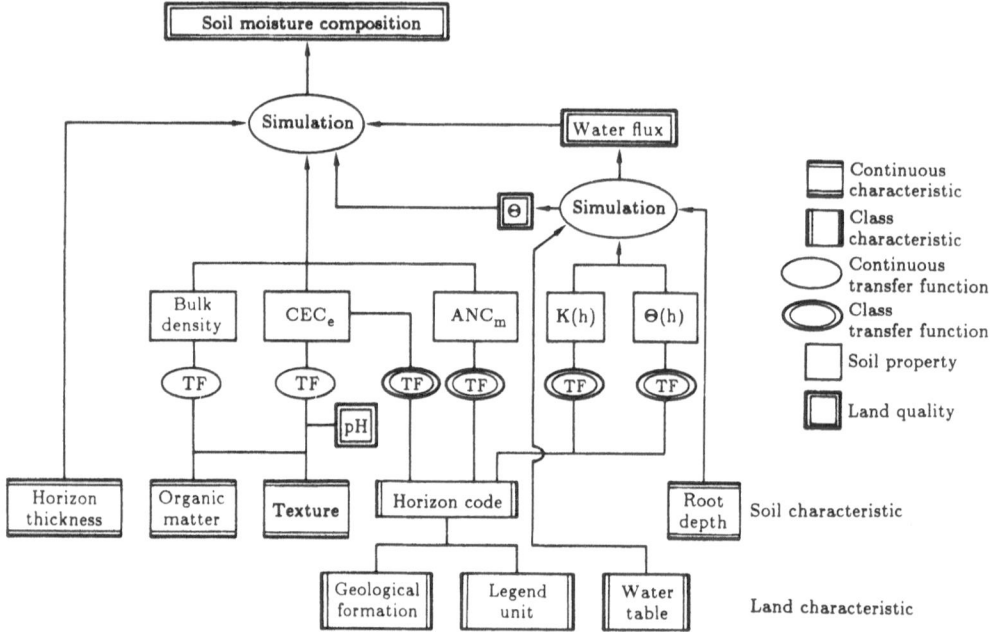

Figure 1.5. The use of soil survey information in RESAM.

A subdivision is made between continuous characteristics that have a continuous range of values (e.g., % clay) and class characteristics that are characterized either by a specified range of values (e.g., texture class) or by a symbol (e.g., soil type, horizon designation). Similarly, transfer functions relating soil properties to soil and land characteristics are divided into continuous and class transfer functions (Bouma *et al.*, 1986).

Bulk density has been related to organic matter content and texture with a continuous transfer function using regression analysis (Hoekstra and Poelman, 1982). The same method has been used to derive cation exchange capacity (CEC) values, measured at pH 6.5, from the clay and organic matter contents of the soil (Breeuwsma et al., 1986). Further investigation is needed to obtain continuous transfer functions deriving the effective CEC ($CECe$) at field pH, because pH influences the exchange capacity of organic matter. This information is needed in the final model when cation exchange is included.

The physical soil properties $\Theta - h$ and $K - h$ required in the hydrological submodel SWATRE have been related to well-defined horizon designations for soil types distinguished in soil surveys (class transfer functions; Wösten et al., 1986). In a similar way, class transfer functions have also been derived for the chemical soil property ANC (including weatherable cations) for A, B, and C horizons of several podzolic soils in the Netherlands (de Vries, 1987).

The class transfer functions derived for these sandy soils illustrate the important role of soil type and soil horizons as carriers of physical and chemical soil properties that are important input data in RESAM (and SWATRE). The soil map, which shows where the various soil types can be found, allows the results to be presented geographically. To our knowledge, the use of model parameters that can be directly or indirectly related to soil survey data is a new approach.

Finally, the most difficult problem is the variability in amounts of elements in biomass – leaves, needles, roots, and litter. There is relatively little information on this factor, and data show an enormous variation. However, using literature data, we are now deriving transfer functions between element amounts in the various compartments and soil-use characteristics, such as tree species, yield class, and age.

It can be concluded that regional application of RESAM is feasible, although the variation in the amounts of biomass elements is an important bottleneck. However, valid application requires the following future activities:

(1) Literature research to derive transfer functions between amounts of elements in biomass compartments and soil-use characteristics.
(2) Process research to derive parameter values for fractions and constants influencing ion fluxes in the soil.
(3) Descriptive research to derive transfer functions for physical and chemical soil properties, determining the transport of water and ions in the soil, from soil survey data.
(4) Geographic research (e.g., remote sensing) to develop a geographical information system (GIS) on atmospheric input, soil use, and soil type.

Acknowledgments

I gratefully acknowledge Ir. M. Waltmans and Ir. R. Verzendaal for implementing the preliminary version of the model on the computer; and Ir. J.J.M. van Grinsven, Ir. P.K. Koster, Dr. J. Hoeks, Dr. N.M. de Rooy, and Ir. L.J.M. Boumans for helpful suggestions during the development of the model. This work was funded by the Dutch Priority Program on Acidification.

CHAPTER 2

Effects of Acid Deposition on Agricultural Soils During and After Use

K. Brodersen, H. Christiansen, B.R. Larsen, and T. Petersen

2.1. Introduction

Denmark is a small and rather densely populated country with little heavy industry. About 70% of the area is used for agriculture. The main sources of acidifying air pollution are domestic heating (oil, gas), regional heating systems (oil, gas, refuse, coal), electricity generation (coal), and transportation. Politically it has been agreed that SO_2 and NO_x emissions should be reduced, and gradually the necessary technological changes are being made. The small size of the country makes the transboundary aspects of air pollution very important, and for this – among other reasons – the development of good, internationally accepted models of the behavior and effects of such pollutants should be of considerable interest to the country.

Back in 1979 an attempt was made to start an inter-Scandinavian project to compare environmental effects of various energy production options. The project was ambitious, and was met with considerable skepticism from the responsible authorities. Although contacts between the interested institutes (mainly, Risø in Denmark; Studsvik in Sweden; VTT in Finland; and IFE in Norway) continued for some years, most of the later developments have been on a national basis. In Denmark the main result has been the creation of the ECCES codes financed by special research grants from the Ministry of Energy.

2.2. The ECCES Model

ECCES stands for Environmental Calculations of Consequences from Energy Systems. It is a general purpose model system composed of a number of

J. Kämäri (ed.), Impact Models to Assess Regional Acidification, 23–31.

submodels that calculate transport, conversion, and deposition of air-polluting materials emitted from point and area sources, and some effects of the deposited material on the soil, crop, and percolating water in the exposed areas.

ECCES can describe the combined effects of a large number of independent sources in and outside a regional area. Subdividing the region according to general administrative practice is desirable to ease acquisition of data concerning population, area use, etc. Test cases using the island of New Zealand, for example, have been run, but only for the atmospheric transport part of the system.

In the following sections, submodels dealing with soil chemistry and plant growth are applied in a special scenario. The models calculate equilibrium between soil and soil pore water for a compartment system of from one to five soil layers on top of each other. The layers are connected by a simple hydraulic model, taking into account vertical movements of soil water and water extraction by plants growing on the area. Ion exchange equilibria for the major elements and selected minor elements are calculated. The interdependence of pH, Ca + Mg concentration, and the base saturation of the ion exchange capacity is modeled using lime potential concept (Clark and Hill, 1964). The ion exchange capacity is divided into a permanent and a variable (i.e., pH-dependent) capacity.

CO_2 equilibria and calcium carbonate dissolution are modeled using temperature-dependent constants, but the soil temperature and the CO_2 concentration in the soil air are introduced as independent parameters – that is, we do not attempt to model the dependence on plant growth. Brodersen *et al.* (1986) recently summarized the status of the model system; various extensions and refinements are planned.

2.3. Scenario

The increasing efficiency of modern agricultural practices has over the past 30 years led to much larger yields. This is demonstrated in *Figure 2.1* for grass produced as fodder in Denmark. A similar, even more pronounced trend line applies to cereals. This generally positive phenomenon is not without certain drawbacks, because for some of the developed countries it has led to overproduction and associated storage, distribution, and economic problems and probably also to unnecessary environmental stress. For many years the total arable area has been declining slightly but steadily in Denmark, reflecting increasing urbanization, as illustrated in *Figure 2.2*. Cereals production since 1980 has declined in favor of other crop types, while the area used for grass has steadily shrunk during the whole period.

The rational response to overproduction would be to decrease further the area used for agriculture, especially marginally productive land. This policy is now under discussion in Denmark. Such areas could be turned into forests or reserved for less intensive use, such as grazing. The scenario described in the following paragraphs was selected to illustrate some phenomena that may be caused by acid rain falling on such old agricultural grounds.

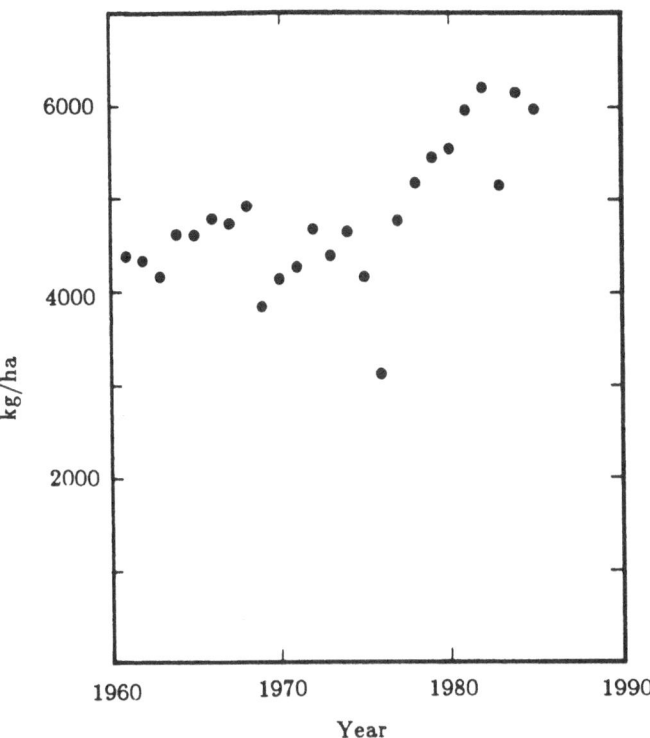

Figure 2.1. The productivity in equivalent kg of barley per hectare of grass used as fodder combined for areas in the rotation and for permanent grass. (Source: *Statistical Yearbooks for Denmark, 1962–1986.*)

For the scenario calculations, it is assumed (for the sake of simplicity) that the area is used exclusively for grass production during a 10-year period. The grass is cropped five times each year and used as fodder. The area is supposed to be tended by a farmer, who maintains a reasonable balance, as far as macro-nutrients (N, P, K) and acidification are concerned, by applying a mixture of manure, NPK fertilizer, and chalk, as appropriate.

A constant yearly rainfall with a distribution typical of Western Jutland is assumed. The rain is supposed to have a composition corresponding to mean values of measurements obtained under present Danish conditions. Typical dry deposition of SO_2 and N components are also taken into account (Hovmand and Petersen, 1985). The main features of the mass balances are given in *Table 2.1*.

The topsoil is divided into four layers of 25 cm thickness, each regarded as a separate compartment. Data for the soil content, the air-filled pore volume, and the maximum and minimum water contents are indicated for each layer in *Figure 2.3*. The actual calculated water contents during the year for the four layers are also shown. Some percolation takes place in autumn, winter, and early spring. The assumed temperature profiles vary from 1° to 19°C in the

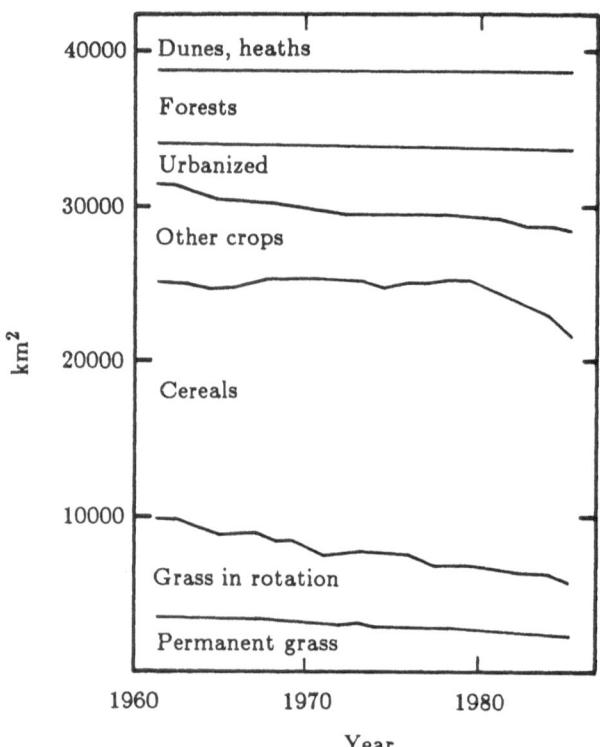

Figure 2.2. Land use in Denmark. (Source: *Statistical Yearbooks for Denmark, 1962–1986.*)

Table 2.1. Yearly amounts of nutrients added or removed (during harvesting) from the area during the initial period with intensive farming. (Values in parentheses were not or only partially taken into account in the calculations.)

Dry mat. g m^{-2}	Fertilizer 50	Manure 440	Lime 6	Rain 27	Dry dep. 2	Harvest 800
Cl		2.0	0.06	4.7		3.8
P	(0.8)	(1.7)				(2.5)
S		1.5		1.34	1.18	2.1
NO_3-N	8.3	4.9		0.45	0.44	13.2
NH_4-N				0.66	0.20	
K	12.6	7.1		0.27		13.9
Na		0.8		3.0		0.8
Ca	(6.6)	3.5	2.5	0.85		4.7
Mg		0.6	0.02	0.29		1.5
Cd mg/m^2	0.34	0.04		0.20		0.06
pH				4.2		

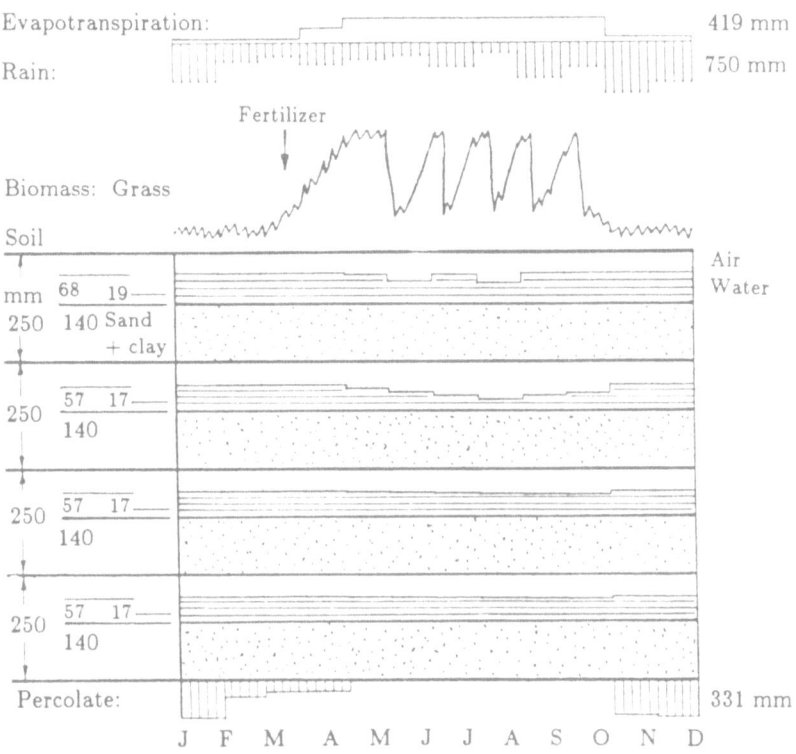

Figure 2.3. Schematic presentation of the yearly cycle of the water contents in the four soil layers used to simulate 10 years of intensive farming and influence from acid rain. In the next 20 years without farming, evaporation is assumed to decline, resulting in higher percolation.

upper layer (3° to 16°C in the bottom one), and the CO_2 content in the soil air from 0.1% in winter for the upper layer to 2% in summer for the deeper layers.

The land has poor-quality, sandy soil with low ion exchange capacity. The permanent capacity $(CECP)$ is assumed to be 3.0 meq/100 g soil, and the pH dependent capacity $(CECV)$ is 2.0 meq/100 g in the neutral soil declining to zero at pH 4.

After 10 years, agricultural production is assumed to stop, but the soil development is calculated for the following 20 years. In this period the area is supposed to be covered by permanent grass growing and decaying in a closed cycle, i.e., no harvesting, fertilizing, or liming takes place. Compared with the previous period, the biomass will be considerably reduced and therefore also the evaporation from the area. The CO_2 content in the soil air may also be reduced, but this is not taken into account. Otherwise, the situation is supposed to be unchanged, i.e., the same yearly amounts of rain with the same composition and the same amounts of dry deposited materials influence the system in the last 20 as in the first 10 years. Selected results from these calculations are given in *Figure 2.4* and *Table 2.2*.

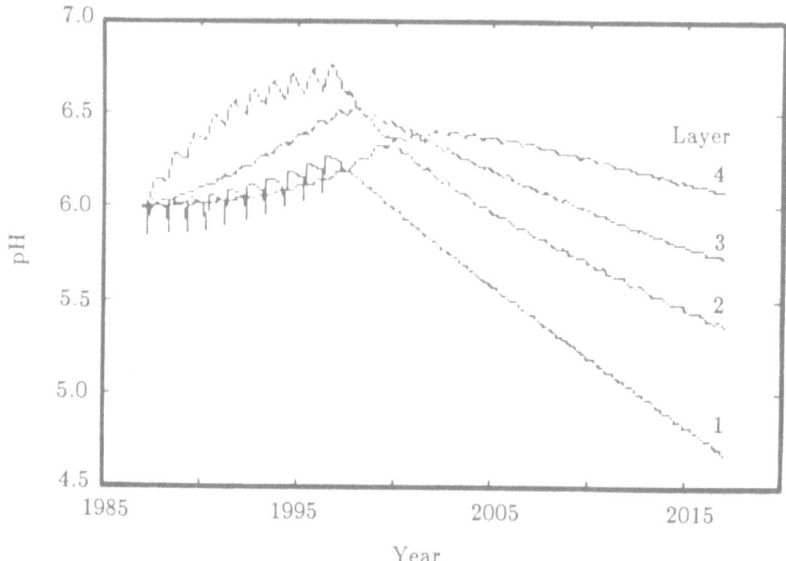

Figure 2.4. Development of soil pH during the 30-year simulation period.

Table 2.2. Mass balance for calcium. The cumulative amounts of Ca going into or leaving an area are given for 10-years periods. From 1987 to 1997, intensive agricultural grass production is assumed. In the following period (1998 to 2017), no further harvesting or fertilizing takes place. The fluxes are compared with the calculated calcium concentrations in the grass and the percolate at the ends of the 10-year periods, and with the change in the calcium magazine in the four soil layers.

Ca ($g\ m^{-2}\ 10yr^{-1}$)[a]	1987	1997	2007	2017
Added	134	8	8	
Removed with harvest	44	0	0	
Concentration in biomass (%)	0.44	$0.6/1.3^{b}$	1.2	1.1
Content (g/m^2) in soil layer:				
No. 1	254	283	250	222
No. 2	225	244	233	226
No. 3	226	235	232	228
No. 4	226	229	233	230
Change in content	60	−43	−42	
Removed with percolate	30	51	50	
Concentration in percolate (ppm)	9.6	13.2	13.8	13.5

[a] Except as noted in parentheses.
[b] These values are for 1997 and 1998, respectively, and illustrate the effect the decrease in biomass production has on the concentration in the plant material.

The general trends are as expected. In the first 10 years, a slight improvement of the pH is obtained – from the originally assumed 6.0 to about 6.6 – as a result of adding lime, etc. In the same period the Cd content of the upper soil layer increases considerably, mainly owing to the Cd present as an impurity in the fertilizers. A one-year transition period is assumed after intensive farming is abandoned, followed by 19 years when the area is used, e.g., for grazing of cattle, without compensation for the various acidifying processes. This results, as shown in *Figure 2.4*, in a practically linear decrease of the pH in the upper soil layer from 6.2 to 4.7 over the 20 years. In the second soil layer, the acidification is considerably less. The general shape of the curve is also different, probably due in part to dry-out in the summer periods (see *Figure 2.3*), an effect that is especially pronounced in this layer, and partly as an artifact of the simple hydraulic model employed. The pH decreases in the two lower layers are even less pronounced. It should be mentioned that the model in its present stage does not take into account pH compensation through weathering of silicate minerals.

It can be seen from the mass balance figures in *Table 2.2* that the soil is depleted in calcium content at the end of the 20-year period without farming, but only to values similar to – or for the upper layer slightly less than – the value used at the start of the simulation. Under unchanged conditions the leaching of Ca will continue at more or less the same rate, and will eventually result in significant degradation of the soil. However, even given poor soil with low ion exchange capacity, it takes quite some time before significant depletion occurs, provided the soil starts in a good condition. Also, even without the acid rain impact, a considerable leaching of Ca will take place owing to CO_2 in the soil and other effects of plant growth.

Similar calculations were made for the other elements. Cadmium, selected as a polluting micro-component, was demonstrated to accumulate in the upper soil layer. A slight downward movement takes place with percolating water, but this is not able to compensate for the influx with fertilizers or from air pollution.

The concentration in the dry material of grass grown on the area increases sharply when the intensive farming is given up. This is mainly an effect of less dilution owing to the decrease in biomass production, but the increasing acidification will also make the Cd more available for plant uptake. This phenomenon may be important for cattle grazing in the area.

2.4. Discussion

2.4.1. Regional aspects

The scenario calculations presented above were made for a single area with only one soil type. Furthermore, simplified assumptions were used to describe agricultural practice and the exposure to polluting materials.

In principle, the facilities in the ECCES code can be applied also to the modeling of much more complex situations involving areas with different exposure, rainfall, hydrological behavior, soil and crop types, use of fertilizers, etc. Exposures for different emission scenarios and source configurations can be calculated and used in comparative evaluations for decision making, etc.

In practice, of course, this is not simple. Questions about the completeness and correctness of the models and the data used in the simulations must be answered. Credibility of the scenarios and how to present and compare the results are also problematic. Computation time may be a limiting factor, etc. A few general comments are addressed to these types of problems in Sections 2.4.2 and 2.4.3.

2.4.2. Completeness and correctness

All models are simplifications compared to reality, and submodels in codes able to handle complex regional phenomena must often be simplified even further, to enable the code to operate within reasonable computation time.

The soil model in the ECCES system is based on a structure where each separate layer is regarded as a completely mixed compartment. This is, of course, unrealistic, but the alternative – some sort of column simulation – would probably not do much better because inhomogeneities, short circuits in water flow, and other deviations from ideal behavior in real soil layers are difficult to take into account.

In ECCES a very simple model is used to simulate the water movement through the unsaturated soil: water is supposed to percolate from one compartment to the next below, only when the maximum water retention capacity is exceeded. Below a certain limit, no further water can be extracted by plants growing on the soil.

As far as chemistry is concerned, it is of special interest to be able to model synergistic effects that changes in the pH may have on other aspects of the soil/pore water system. The soil model in the ECCES code has been developed with this in mind. The set of equations describing ion exchange, calcium carbonate dissolution, and carbon dioxide behavior is restricted to equilibrium calculations, but these have been found to give reasonable reproductions of the macro-chemistry in laboratory-scale extraction experiments with well-mixed samples under controlled CO_2 conditions.

The variable cation exchange capacity – e.g., associated with the weak ion exchange sites in humic acid – is found to be a major pH-stabilizing factor in high-pH soil. Dissolution and speciation of Al are important in acid soil, but are not yet represented in a satisfactory way in the model. Slow dissolution or conversion of other minerals – primarily silicates – are also important as long-term pH buffer mechanisms and as sources for nutrients such as K and Mg. These processes are not included in the present version of the model, and the pH development in the soil will therefore in reality be less extreme than indicated in *Figure 2.4*.

In general, a model simulating reasonably closely the mechanisms of the real reactions taking place is preferable to the black-box type of model, which may be tuned to fit available series of measurements, but is rather problematic for use under circumstances deviating from the measurements.

ECCES attempts to serve as a mechanistic model for the macro-chemistry in a soil compartment. The model is by no means complete, but it should be relatively robust over a large application field. The modeling of fertilization and the effects of plant growth are included because they have a major influence on soil chemistry.

As far as micro-chemistry is concerned (concentrations of heavy metals, etc.), it has been necessary to introduce some features of the tunable type to obtain agreement, for example, with measurements of Cd concentrations in pore water, as functions of pH and total Cd.

Along with questions about the correctness and completeness of the models, questions may arise about the data going into the models. The obvious answer to all three problems is to test the models and their sensitivities against well-documented field experiments. This is a complex and time-consuming task, but some work of this kind is planned for the ECCES model.

2.4.3. The role of sensitivity studies

For regional applications, data may be difficult to obtain in sufficient detail, and it may be necessary to aggregate them to construct manageable systems. The influence of this type of uncertainty can probably be documented by sensitivity studies.

Other types of uncertainties may also enter into regional calculations. For instance, environmental models tend to be used in a deterministic mode, but reality contains strong probabilistic elements. This is, of course, especially true of the atmospheric parameters, such as wind direction and rainfall; but it is also the case for individual land areas with regard to crop rotation, fertilization, etc. The problem always arises in making projections, but it also applies to historical data when sufficiently detailed information may not be available. Some studies of this type may also benefit from sensitivity analysis.

Another type of uncertainty is associated with the emission sources. Some emission is certain to take place, but the details in projections are mainly estimates that depend heavily on political decisions, economic conditions, and technical developments. Although the absolute values are important for planning purposes, the motivation for further development of modeling tools to quantify pollution effects is strong, regardless of even relatively large changes in projected emissions.

Finally, a comment about an especially undesirable type of uncertainty. Calculations – even quite accurate calculations – on complex systems involving many parameters and data must be digested and understood without distortion. Presentation of the results and evaluation of their significance are just as important as development of reliable calculation tools. Within this area one typically encounters problems with low-level dose/effects relationships; estimating costs of damage to nature, to health, and to future human possibilities; extrapolations to foreign countries and under some future circumstances, etc.

However, the basis for any management model must be a well-documented description of the physical and chemical changes that pollution can be expected to cause, supplemented with an estimate of the reliability of the results. There is scope for much future work in these areas.

CHAPTER 3

Analysis of Regional Changes in Forest Growth

P. Hari and M. Holmberg

3.1. Introduction

There are two strategies for constructing models to deal with environmental change and its consequences for forest growth: either we can model the properties of available data, or we can construct theoretically based models. The data-oriented approach has the advantage of producing rapid projections. On the other hand, the theoretical approach, although it usually proceeds slowly, has, at least in the long run, wider applicability. Theory-based models can be generalized even outside the data range. In this chapter, we emphasize the theoretical approach.

The analysis is based on flows of material in the system formed by the atmosphere, trees, forest soil, soil solution, and groundwater. The concentrations of nutrients and toxic compounds are environmental factors for forest growth. Owing to the anthropogenic input into the atmosphere, several environmental factors are changing. The aim of our modeling work is to analyze these changes in environmental factors and produce scenarios of forest growth in Finland.

3.2. General Structure of Models

There are usually growing stands of different age in a forest region, and most trees live less than 200 years. If forests are treated according to the principle of sustained yield, the age structure of the stands in the region does not change with time: old stands are replaced by young ones, but the area occupied by each class remains the same size. In our model we adopt this assumption.

J. Kämäri (ed.), Impact Models to Assess Regional Acidification, 33–47.

A region as large as a nation usually consists of a wide range of environments. We divide Finland into three subregions of constant deposition. Within each subregion, we consider four classes of soil fertility. By treating the forests according to the principle of sustained yield, it is sufficient to analyze the time development of the impact of the changing environment on the mean growth per hectare in each subregion. Let u_{jk} denote the environment in the region j and the fertility class k, and g_{jk} the corresponding impact on growth.

The concentrations of nutrients and toxic substances are environmental factors. Anthropogenic emissions into the atmosphere change the flows of material in the system formed by the atmosphere, vegetation, forest soil, soil solution, and groundwater. Concentrations of several elements and compounds, such as nutrients and toxic substances, are changing due to the emissions. Thus, the environment varies with time:

$$u_{jk} = u_{jk}(t) \quad . \tag{3.1}$$

The environment consists of several factors ($i = 1, \ldots, n$), so we use a vector to describe the environment in subregion j, fertility class k:

$$u_{jk}(t) = [u_{ijk}(t)] \quad . \tag{3.2}$$

The forests are responding to changes in the environment, and the impact on growth g_{jk} of the environment in fertility class k of subregion j is thus indirectly a function of time:

$$g_{jk}(t) = g[u_{jk}(t)] \quad . \tag{3.3}$$

We need two types of models for analyzing regional forest growth. One for simulating the time development of environmental factors, and another for linking forest growth to environmental factors.

Before anthropogenic influences came to dominate, environmental factors changed in most cases so slowly that we can assume that they were in a steady state, fluctuating around some constant mean value. Currently, several environmental factors are changing, so that this steady state assumption is no longer justified. The most important environmental changes in our forests are increased atmospheric CO_2, soil acidification, and increased nitrogen in soil. The effects of soil acidification include changing concentrations of nutrients (both adsorbed and in soil solution). Let u_{1jk} denote CO_2 concentration. Let u_{2jk} denote the amount of excess nitrogen in soil due to atmospheric deposition, and u_{3jk} the amount of nutrient cations available for uptake by plants.

We assume that a multiplicative model can be applied in calculating the total impact of the three environmental factors on growth in fertility class k of subregion j. The multiplicative approach is chosen because it allows for a simple mathematical formulation of the principle that environmental factors interact without compensating for each other. We get

$$g_{jk}(t) = \mathop{\pi}_{i=1}^{3} f_i[u_{ijk}(t)]$$ (3.4)

where f_i, $i = 1, 2, 3$, is the impact on forest growth of CO_2, excess nitrogen, and depletion of nutrient cations, respectively.

3.3. Linkage between the Environment and Regional Forest Growth

3.3.1. Carbon dioxide

The impacts of CO_2, cation availability, and nitrogen concentration in soil on regional forest growth are poorly understood. Thus, rather rough estimates have to be used. Process-based causal stand models, such as those proposed by Hari et al. (1985) and Mohren (1987), may provide better information in the future. Increasing CO_2 concentration in the atmosphere accelerates photosynthesis (Thornley, 1974; and Kaitala, et al., 1982). The dependence of photosynthetic rate on carbon dioxide concentration is linear at the present level of CO_2. Assume that the response of regional forests to increasing CO_2 is the same as the dependence of photosynthetic rate on CO_2. The impact f_1 is formulated as a linear function of the CO_2 concentration u_1 (ppm):

$$f_1(u_1) = a_1 u_1 + b_1$$ (3.5)

where $a = 0.00432$ ppm^{-1} and $b = -0.259$. The factor has no regional characteristic, because the CO_2 concentration is assumed to vary globally with time.

3.3.2. Excess nitrogen

Nitrogen fertilization experiments allow the approximation of the effect of excess nitrogen on forest growth (Kukkola and Saramäki, 1983). We assume that the effect of excess nitrogen due to deposition is similar to that obtained in fertilization experiments, and we approximate the impact f_2 by a linear function:

$$f_2(u_2) = a_2 u_2 + b_2$$ (3.6)

where $a_2 = 0.02$ kg^{-1} ha a, and $b_2 = 1$.

3.3.3. Nutrient cations

The nutrients calcium, potassium, and magnesium are taken up by the roots. When roots grow, they exchange cations from the soil solution for hydrogen ions, and anions for hydroxyl ions, thus changing their microenvironment. The ability of roots to collect nutrients, by growing into yet undepleted layers of soil, can be accounted for by describing root nutrient uptake as if it happened both from soil solution and from the pool of exchangeable ions, adsorbed to the solid phase of the soil. The amount of nutrients available for uptake is calculated as a weighted mean of the base cations in the soil solution and on the exchange sites, since the exchangeable pool in absolute terms is much larger, although relatively of the same importance as the dissolved nutrients:

$$u_3 = (a_3 \, M_{sol} + b_3 \, M_{ex})/(a_3 + b_3) \quad . \tag{3.7}$$

The weighting coefficients a_3 and b_3 are determined from the ratio of adsorbed to solved cations in undisturbed soil, and from the ratio of the effective volume in which the exchangeable cations reside.

We assume that a decrease in the amount of cations available for uptake, u_3, has a detrimental effect on growth. The impact f_3 is computed by comparing the value of u_3 at each instant with the initial amount of cations available u_3^0. As long as the availability exceeds 90% of the initial availability, no impact is assumed. Below that level, we assume the growth to decrease linearly.

$$f_3(u_3) = \begin{cases} \dfrac{u_3}{0.9 \, u_3^0} & \text{if } u_3 < 0.9 \, u_3^0, \\[2mm] 1 & \text{if } u_3 \geq 0.9 \, u_3^0 \quad . \end{cases} \tag{3.8}$$

3.4. Time Development of Environmental Factors

The environmental factors, i.e., concentrations of nutrients and carbon dioxide, are changing due to anthropogenic input into the atmosphere. The concentrations in a volume element are determined by the inflows and outflows and by the production and consumption of the compound in the volume element. The model for concentrations in a volume element can be constructed using the material balance for each compound under consideration. The transport processes have to be modeled separately.

3.4.1. Carbon dioxide

Carbon dioxide concentration is increasing in global scale. The analysis of the time development of CO_2 can be based on the carbon balance of the entire atmosphere. Models are available to produce the time development of CO_2. Not knowing the annual amount of fossil fuels to be used in the future causes uncertain predictions of atmospheric CO_2.

3.4.2. Excess nitrogen

Anthropogenic nitrogen deposition has a clear geographical pattern: in southern Finland nearly 10 kg ha^{-1}, in northern Finland only 1–2 kg ha^{-1}, are deposited yearly (Järvinen and Haapala, 1980). Nitrogen is a key nutrient in the northern forests. Deposited nitrogen takes part in the metabolism of trees and microorganisms, and is accumulated in the forest ecosystem.

3.4.3. Nutrient cations

The deposition of hydrogen ions and base cations alter the composition of soil solution, and therefore also the proportion of base cations and hydrogen ions on the exchange sites. The exchange of ions between the soil solution and the solid phase dominates the short-term response of the soil to a changing deposition. Weathering of the mineral structure contributes to the long-term pattern of the response. In an earlier study we have modeled the dynamics of dissolved and adsorbed hydrogen ions and base cations, weathering, cation uptake, decomposition, transpiration, and leaching (Holmberg *et al.*, 1985; Holmberg and Hari, 1987).

3.5. Model of Soil Processes

3.5.1. Model structure

A dynamic model of cation exchange, mineral weathering, and mass flow of water in forest soil was developed to study the impact of external inputs on the temporal development of the concentrations of ions in the soil. The time span 1900 to 2060 has been simulated. The ions considered are hydrogen, potassium, calcium, and magnesium ions. The model is based on the following assumptions:

A1   The rate of cation exchange depends on the equivalent concentrations of ions in soil solution and on the exchange sites. The amount of exchangeable base cations is increased by physical disintegration of the minerals.
A2   The rate of physical weathering is constant, independent of ion concentrations in soil solution.
A3   The rate of water percolating through the soil equals precipitation minus evapotranspiration – that is, no surface runoff is assumed to take place.
A4   The water content of the simulated soil layer remains constant.
A5   The organic cycle is in equilibrium – that is, annual net decomposition equals annual net uptake.

The temporal development of the concentrations of hydrogen ions and base cations (potassium, calcium, and magnesium) in the uppermost 0.3 m of the soil is simulated using differential equations (*Table 3.1*). The equations are mathematical formulations of assumptions A1–A5. The input variables are the pattern of the deposition of hydrogen ions and base cations. The state variables are the equivalent concentrations of hydrogen ions and base cations in soil solution and adsorbed on soil particles. The water content in the soil, the thickness of the simulated layer, the rates of ion exchange and mineral weathering, and the rates of precipitation and evapotranspiration are the parameters of the model.

Table 3.1. Model equations, variables, and parameters.

Model equations

$$dH/dt = H_{in} - (P-ET)H/(\Theta L) - v_1(M_a/CEC)\,[H/(\Theta L)]^2$$
$$\qquad\qquad + v_2[M/(\Theta L)]\,(H_a/CEC)^2$$
$$dH_a/dt = v_1(M_a/CEC)\,[H/(\Theta L)]^2 - v_2[M/(\Theta L)]\,(H_a/CEC)^2$$
$$dM/dt = M_{in} - (P-ET)M/(\Theta L) + v_1(M_a/CEC)\,[H/(\Theta L)]^2$$
$$\qquad\qquad - v_2[M/(\Theta L)]\,(H_a/CEC)^2$$
$$dM_a/dt = -\,v_1(M_a/CEC)\,[H/(\Theta L)]^2 + v_2[M/(\Theta L)]\,(H_a/CEC)^2 + w$$

Variables

H, H_a (meq m^{-2}): hydrogen ions in soil solution and adsorbed
M, M_a (meq m^{-2}): Ca + Mg + K in soil solution and adsorbed
H_{in}, M_{in} (meq m^{-2}a^{-1}): total deposition of ions
$CEC = H_a + M_a$: total cation exchange capacity

Parameters

Soil water content $\Theta = 0.3$ m^3 m^{-3}
Layer thickness $L = 0.3$ m
Ion exchange parameters $v_{1,2}$
Weathering rate $w = 0.002$ meq m^{-2} a^{-1}
Precipitation $P = 600$ mm a^{-1}
Evapotranspiration $ET = 360, 420, 480,$ or 540 mm a^{-1}

3.5.2. Initial states of the soil model

The initial values were chosen to reflect four different soil types (*Table 3.2*). The types range from a fertile soil – developed on sites with low net throughflow, showing large cation exchange capacity, coupled with high base saturation – to the other extreme – a poor soil with small cation exchange capacity, high percolation, and low base saturation. The spatial distribution of these soil types over Finland (*Figure 3.1*) is approximated on the basis of a classification of forestland

Table 3.2. Initial values used in the model.[a]

Soil type	CEC	M_a	M_a/CEC	ET	P–ET/P
1	10,000	8,000	80%	540	10%
2	6,000	2,400	40%	480	20%
3	4,000	800	20%	420	30%
4	2,000	200	10%	360	40%

[a] Abbreviations: CEC = cation exchange capacity in meq m^{-2}; M_a = adsorbed bases in meq m^{-2}; M_a/CEC = base saturation; ET = evapotranspiration in mm a^{-1}; $P - ET$ = percolation.

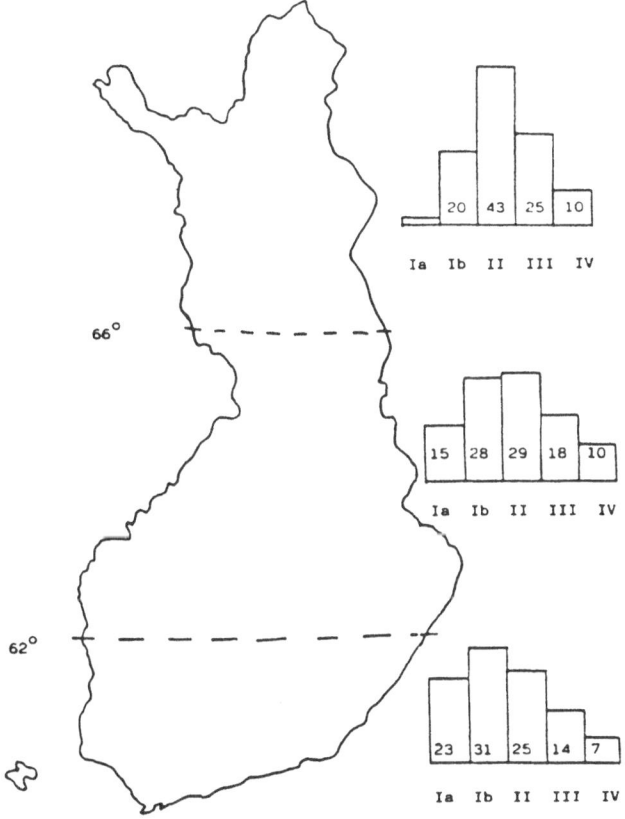

Ia = Rich forest sites (*Oxalis–Myrtillus* type)
Ib = Moist forest sites (*Myrtillus* type)
II = Dry forest sites (*Vaccinium* type)
III = Lichen sites (*Calluna* and *Cladina* type)
IV = Peatland sites

Figure 3.1. The spatial distribution of forestland by taxation class. (From *The Yearbook of Forest Statistics*, 1979.)

into taxation classes, according to the National Forest Inventory taken in 1971–1976 (*Yearbook of Forest Statistics*, 1979).

The initial states were chosen to represent conditions from extremely rich to extremely poor soil. For the rich soils, the chosen values for cation exchange capacity and base saturation agree with the values given by Urvas and Erviö (1974); whereas the poorest soils they surveyed were more fertile than our approximation of a poor soil.

3.5.3. Input to the soil model

The driving force of the changes of ion concentrations in the soil is the deposition of hydrogen ions and base cations (calcium, magnesium, and potassium). The history of this deposition was generated from energy production data for the period 1900 to 1970. A peak in the deposition is assumed to have occurred in the 1970s, followed by a 30% decrease by the year 1990. The country is divided into three areas, separated by the latitudes 62° and 66° (*Figure 3.2*).

The peak value of the hydrogen ion deposition is approximated on the basis of measurements performed in the 1970s by the National Board of Waters (Järvinen and Haapala, 1980). The peak amounts to 23.8 meq $m^{-2} a^{-1}$ in southern, 17.5 meq $m^{-2} a^{-1}$ in central, and 6.7 meq $m^{-2} a^{-1}$ in northern Finland. The deposition of base cations is assumed to have reached a peak value of 2.3 meq $m^{-2} a^{-1}$ in all three areas in the 1970s.

3.5.4. Sensitivity of the soil model

The model structure implies that the rate of the change occurring in the ion concentrations of the soil is primarily determined by the ratio of the ion exchange parameters, v_1/v_2. The rate of mineral weathering, w, on the other hand, determines the level of the long-term change. Besides these geochemical parameters, the hydrological parameter, rate of percolation $(P - ET)/P$, influences the rate and the level of the response.

In modeling the dynamics of ion exchange, the soil solution is assumed to be diluted enough to allow the use of equivalent concentrations instead of ion activities. The ratio v_1/v_2 of the ion exchange parameters corresponds to Ericsson's coefficient (Bolt *et al.*, 1976). This ratio influences the rate of the system's response. The larger the ratio v_1/v_2, the faster the transfer of base cations from the exchange sites into soil solution. In these simulation runs, the ratio has been calculated from the assumption that the four initial states of varying fertility all represent steady state conditions.

The constant weathering rate (0.07 meq $m^{-3} a^{-1}$) used in these runs is an approximation of the mineral disintegration rate in the uppermost layers of soil. If cation exchange is included, or deeper horizons are considered, a higher rate is obtained; for example, von Brömssen (1986) estimates the rate of neutralization due to weathering as 10 meq $m^{-3} a^{-1}$ in the C-horizon.

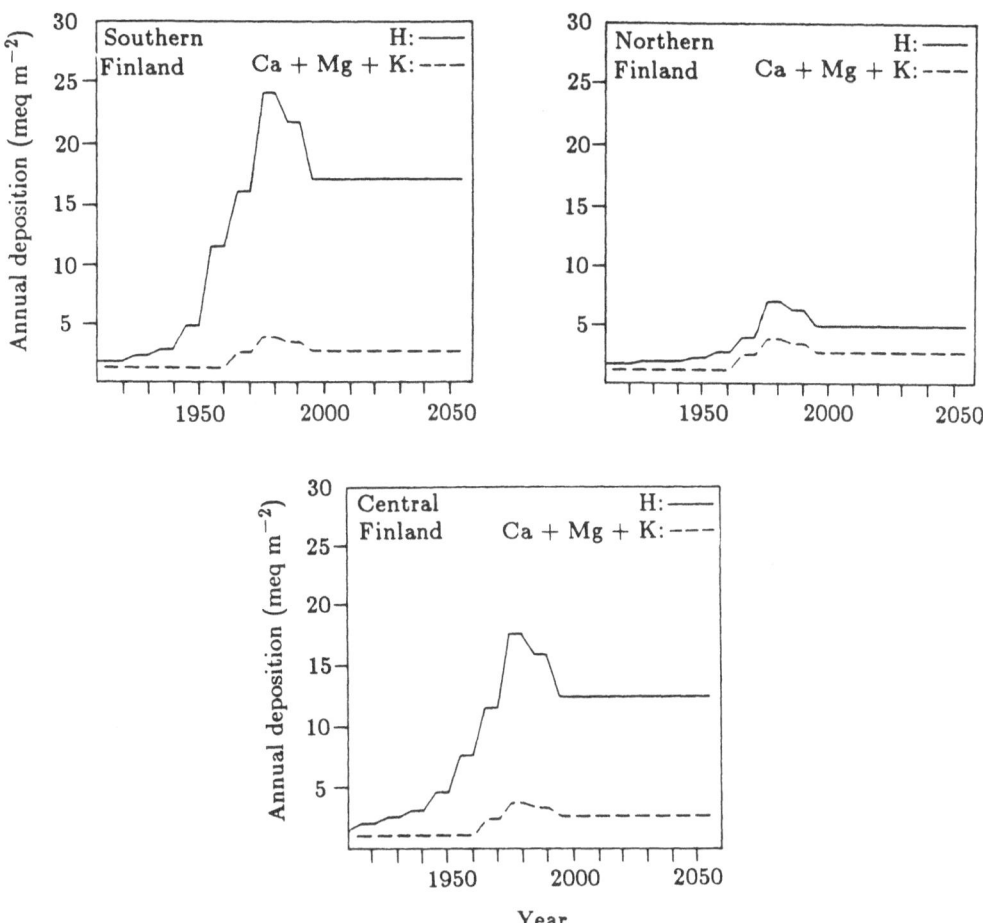

Figure 3.2. Temporal development of the deposition of hydrogen ions and base cations in southern, central, and northern Finland.

Empirical sensitivity analysis of the model showed that the response is not a monotonic function of the percolation rate (Holmberg and Hari, 1987). The minimum of the response time versus percolation rate graph was attained for an evapotranspiration rate of 480 mm a^{-1}, resulting in a percolation rate of 20% of the annual precipitation. In that case, only 40 years are needed to produce a decrease of 30% in the amount of base cations on the exchange sites, whereas nearly 100 years are needed for the same response when the percolation rate is 10% or 40%. This is because a high percolation rate gives rise to low concentrations in soil solution, slowing down the exchange process. A low percolation rate, on the other hand, may cause an intensive exchange, while only little is lost due to the slow transport out of the system. Correspondingly, a maximum in the response volume versus percolation rate graph was attained for $ET = 480$ mm a^{-1} and percolation rate = 20%.

3.5.5. Soil simulation results

The amounts of hydrogen ions and base cations in soil solution [*Figures 3.3(b)*
and *3.3(d)*] mirror the pattern of deposition (*Figure 3.2*). The peak in the input
of hydrogen ions in the 1970s caused the concentrations of the soil solution to
rise correspondingly. This is reflected, by the process of ion exchange, as an
increase in hydrogen ions and a decrease in base cations adsorbed to the solid
phase of the soil [*Figures 3.3(a)* and *3.3(c)*]. The changes are not as drastic in
the north (*Figure 3.4*), because of a smaller deposition (*Figure 3.2*).
    The amount of potassium, calcium, and magnesium available for uptake by
the tree roots increases slightly, at first, owing to the rising soil solution

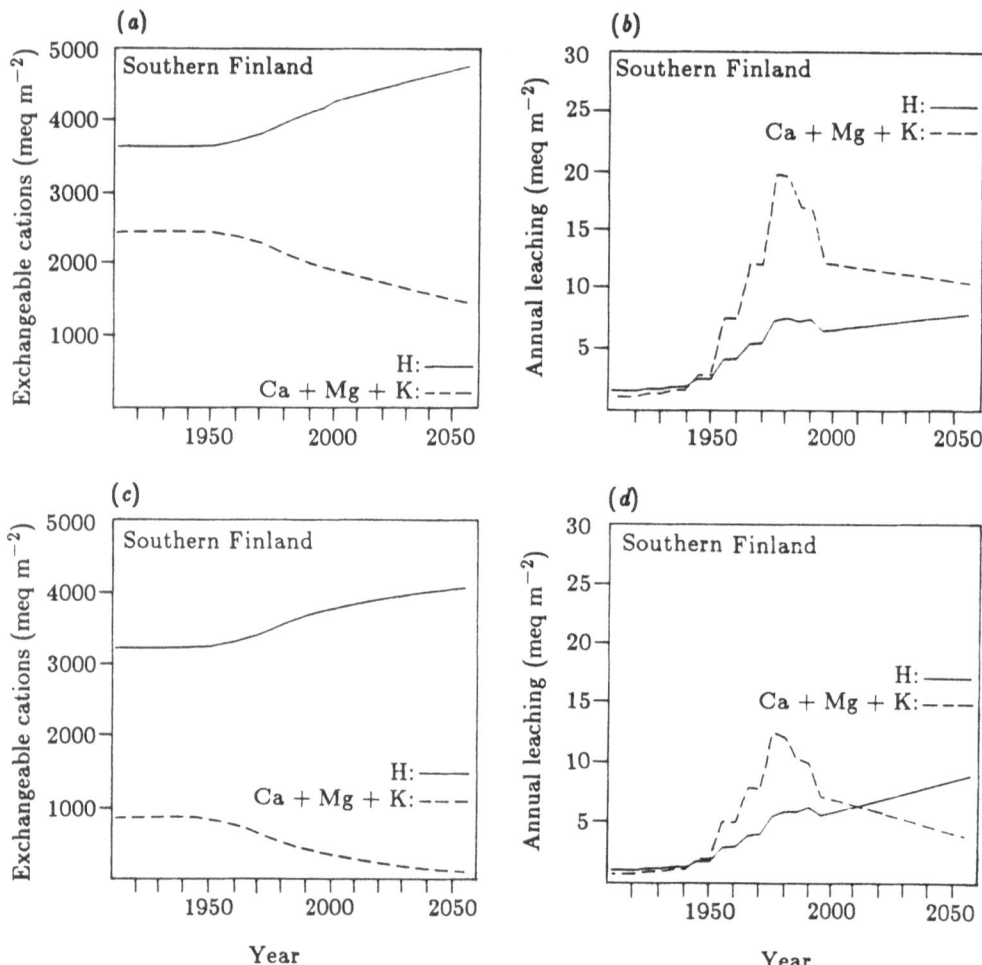

Figure 3.3. Temporal development of the amount of exchangeable cations adsorbed to
the solid phase and of the leaching of cations from the uppermost 0.3 m of mineral soil.
Graphs (*a*) and (*b*) show the simulation results for a moderately rich forest site (initial
value 2), and graphs (*c*) and (*d*) those of a moderately poor site (initial value 3).

concentrations. The leaching of cations, however, changes this development, and, for a rich soil, the availability returns to its 1900 value as the deposition stabilizes. If the amount of exchangeable base cations is low to start with (initial state 4), the availability continues to decrease (*Figure 3.4*), although the input does not change after the year 1990 (*Figure 3.2*).

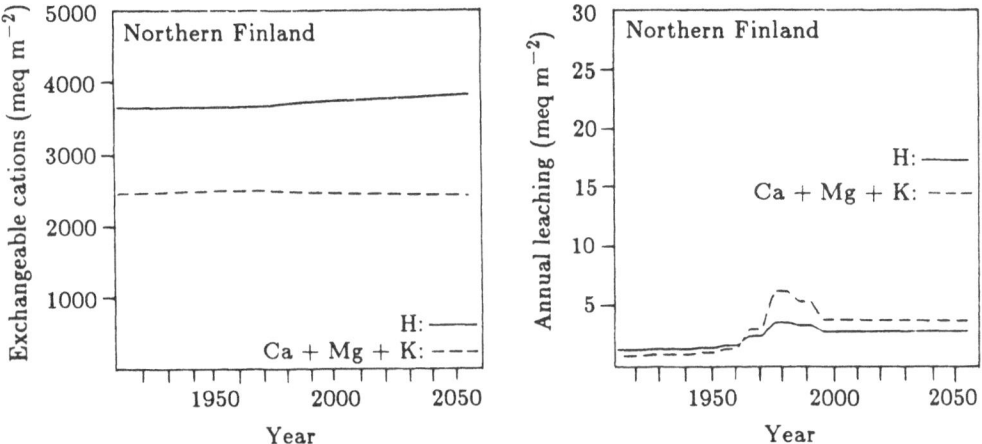

Figure 3.4. Temporal development of the amount of exchangeable cations and the leaching of cations, simulated for a moderately rich site in northern Finland.

The smallest changes occurred with the most fertile soil, initial state 1. The amount of base cations decreased by the year 2060 from 8,000 meq m^{-2} to 7,042 meq m^{-2} in southern Finland, which corresponds to a change of 12%. In central Finland, the change was 8% and 2% in the northern part. The least fertile soil type, initial state 4, showed the largest changes. The amount of base cations decreased from 200 meq m^{-2} to 10 meq m^{-2} in the south — that is, by 95%. The changes in central Finland and in the north were 91% and 7%, respectively.

The simulation results can be interpreted such that a shift from fertile soils to poor soils occurs by the year 2025 in the south of Finland. A strong leaching of base cations from the least fertile soils takes place. The results indicate a potential deterioration within the next 50 years of forest soil, which in the 1970s had a low content of base cations. Medium fertile soils may be strongly leached in areas with high deposition. Changes in the ionic balance of forest soil are expected to take place about 20 years earlier in southern than in central Finland.

3.6. Regional Aggregation

The deposition of hydrogen ions and cations to forest soil has a strong regional pattern. In addition, the properties of soil are characteristically regional. Carbon dioxide concentration is an exception: it changes on a global scale.

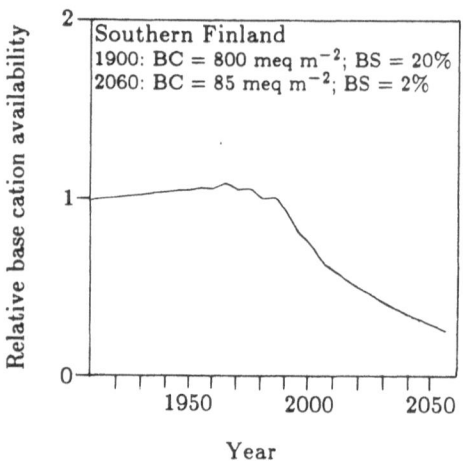
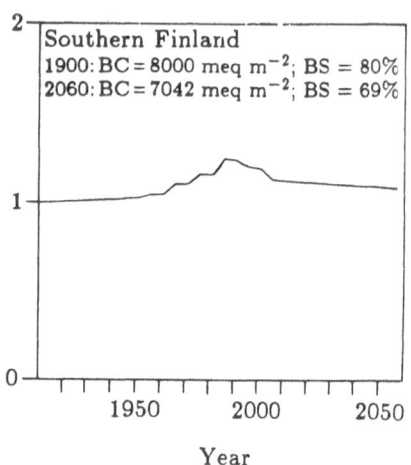

Figure 3.5. Temporal development of the relative amount of nutrients available for up-take, compared with the amount in 1900. The graph on the left shows the simulation results for a moderately rich soil (initial value 2), and that on the right those for a rich soil (initial value 1).

The changes in the status of soil are converted to forest productivity using cation availability, which is calculated as the weighted mean of the amount of cations in soil solution and on soil particles (*Figure 3.5*). It declines rapidly after the onset of acid load on poor sites in the south. The changes are smaller, and the response time longer, on richer sites and on sites further north.

The growth-decreasing effect of the declining nutrient availability determines the impact on the poorer sites, whereas the growth-increasing effect of the rising CO_2 concentrations predominates on the richer sites (*Figure 3.6*). The impact of the changing environment on regional forest growth is obtained as an areally weighted sum of the impact on the growth in the four soil fertility classes in each region (*Figure 3.7*).

When changes in the status of the environment are fed into the forest growth model, an estimate for the changes in forest productivity is obtained for each region and each fertility class. The increasing effect of the concentration of CO_2 is dominating in the north and on rich sites in the south. The effect of decreasing cation availability determines the response pattern on poor sites in the south.

Let A_{jk} denote the area of fertility class k in the subregion j, $V_{jk}(t)$ the mean annual forest production per hectare and $g_{jk}(t)$ the total impact of the environment on the growth. The estimate of regional forest growth, g, is obtained as the weighted mean of the growth in the four fertility classes in the three subregions:

$$g(t) = \frac{\sum\limits_{j=1}^{3}\left\{\sum\limits_{k=1}^{4}\left[A_{jk}\,V_{jk}\,g_{jk}(t)\right]\right\}}{\sum\limits_{j=1}^{3}\left[\sum\limits_{k=1}^{4}\left(A_{jk}\,V_{jk}\right)\right]}. \tag{3.9}$$

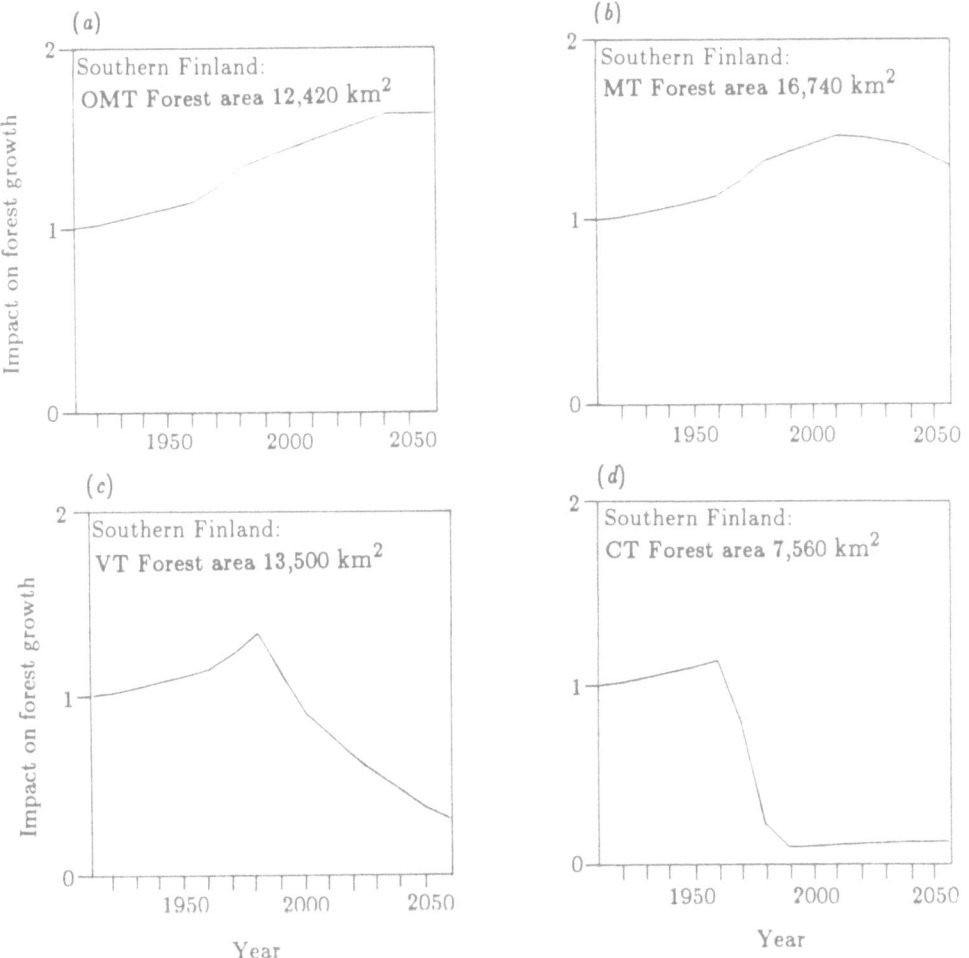

Figure 3.6. The temporal development of the impact on growth in the four different soil fertility classes: (a) rich, (b) moderately rich, (c) moderately poor, and (d) poor.

The increasing and decreasing effects compensate for each other on the national level, and the impact on growth is slightly positive in the 1980s.

3.7. Concluding Remarks

There are no time series of monitored impacts of a changing environment on forest growth that could supply the necessary information to draw conclusions regarding future development. Therefore, predictions have to rely on well-formed theoretical models.

The analysis of changes in the flow rates and concentrations of nutrients and toxic compounds, in the system formed by the atmosphere, forest soil, trees,

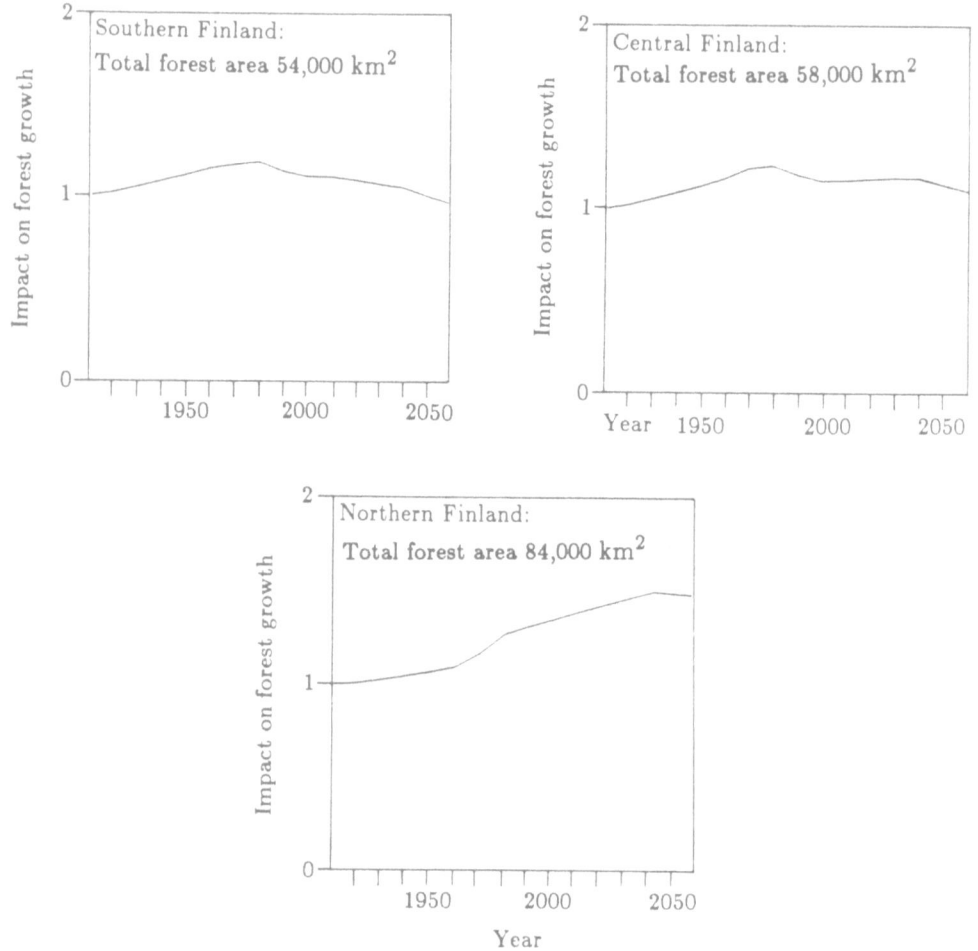

Figure 3.7. The temporal development of the impact on growth in each region.

and groundwater, forms a sound basis to construct models. In addition, this theoretical approach determines the structure of submodels to be applied. Thus, the analysis of changes in these flows has a practical research benefit.

   More research is needed on the nature of the response of forest growth to cation nutrient availability and nitrogen saturation. The model presented here forms an operational framework for the analysis of the impact of changing mass fluxes into the forest ecosystem. The framework is being refined iteratively, as more empirical and theoretical knowledge is gathered concerning the effect of nutrient availability on forest growth.

   The applied method requires rather specific data, which are not always available. The flow of external cations into the forest is weakly known. All the measurements available include cations circulating in the vegetation. This gives rise to a biased estimate of external cation input. The model is sensitive to the ratio of hydrogen ions to base cations in deposition. This ratio is probably higher in remote and mountainous areas than near large emittors.

A major change in environmental factors and in forest growth is taking place. The symptoms observed are the first signs of these significant change processes.

PART II

Aquatic Effects

CHAPTER 4

Mapping Groundwater Sensitivity to Acidification in Europe

M. Holmberg, J. Johnston, and L. Maxe

4.1. Introduction

Acid deposition alters the properties of topsoil and the composition of the fluxes in forested soil–water systems. Recent changes in the chemical properties of soils and surface waters in small watersheds have been reported in northern Europe and in the northeastern United States (e.g., Likens *et al.*, 1977; Wright *et al.*, 1980b). Nutrient cycling in soil appears to be disturbed, and mobilization of metals from mineral structure seems to accelerate in certain areas owing to acidic precipitation (e.g., Norton *et al.*, 1980; Ulrich, 1983b). Ultimately, these changes may affect the chemical composition of groundwater.

Historical records of groundwater quality are scarce. However, recent monitoring of groundwater chemistry has uncovered changes occurring in some shallow, noncalcareous, sandy aquifers. An increased content of leached base cations and sulfate, accompanied by a decrease in bicarbonate, has been documented by Jacks *et al.* (1984) in parts of Sweden, by Soveri (1985) in Finland, and by Wieting (1986) in the Federal Republic of Germany.

Groundwater is an important source of Europe's water supply. The proportion of drinking water extracted from groundwater ranges from 30% to 70% in most of Europe. Denmark, which uses only groundwater, and Norway, where less than 10% of the waterworks use groundwater, form exceptions. A high sulfate-to-bicarbonate ratio in groundwater extracted from private wells causes the corrosion of supply pipes, which becomes an economic issue in regions with many private wells. Acidic drinking water may also dissolve copper from the pipes.

J. Kämäri (ed.), Impact Models to Assess Regional Acidification, 51–64.

At the International Institute for Applied Systems Analysis (IIASA), a methodology has been developed for evaluating the sensitivity of European aquifers under forested soil to the deposition of acidifying compounds originating from transboundary atmospheric pollutants. This chapter presents maps of aquifer sensitivity and risk resulting from the application of this methodology on a European regional scale.

4.2. Impact of Acid Decomposition on Aquifer Systems

The impact of acid deposition on the aquifer is determined by the neutralizing properties of the overlying soil and that of the aquifer itself. Cation exchange and mineral weathering are the primary transformation processes that contribute to the neutralization of acid deposition. Hydrogen ions in the matric water participate in cation exchange at the surface of the soil solids, thereby causing the leaching of base cations. The pool of exchangeable base cations is replenished by the weathering of minerals.

Ions in precipitation affect the matric water concentrations. The rates of the soil-forming processes, such as cation exchange, weathering, nutrient mineralization and assimilation, depend on the concentrations of ions in the matric water. If the atmospheric input of hydrogen ions changes over time, the rates of the soil processes change. In the short run (years), this will influence the concentrations of the matric water. In the long run (decades), the changing composition of the matric water is expected to induce changes in the properties of the solid phase of the soil and the composition of groundwater.

The concentrations of elements in groundwater depend on the chemical composition of the recharge from the unsaturated zone, on the mineral composition and the weathering rate in the saturated zone, and on the residence time of groundwater.

The composition of the recharge to groundwater is influenced by the residence time of the water in the unsaturated zone. The ion exchange reactions are almost instantaneous; but the longer the time available for contact between the matric water and the mineral surfaces, the higher the content of weathered cations (calcium, magnesium, potassium, and aluminum) in the recharge to groundwater. The residence time in the unsaturated zone increases with the depth and decreases with the permeability of the soil. Furthermore, the mineral composition of the soil and the organic matter content influence the recharge composition, through their impact on the rates of ion exchange and weathering.

The weathering rate and the chemical composition of the minerals in the saturated zone determine the alkalinity production rate within the aquifer. Calcareous bedrock is easily weathered, whereas silicate bedrock weathers slowly. In the weathering reactions of silicate minerals, base cations and aluminum ions are released and bicarbonate ions are formed. The rate of alkalinity production in noncarbonate soil material of granitic origin has been estimated in the range of 1 to 40 meq m^{-3} a^{-1} (Nilsson, 1986).

The residence time of the groundwater in reservoirs determines the time available for the weathering reactions. The residence time of groundwater is determined by the rates of recharge, discharge, and extraction, as well as by the size of the aquifer. The climatic regime influences the rate of recharge through the rate of precipitation and the variables of temperature and vegetation, which together affect the rate of evapotranspiration. If there is no surface runoff, the rate of recharge can be approximated by precipitation minus evapotranspiration. The rate of recharge increases also with the hydraulic conductivity of the soil, which depends on soil texture and water content. The rates of recharge and discharge, furthermore, depend on the relief of the region, on the physical location of the aquifer, and on whether it is confined or unconfined.

4.3. Assessing the Impact on a Regional Scale

Few countries in Europe monitor environmental response to diffuse pollution. Data from regional monitoring of several environmental variables would obviously be the logical starting point for any appropriate assessment of regional acidification. Efforts have been made to intensify the regional surveillance of air quality and forest health. Few networks, however, have been established to monitor soil and groundwater quality on a national level, much less on the regional level. Sweden is an exception; a national network for monitoring groundwater quality was established there in 1968.

The lack of data regarding present and past chemical characteristics of soils and aquifers in Europe complicates the assessment of future acidification. Despite the empirical shortcomings, an evaluation has to be made. There are basically two possible approaches.

First, the geochemical processes that determine the impact of acid deposition on soils and aquifers may be described in a structural simulation model. The mechanisms of the neutralizing processes are mostly well known, and the main modeling problems lie in the complexity of the groundwater system and in the quantification of the process rates. The structural models are expressed as differential equations, describing the change per unit time in the state variables, or as algebraic equations, describing equilibrium reactions. A structural model of soil acidification on a regional scale has been developed by Kauppi *et al.* (1986). Extending Kauppi's soil model or an alternative structural model of the neutralizing processes in the unsaturated zone to include the dynamics of the saturated zone would require a complex hydrological model to account for the seasonal and regional variations in European hydrology. In addition, quantitative estimates of elemental transport, release, and accumulation in the unsaturated zone below the rooting zone are not available on a regional scale.

The second approach to assessing the impact of acid deposition on aquifer systems is to evaluate the regional potential for acidification on the basis of sensitivity analysis. An ecological system responds to its environment by variations in certain state variables, incited by changes in some driving functions. Apart from the driving functions, certain physical system characteristics, or sensitivity indicators, determine the amplitude of the response and the response time of the system.

If a system reacts very slowly to the driving functions, i.e., if the system is highly inert, changes that occur in the state variables are only slowly reversible. In assessing the anthropogenic impact on ecological systems, both the sensitivity and the inertia of the systems should be taken into account. In the long run, a highly sensitive system is worse off if it is also very inert, whereas a quickly reacting system may represent the worst case in the short run. The best case with respect to environmental damage is an inert system with low sensitivity to changes in the driving functions.

4.4. Assessment Method

In indicator or sensitivity analysis, the basic geochemical and physical characteristics that influence the chemical behavior of soil, groundwater, and surface waters are lumped into a number of indicators. The indicators are discrete variables that correspond either to a classification of the original continuous physical characteristics, such as soil depth and texture, or to a derived entity, such as neutralizing capacity. An indicator reflects certain properties that contribute to the overall sensitivity of the system. The overall sensitivity is obtained by aggregating the individual indicators.

The aggregation may be done, for example, by computing the weighted average of the individual indicators. Jacks and Knutsson (1982) followed this approach in evaluating the sensitivity of Swedish soil and groundwater to acidification. The disadvantage of the linear model is that the determination of the constant weighting coefficients is difficult. The mineral weathering rate, for instance, is an exponential rather than a linear function of the chemical composition of the minerals.

The difficulty of constant weathering coefficients can be overcome by using piecewise linear functions, which allow for the coefficients to vary with the indicators. Piecewise linear functions can be implemented by two-dimensional matrices for the stepwise aggregation of pairs of indicators. A method for evaluating the sensitivity of European groundwater to acidification, using aggregation matrices, was developed at IIASA (Holmberg *et al.*, 1987). Carter and co-workers (1987) used a similar method for mapping the vulnerability of groundwater to pollution from agriculture.

4.5. Sensitivity Indicators

In our methodology, the neutralization capability is assessed on the basis of soil depth, texture, and base cation content; the size and the mineral composition of the aquifer; and the annual amount of water potentially available for recharge (*Figure 4.1*). The choice of these indicators was based on knowledge of the geochemical processes involved in neutralizing acid deposition and on the availability of data. The choice was also influenced by sensitivity studies conducted by Jacks *et al.* (1984), Aust (1983), and Edmunds and Kinniburgh (1986a). Data availability was a major consideration because of the regional scale of this model;

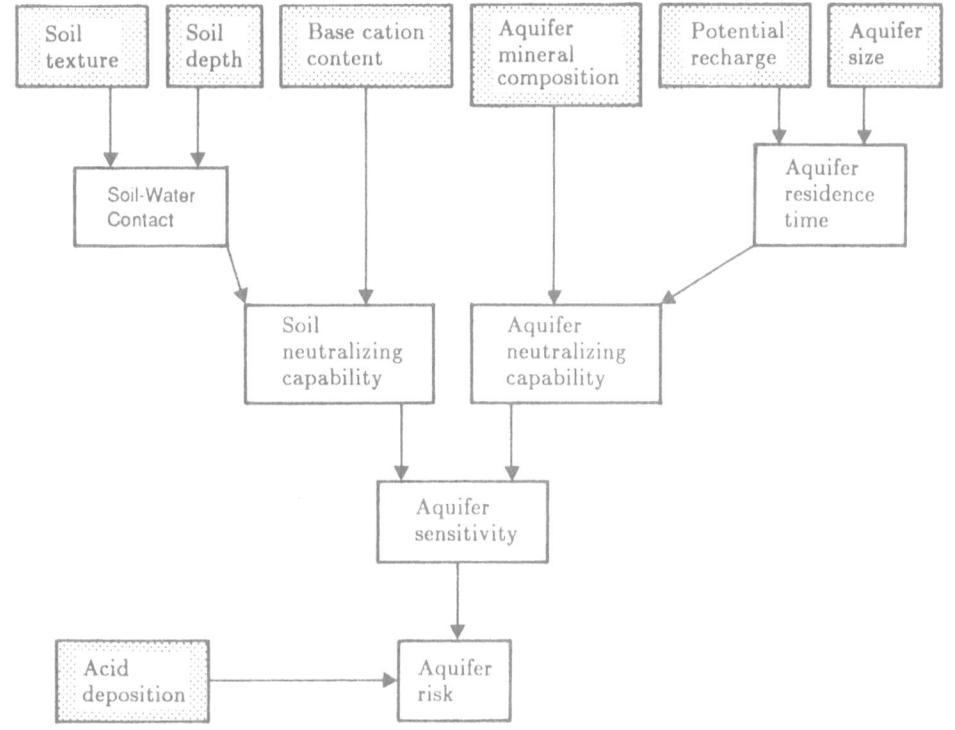

Figure 4.1. Aquifer sensitivity is assessed as the combined neutralizing capability of the aquifer and the overlying soil.

few sources of data are consistent over large areas, yet detailed enough for this purpose. Soil depth, texture, and base cation content were compiled from the Food and Agriculture Organization (FAO) Soil Map of the World (1974). Aquifer size and mineral composition were taken from the International Hydrological Map of Europe currently being compiled (UAH-UNESCO, 1970–1985).

The efficiency of neutralization by cation exchange in the soil depends upon the exchanging surface available for contact with the matric water. This factor is introduced through an indicator called soil–water contact, the value of which is determined on the basis of soil depth and soil texture. Soil texture is divided into five classes, ranging from coarse to fine. Soil depths are classified as either shallow or deep (less than or greater than 0.5 m, as defined in the soil types of the FAO soil map).

$$\text{soil–water content} = f_1 (\text{depth, texture}) \quad . \tag{4.1}$$

The resulting indicator soil–water contact, is assigned the rating poor (1), moderate (2), or good (3).

The total neutralizing capability of the soil is evaluated from the soil–water contact and the content of base cations in the top 0.5 m of soil. The base cation content was calculated from data on cation exchange capacity and percentage of base saturation for 80 European soil types. The indicator was divided into six classes: class 1 < 100 keq/ha, class 2 100–200, class 3 200–400, class 4 400–800, class 5 800–1,600, and class 6 > 1,600 keq/ha. We assume that the base cation content reflects the weatherability of the minerals in the unsaturated zone:

$$\text{soil-neutr. capability} = f_2 \text{ (soil–water contact, base cation content).} \qquad (4.2)$$

The neutralizing capability of the soil is assigned the rating poor (1), moderate (2), or good (3).

Rather than create a comprehensive hydrological model of Europe, potential annual recharge, defined as annual precipitation minus annual evapotranspiration, has been chosen to account for the climatic impact. The potential recharge is divided into three classes: (1) > 400 mm, (2) 400–100 mm, and (3) < 100 mm. The International Hydrological Map of Europe defines four aquifer productivity classes based on permeability and aquifer size. This makes the estimation of residence time difficult. High productivity results from high permeability in extensive aquifers. However, low productivity may be caused by low permeability, small discontinuous water-bearing formations, or both conditions.

For the purpose of assessing aquifer sensitivity, highly productive aquifers were assumed to possess long residence times because of their sizes and depths. Low or nonproductive aquifers were assigned shorter times because groundwater moves primarily through fractures in these aquifers. The main aquifer of an area, given by the hydrogeological map, is considered. This implies that different types of aquifers are encountered – open as well as confined, and rock aquifers as well as aquifers in unconsolidated material. The aquifers were classified into four size categories ranging from small (1) to extensive (4). The potential recharge combined with the aquifer size gives an estimate of the residence time:

$$\text{aquifer residence time} = f_3 \text{ (potential recharge, aquifer size).} \qquad (4.3)$$

The resulting residence time of the aquifer is classified in three classes from short (1) to long (3). The residence time is, on the other hand, also a measure of how easily any changes in the chemical composition may be reversed.

Aquifers have been classified into four mineral composition categories according to their mineral weathering rates. Class 1 contains acid or intermediate silicate rocks, sandstones, sands, gravels, and silts (i.e., no easily weatherable materials). Class 2 includes clays, claystones, slates, shales, graywackes, phyllites, and undefined silicate rocks. Class 3 contains the basic silicate rocks. Class 4 includes carbonate rocks such as limestone, dolomite, and marl (i.e., easily weatherable materials). The neutralizing capability of the aquifer was evaluated on the basis of the residence time of the groundwater and on the weatherability of the minerals in the aquifer:

aquifer-neutr. capability $= f_4$ (residence time, mineral composition).   (4.4)

Finally, the sensitivity of the aquifer is evaluated as the inverse of the combined neutralizing capability of the soil and the aquifer. The sensitivity may in turn be combined with an estimate of regional deposition to yield the risk of groundwater acidification. The resulting sensitivity and risk values are ranked into three classes, ranging from the rating 1 for low to 3 for high sensitivity or risk (*Figure 4.2*).

4.6. Sensitivity

In the computer implementation of this methodology, Europe is subdivided by a grid system with individual cells of 1.0 degree longitude by 0.5 degree latitude. Data for each of the 1,844 grid cells are passed through the assessment algorithm and assigned a sensitivity class.

The resulting sensitivity class assignments can be mapped as shown in *Figure 4.3*. This map is based on the indicator data for the dominant (largest area) soil type and aquifer in each grid cell. It contains 528 grid cells of class 1 (low), 815 cells of class 2 (medium), and 501 cells of class 3 (high) sensitivity. As can be seen, nearly all of the aquifers in the Nordic countries are categorized as highly sensitive. Other regions of high sensitivity include northern Scotland, northwestern Spain, and parts of Central Europe.

4.7. Risk

This process can be carried one step further to assess aquifer risk – a combination of aquifer sensitivity and acid deposition. Regions with little ability to resist acidification (highly sensitive) and exposed to the highest degree of deposition carry the greatest risk of change. Conversely, those areas with large buffering capabilities and/or not subject to significant acid deposition carry little risk of change.

A sample risk map is shown in *Figure 4.4*. This map results from combining the sensitivity map in *Figure 4.3* with the estimated 1980 sulfur deposition pattern (Alcamo *et al.*, 1985). The deposition has been divided into three classes for this purpose (0–1, 1–5, and > 5 g $m^{-2}a^{-1}$). As can be seen, the majority of the aquifers at risk lie in Central Europe, the Nordic countries, the United Kingdom, and northern Italy. For some parts of the Continent, the Hydrogeological Map of Europe is not yet available. These areas are left unshaded in the maps.

In interpreting these maps, it is important to realize that intensive fertilization in some agricultural areas contributes more to groundwater acidification than transboundary atmospheric pollutants. For this reason, the results of this method are strictly applicable only to uncultivated areas, such as forests. Consequently, soil-neutralizing capability was evaluated on the basis of geochemical data for forest soils. Nevertheless, the same technique could be expanded to

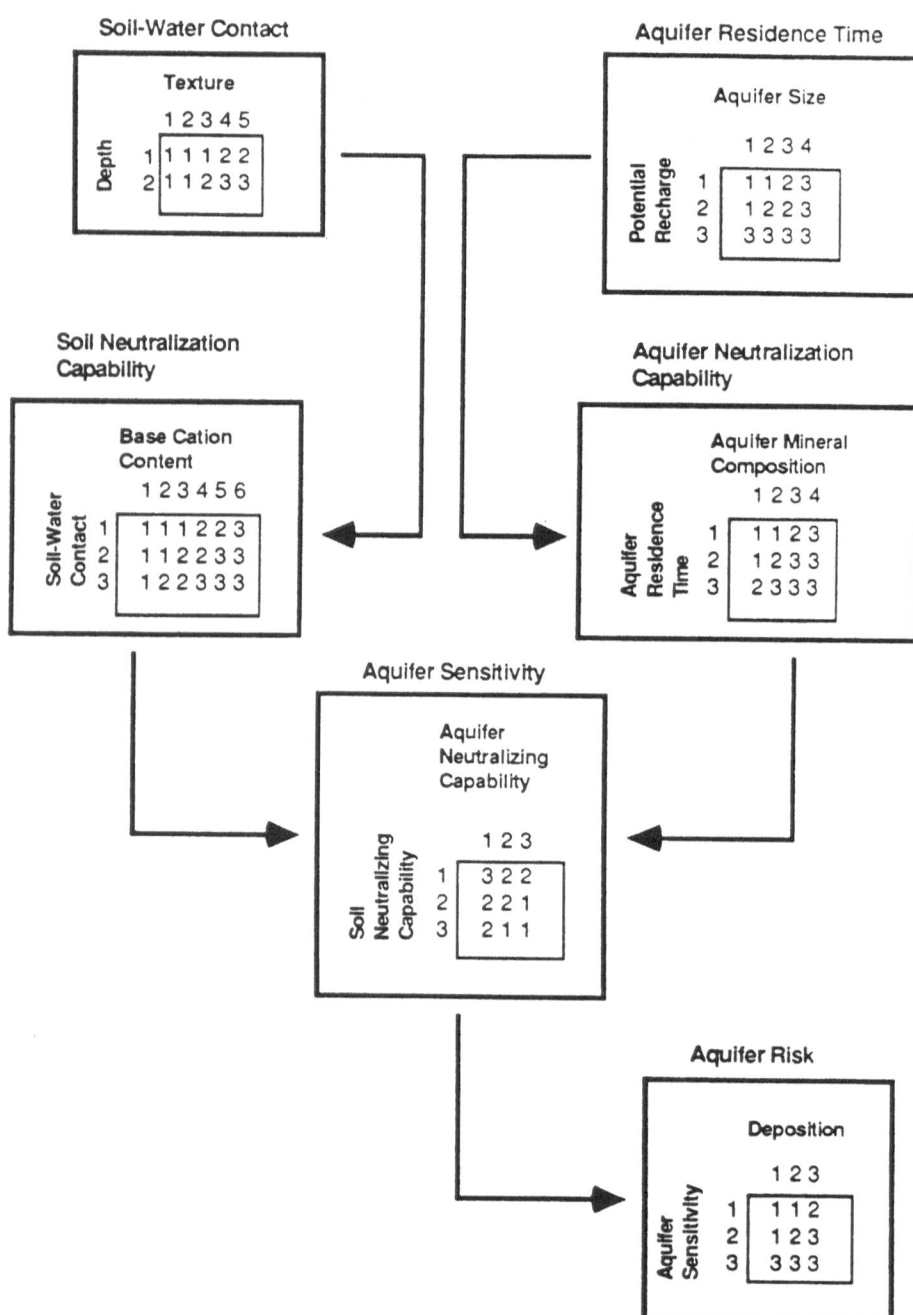

Figure 4.2. Aquifer sensitivity is assessed using combination matrices. The rating 1 stands for low and 3 for high neutralizing capability, sensitivity, or risk.

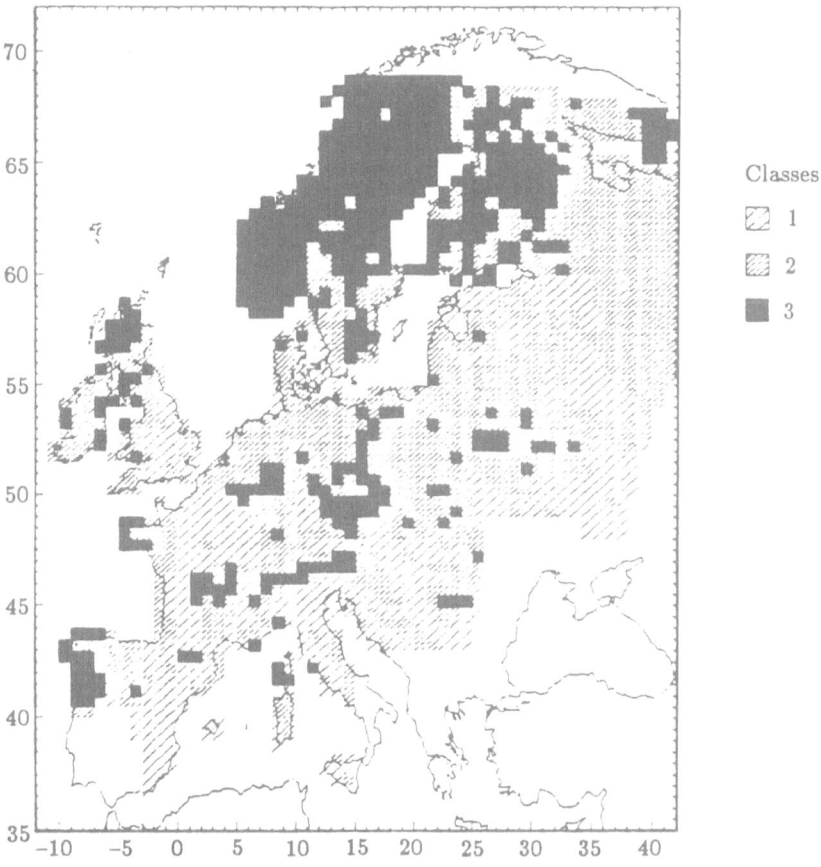

Figure 4.3. Aquifer sensitivity to acidification. The lightest shading stands for low sensitivity, while the darkest stands for high sensitivity, evaluated from the areally dominant indicators.

include agricultural soils. In that case, the risk would be evaluated by combining aquifer sensitivity with indicators representing land use and atmospheric deposition.

4.8. Uncertainty

Examination of the assessment technique and output variations under different input conditions suggests three sources of uncertainty: the input data, the function tables, and the spatial resolution of the data. There is no denying that the realities of regional environmental databases have forced some compromise in how the input data are included in the assessment technique. The indirect method for estimating aquifer size has already been described. Potential recharge may or may not be a good substitute for actual recharge because the

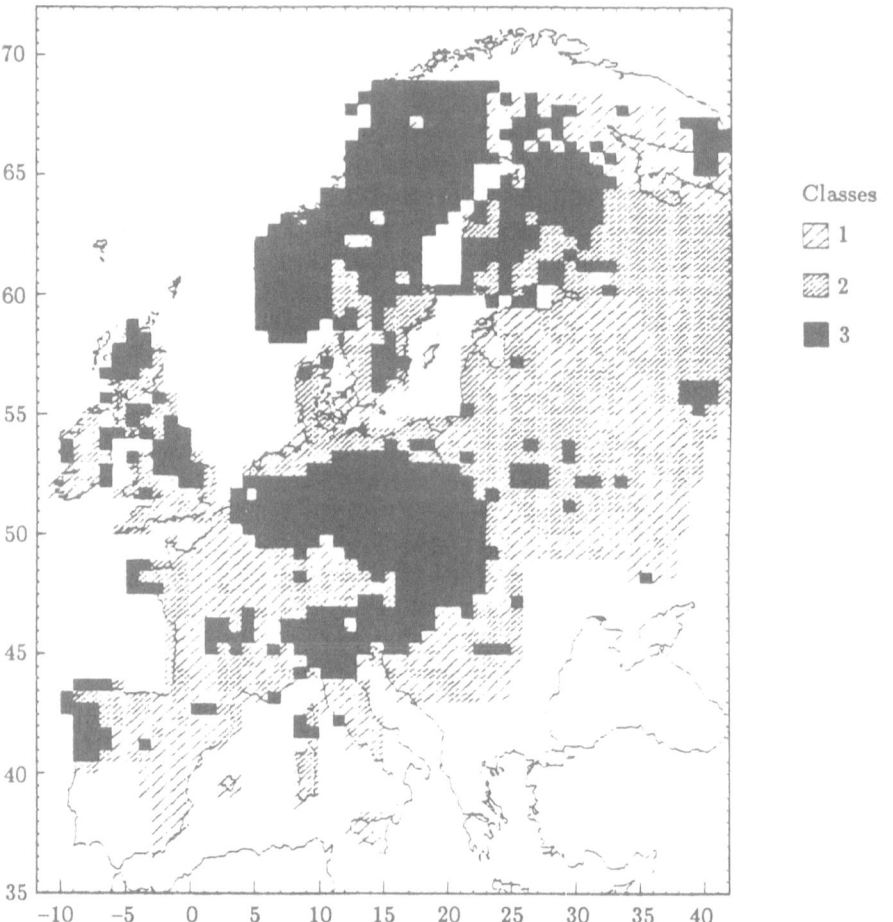

Figure 4.4. Risk of aquifer acidification. Light shading stands for low risk; dark shading represents high risk. The risk is evaluated as a combination of the sensitivity map in *Figure 4.3* and the sulfur deposition pattern for the year 1980 (Alcamo *et al.*, 1985).

recharge locations of many aquifers are unknown and may be some distance from the main body of the aquifer itself. Again, for lack of data, the neutralization potential of the soil and rocks between the root zone and the phreatic surface has not been incorporated into the methodology. As presently constructed, the methodology assumes no neutralization from this source. At present, it is not clear how to estimate the uncertainty arising from these factors.

A factor that is easier to evaluate is whether the data have been correctly divided into classes for use by the combinational algorithm. There is some judgment involved in this process with the attendant possibility of error. For any particular grid cell, changing the class assignment of one indicator may or may not cause a change in the aquifer sensitivity, depending on the values of the other indicators. For instance, if all the other indicator values are equally likely, the probability that the sensitivity will be changed by a 1-unit error in the potential recharge indicator is only 13%. In fact, because all other indicator values are not equally likely, but are determined by the data set, the probability of changing the sensitivity is actually about 25%. This value is the maximum probability of change associated with any indicator. The probability associated with aquifer mineral composition is also 25%, followed in decreasing order by base cation content (20%), aquifer size (18%), soil depth (13%), and soil texture (9%). In contrast, the probability of changing the risk assessment by changing the deposition class of a grid cell is 69%.

The second source of uncertainty is the choice of ratings used in the matrices, reflecting the coefficients of piecewise linear functions. These are supplied by the model user, and bias may be unintentionally introduced. Test runs show that that results are relatively unaffected by reasonable changes in every functional table except the sensitivity and risk tables. The position of the table in the algorithm determines its importance to the final result (i.e., those tables closest to the sensitivity table exert greater influence on the final result than tables located earlier in the flowchart, *Figure 4.2*).

The third source of uncertainty is the spatial resolution of the data. The individual grid cells are larger than some essential map features, such as individual soil types and aquifers. As a result, up to seven soil types and six aquifers per grid are included in the input data set. Unfortunately, because these data were taken from different sources, they are not related spatially within the grid cell. Given these multiple readings, which soil and aquifer data should be used to represent the grid cell?

Several approaches have been tried. One is to use the dominant soil type and aquifer in each grid and ignore the rest of the data. This is the approach used in creating *Figures 4.3* and *4.4*. Another strategy is to combine systematically the soil types and aquifers within each grid cell, calculate the sensitivity, and choose the best or worst cases (lowest or highest sensitivity values) subject to a minimum area criterion to weed out very small, and possibly anomalous, aquifers and soils. Examples of best-case and worst-case maps are shown in *Figures 4.5* and *4.6*. These maps, based on the tables shown in *Figure 4.2*, consider only those soils and aquifers that cover at least 15% of the grid cell. The differences among the three cases can be seen if the average sensitivities and risks are compared. The average sensitivities for best, dominant, and worst cases are 1.62, 1.93, and 2.23, respectively. The average risks are 1.79, 2.13, and 2.41 for the best-, dominant-, and worst-case assessments.

A related spatial resolution problem is associated with the recharge and deposition data. Choosing a single value for each of these indicators to represent an area of 0.5 degrees latitude by 1.0 degrees longitude is a very coarse simplification, considering the variability of European climates, especially in mountainous regions.

Finally, there is the problem of method verification. To date, the results of these analyses have not been validated by comparison with field data, since such data are scarce. Point observations (Jacks *et al.*, 1984; Soveri, 1985; and Wieting, 1986) do not contradict the assessments presented here. However, future research efforts need to focus on exploring these uncertainties and developing

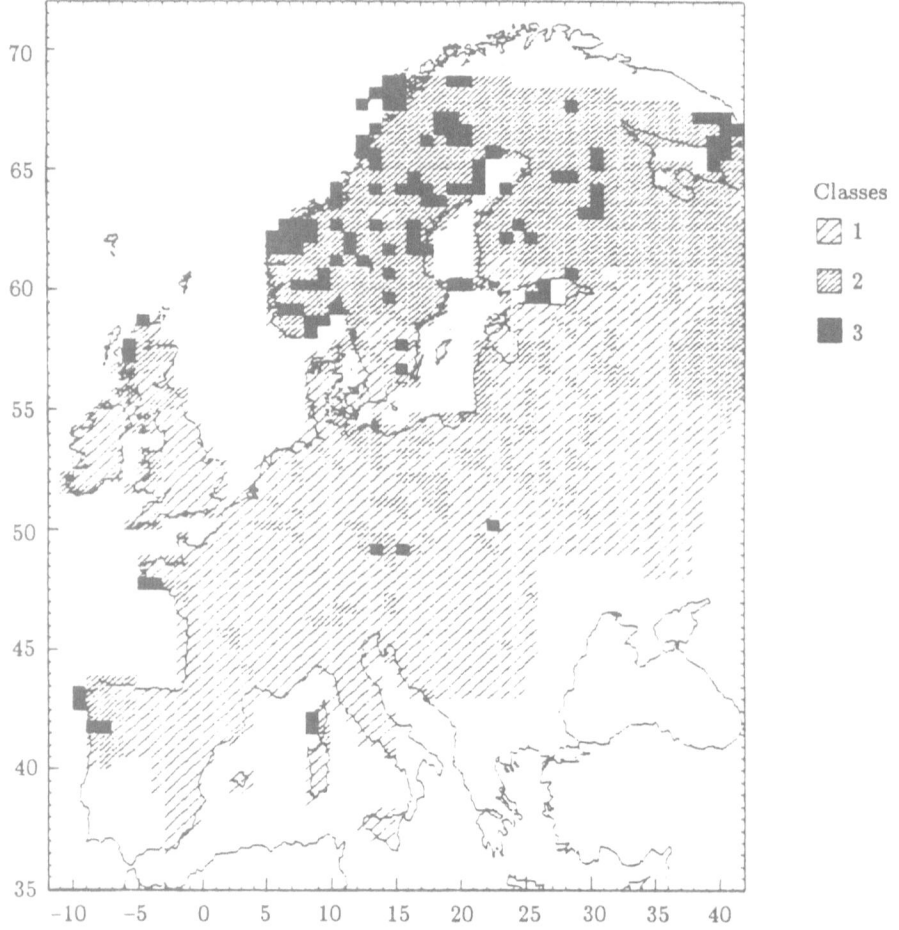

Figure 4.5. Best-case analysis of aquifer sensitivity. Within each grid cell, only those indicators are considered that combine to produce the least sensitive result.

confidence that the results of this technique are an accurate representation of reality.

4.9. Uses

This technique is not intended to predict future groundwater quality in response to continued acid inputs. The maps reflect only the sensitivities and risks of grid cells relative to each other. Nevertheless, this procedure can be used as a screening tool to locate those aquifers that are liable to experience changes first. In this way, the sensitivity and risk maps can aid in planning soil and groundwater monitoring networks. They may also guide decisions on where more detailed site investigations are warranted.

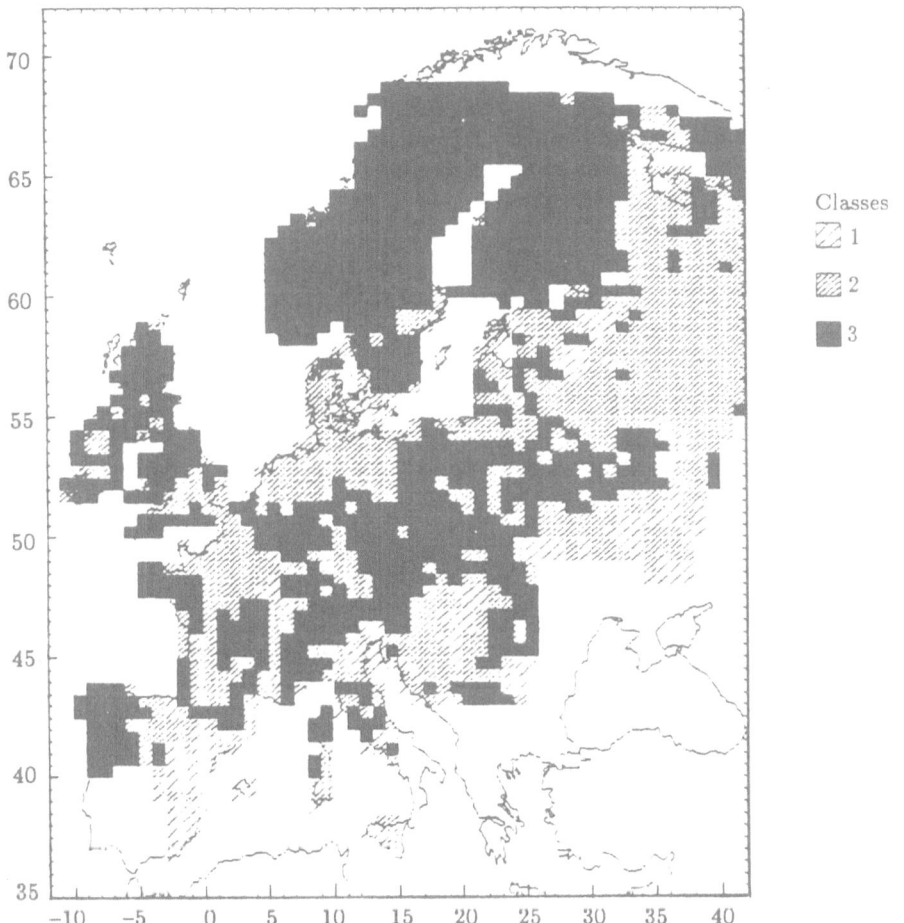

Figure 4.6. Worst-case analysis of aquifer sensitivity. Within each grid cell, only those indicators are considered that combine to produce the most sensitive result.

Although the risk assessment is ultimately based on a static database, it may be possible to use it within an overall framework of providing feedback on the future environmental effects of different air pollution control strategies. By using cumulative or time-averaged acid inputs resulting from different control strategies, maps could be produced showing the changes in aquifer risk compared with some base year.

The choice of spatial resolution strategy would depend upon the use intended for the maps. For example, planning monitoring networks may be best served by using worst-case data, while evaluating general environmental degradation in response to acid deposition might be better served by using the dominant-case calculations.

4.10. Summary

This chapter has presented an analytical technique for estimating the geochemical sensitivity of aquifers and the associated risk that groundwater quality changes may occur as a result of atmospheric acid deposition. European-scale maps prepared by this technique are also presented. Uncertainties in the resulting sensitivity and risk assessments are connected with the compilation of the sensitivity indicators, with the construction of the function matrices, and with the spatial resolution of the data. Nevertheless, the results of this technique can aid in the design of groundwater monitoring networks, and may contribute to the preparation of damage assessments associated with different air pollution control strategies.

CHAPTER 5

Modeling the Impacts of Acidification on Fisheries in Ontario, Canada

C.K. Minns

5.1. Introduction

Initial attempts to assess the regional impact of acidic deposition on fisheries resources showed that many elements were either poorly developed or entirely missing (Minns *et al.*, 1986). The Canadian Department of Fisheries and Oceans (DFO) and the Ontario Ministry of Natural Resources (OMNR) both held adaptive environmental assessment and management workshops (Holling, 1978), which provided much guidance for the evolution of programs to identify mechanisms of acidification effects on fisheries and to predict regional-scale impacts. Since then we, at DFO, have iteratively developed a regional impact model to predict impacts and refine information needs. At each stage, we have taken advantage of the newest data sources and any conceptual gains.

The purposes of this chapter are to describe the development and structure of the modeling framework, to outline in brief the modeling of the watershed and lake components, to present some methods for assessing fisheries impacts on a regional scale, to report some of the fisheries results obtained using an integrated regional model, and to offer an assessment of the modeling process and the results.

5.2. Modeling Framework

Canadian modeling studies have a framework covering the steps from emissions to deposition via long-range transport and transformation, from deposition to its effect on lake chemistry via lake watershed biogeochemical processes, from lake

J. Kämäri (ed.), Impact Models to Assess Regional Acidification, 65–81.
© 1990 *International Institute for Applied Systems Analysis.*

chemistry to its effect on fish via toxicity and changes in populations, and the relation between fish and man via socioeconomic effects (*Figure 5.1*). In my modeling, I have concentrated on the deposition through fish linkages, relying on colleagues to deal with the other components.

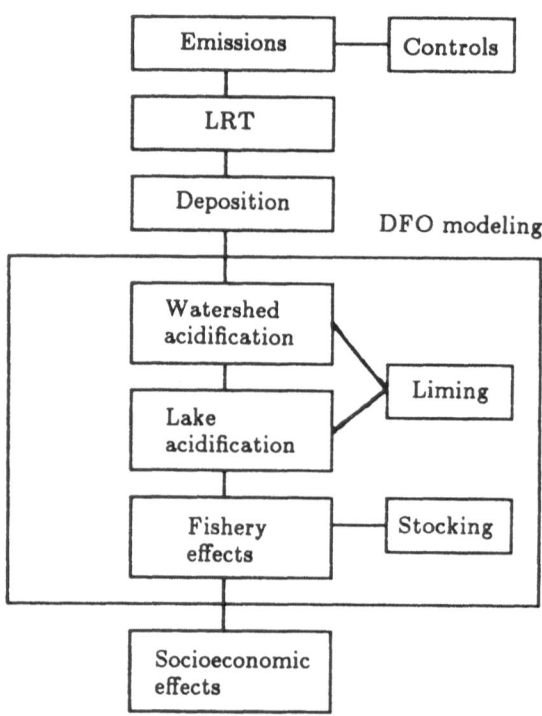

Figure 5.1. A schematic diagram showing the overall framework for modeling the impacts of sulfur emissions on fishery resources.

My modeling philosophy stresses simplicity. Certainly, more complex models of lake acidification have been formulated, e.g., ILWAS and MAGIC (cf. Reuss *et al.*, 1986), but I am constrained by the objective of formulating regional predictions of impact. Models that require large quantities of data and extensive conceptual constructs are limited for the most part to single-site applications. My objective is to capture the essential features of the acidification process, but to minimize the data requirements for individual sites. Thus, I have emphasized the use of regional survey data for lake chemistry and fish as a reference point for model construction.

Many colleagues with the Atmospheric Environment Service–DOE (Department of the Environment) and other agencies are involved in extensive efforts to model the emission–deposition link. Below, I have used a published emission–deposition transfer matrix (Olson *et al.*, 1983) to link changes in regional deposition fields to emission control strategies.

Modeling the fish–man linkage is in its infancy. Most modeling approaches are based on estimates of use cost or willingness to pay rather than the "existence" value of unexploited fish. Market-value assessments of recreational fishing suggest that the costs of impact are marginal compared with emission control costs (Minns and Kelso, 1986). DFO has conducted modeling studies (unpublished) to estimate the economic and employment impact of lake and river acidification in eastern Canada.

5.3. Regional Impact Modeling

5.3.1. The lake model

DFO has used a model derived from those developed by Henriksen (1979, 1980), Wright (1983), and Thompson (1982). The model assumes that a lake's alkalinity ($[\text{Alk}]$) is determined by the difference of the sum of base cation ($C_B = Ca + Mg + K + Na$) concentrations and the sum of sulfate (SO_4) and chloride $[\text{Cl}]$ concentrations. Nitrate and ammonia are assumed to balance one another and so do not contribute to the acidification process. Steady state conditions are estimated for a given level of acidic deposition.

After Henriksen (1980), the eventual steady state lake alkalinity ($[\text{Alk}]_\infty$) is the sum of original alkalinity ($[\text{Alk}]_0$) and the change in lake alkalinity owing to acid deposition ($\delta[\text{Alk}]$):

$$[\text{Alk}]_\infty = [\text{Alk}]_0 - \delta[\text{Alk}] \quad . \tag{5.1}$$

The original alkalinity is estimated from current lake conditions adjusted by the expected increase in base cations resulting from cation exchange in the watershed. A portion (F_W) of the acidic deposition falling on the watershed is presumed to be exchanged for base cations, thereby decreasing the acidity of the runoff. Since the increase in lake sulfate from its original state ($[SO_4]_0$) to its current state ($[SO_4]_t$) is assumed to be proportional to acidic deposition, that increase is proportional to the increase in cations corrected for the portion of deposition falling on the watershed multiplied by the F_W factor. This factor is similar to the F-factor by Henriksen (1982a). Chloride levels are assumed to be unchanged:

$$[\text{Alk}]_0 = [C_B]_t - F_W \cdot \frac{r}{r+1} \cdot ([SO_4]_t - [SO_4]_0) - [SO_4]_0 - [\text{Cl}] \tag{5.2}$$

where $[C_B]_t$ is the current sum of base cations concentrations (eq m^{-3}); $[SO_4]_t$ and $[SO_4]_0$ are the current and original sulfates concentrations (eq m^{-3}); $[\text{Cl}]$ is the chloride concentrations (eq m^{-3}); r is the ratio of watershed area to lake area; and F_W is the proportion of hydrogen ion falling on the watershed that is neutralized.

The ratio $r/(r+1)$ apportions the increase in sulfate due to acidic deposition between that applied to the watershed and that applied directly to the lake where F_W is inoperative. We have further assumed that F_W is a function of original alkalinity such that F_W is zero at zero original alkalinity or less, F_W rises linearly to 1 when alkalinity reaches 200 μeq l^{-1}, and remains at 1 for all higher alkalinities. I have also assumed that the background sulfate concentration in lakes is equivalent to the sum of base ions in deposition divided by runoff.

The change in lake alkalinity ($\delta[\text{Alk}]$) is assumed to be a function of the acidic deposition corrected for F_W and for alkalinity production linked to sulfate reduction in the lake:

$$\delta[\text{Alk}] = \frac{D_\infty}{P \cdot WR} \cdot (1 - F_W \cdot \frac{r}{r+1}) \cdot (1 - F_L) \qquad (5.3)$$

where D_∞ is the annual acid deposition (eq m^{-2} yr^{-1}); P is the mean annual precipitation (m yr^{-1}); WR is the ratio of runoff to precipitation; and F_L is the proportion of acid neutralized due to sulfate reduction.

Acid deposition, D_∞, is assumed to be equivalent to total sulfate deposition, wet plus dry, less background sulfate deposition due to marine and dust sources. We assumed that background sulfate is estimated by the sum of base cations in deposition. Some of the sulfate in deposition is derived from soils or sea spray and must be related to levels of base cations in deposition (cf. Brydges and Summers, personal communication).

The sulfate reduction parameter (F_L) is estimated using the models described by Kelly et al. (1987) and Baker et al. (1986):

$$F_L = S_S/[P \cdot WR \cdot (r+1) + S_S] \qquad (5.4)$$

where S_S is the sulfate mass transfer coefficient (0.46 m yr^{-1}). F_L is analogous to the retention coefficient in lake phosphorus models where sedimentation, or in this case sulfate reduction and removal to the sediments, competes with removal by flushing (Minns, 1986b). In-lake alkalinity production is only important in lakes with a low r ratio.

The complete derivation of the lake model is described in Jones et al. (1987). They found that the model's predictions are not overly sensitive to wide variation in the $F_W - [\text{Alk}]_0$ relationship, but very sensitive to wide variations in the level of background sulfate in deposition and in lakes. Neither of these model elements is well described as yet in results of lake acidification research (Reuss et al., 1986).

The lake alkalinity model is embedded in a regional model (Jones et al., 1989). We have included all secondary watersheds south of latitude 52° and east of the Ontario–Manitoba border, the region of Canada receiving the most acid deposition (Minns and Kelso, 1986). Available resource inventories are used to obtain the size–frequency distributions of lakes in each watershed along with the associated distributions of r, e.g., Minns (1984) for Ontario. Mean precipitation and runoff values are obtained from hydrological atlases. All available water chemistry data, such as those described for eastern Canada by Jefferies et al. (1986), are used to develop frequency distributions of $[C_B]_t$, $[SO_4]_t$, and $[Cl]$ from which $[Alk]_0$ distributions can be calculated.

Consequently, given the assumption that F_W reaches a value of 1 for $[Alk]_0 = 200$ μeq l^{-1}, and the total sulfate deposition for a secondary watershed, the steady state number- and area-weighted distributions of $[Alk]_\infty$ can be estimated. Using a simple empirical pH–alkalinity relationship based on carbonate–bicarbonate formulations, pH distributions can also be derived (Jones et al., 1989).

For the purpose of this chapter, I have used the model to predict the alkalinity and pH distributions for all the lakes in watersheds in the province of Ontario south of latitude 52°N. The range of scenarios includes:

(a) Original conditions with no acid deposition.
(b) Baseline 1980 deposition values.
(c) Canadian emissions controls only [42% reduction from (b)].
(d) Canadian and US controls [42% reduction from (b)].
(e) Worse case with increase of 10% over (b) (see *Table 5.1*).

Table 5.1. Estimated sulfur emissions (kt S yr^{-1}) for the model deposition scenarios.

Scenario	Canada	USA
(a) Original	0	0
(b) 1980 baseline	5,306	24,023
(c) Canada only	3,409	24,023
(d) Canada + USA	3,409	13,933
(e) Worse 10%	5,837	26,425

(Source: US–Canada MOI, Working Group 3B Report.)

The 1980 baseline deposition fields were provided by the Atmospheric Environment Service (AES)–DOE (Summers, personal communication). Using the emission-to-deposition transfer matrices developed for AES by Olson et al. (1983), 1980 emissions were used to predict expected 1980 deposition. Differences between observed and predicted 1980 deposition were used empirically as corrections for predicted deposition fields, given a 42% reduction in Canadian emissions [scenario (c)], a 42% reduction in Canadian and US emissions [scenario (d)], and a 10% increase in all emissions [scenario (e)].

A 42% reduction in Canada is required for implementation of the existing Canadian federal–provincial agreement to control sulfur emissions. In scenario (d), I have assumed that the USA implements 42% reduction in all source areas matching the Canadian strategy. A worse case of a 10% increase over 1980 levels was arbitrarily selected to illustrate the potential for further damage had not some emission reduction already taken place both in Canada and the USA.

Under different deposition scenarios, the proportion of Ontario lakes with alkalinities less than zero ranges from 0.0–9.4% by number and 0.0–7.3% by area

Table 5.2. Predicted percentage of Ontario's inland lakes with alkalinities below 0, 60, and 120 μeq l^{-1}, by number and area of lakes, for different deposition scenarios. Estimated total number of lakes = 137,415.

Scenario	Number			Area		
	< 0	< 60	< 120	< 0	< 60	< 120
(a) Original	0.0	1.0	14.5	0.0	0.7	10.9
(b) 1980 baseline	7.1	16.0	27.2	5.5	11.8	21.4
(c) Canada only	4.7	14.0	25.4	3.5	10.2	19.8
(d) Canada + USA	0.9	8.5	22.0	0.7	6.3	16.0
(e) Worse 10%	9.4	18.0	27.8	7.3	13.1	22.1

(Table 5.2). The number and area percentages are different because the size distribution of lakes varies across Ontario. In the 1980 baseline deposition scenario, the model predicts that 7.1%, or nearly 10,000, of Ontario's lake have an alkalinity less than zero (Table 5.2) out of an estimated total of 137,415 lakes (Jones et al., 1987). The Canada only emission control strategy produces a 34% recovery, while a Canada–US strategy produces an 87% recovery. A 10% increase in emissions is predicted to cause a 32% increase in the number of acidic lakes. The emission control strategies do not protect all lakes from the risk of acidification. With the Canada–US strategy, 8.5% by number and 6.3% by area of lakes would have alkalinities less than 60 μeq l^{-1} and would be prone to episodic depression of alkalinity.

Table 5.3. Predicted percentage of Ontario's inland lakes with pH below 5.0, 5.6, and 6.0, by number and area, for different deposition scenarios.

Scenario	Number			Area		
	< 5.0	< 5.6	< 6.0	< 5.0	< 5.6	< 6.0
(a) Original	0.0	0.0	0.0	0.0	0.0	0.0
(b) 1980 baseline	6.3	8.6	13.0	4.8	6.8	9.5
(c) Canada only	3.8	5.3	6.2	2.7	3.8	4.7
(d) Canada + USA	0.7	1.2	3.4	0.5	1.0	2.6
(e) Worse 10%	7.5	11.3	14.0	5.9	8.6	10.2

Conventionally, acid lakes have been considered to be those with pH less than 5. The model predicts that 6.3% (number) or 4.8% (area) of lakes would be in that state, given 1980 deposition levels (Table 5.3). This is equivalent to 8,600 lakes or 290,000 ha. This is greater than previously predicted by Minns (1986a) and Minns and Kelso (1986). Minns (1986a) estimated there are already 1,400 acid lakes in Ontario, while Minns and Kelso (1986) estimated there were 2,000. Thus, the regional model predicts that 6,000–7,000 lakes will be acidified. The difference in the prediction could arise as a result of sulfate levels in lakes predicted from a sulfate deposition field being greater than levels observed in the survey data sets. Current lake sulfate levels may not be at steady state owing to changes in deposition and the flushing time, while the model predicts the eventual steady state for a fixed level of sulfate deposition.

Changes in lake alkalinity and pH levels are predicted for all lakes with original alkalinities up to 200 μeq l^{-1}. Decreases in alkalinity and pH can have a biological impact over a wide range of lake conditions. Lake acidification is more than just the most sensitive lakes becoming acid.

5.4. The Fisheries Models

Minns *et al.* (1986) have outlined several ways that lake fish might be impacted by acidic deposition. Here I have chosen three indices of response: species richness, presence–absence, and production. The three indices each emphasize different aspects of our concern for the fish resource. Species richness represents the total diversity and character of Canadian lake ecosystems. Presence–absence analyses allow us to highlight highly prized species and to identify important thresholds on a pH gradient. Finally, fish production supports the harvests desired by man, be it for sport or commerce. Taken together these indices provide some measure of the importance of freshwater fisheries resources to Canada.

5.4.1. Species richness

Since Barbour and Brown (1974) first showed a positive relationship between fish species richness and area in lakes, a number of authors have further studied the phenomenon, e.g., Mahon and Balon (1977) and Eadie *et al.* (1986). These results are consistent with the theory of island biogeography put forward by MacArthur and Wilson (1967). Indeed, several authors (Eadie *et al.*, 1986; Eadie and Keast, 1984; and Tonn and Magnuson, 1982) have provided evidence for a habitat diversity explanation of species-area curves as put forward by MacArthur and Wilson. In relation to acid deposition, others have shown that pH or related variables are important determinants of fish species richness in lakes (Pauwels and Haines, 1986; Rago and Wiener, 1986; Somers and Harvey, 1984; and Harvey, 1975).

To date, however, these species-area-pH models have not been used to assess the status of regional fisheries resources. I have used the lake inventory database of the Ontario Ministry of Natural Resources for earlier analyses of fish (Minns, 1986a). This database is very large; and most variables, e.g., lake area, lake depth, elevation, and pH, span a considerable range (*Table 5.4*). I developed a multiple regression model of fish species richness with lake area, mean depth, elevation, tertiary watershed richness (the number of species found in all surveys in each tertiary watershed), survey year, and pH as input variables (*Table 5.5*). Rago and Wiener (1986) do not recommend regression analysis as a method of describing species-area-pH effects because the input variables tend to be highly correlated. However in this case, the large sample size and relatively low correlations among the independent variables should allow the resulting model to be fairly reliable.

The input variables used in the model are much the same as those used by Pauwels and Haines (1986) and Eadie *et al.* (1986). The tertiary watershed

Table 5.4. Minimum, mean, and maximum values for selected variables in the Ontario lake inventory database up to 1981 ($N = 8,772$).

Variable	Units	Minimum	Mean	Maximum	ln-mean
Area	ha	0.1	349.5	79,771.6	4.086
Elevation	m	42.0	595.0	335.4	5.788
Tertiary richness[a]		1.0	40.6	57.0	3.669
Survey year[b]		68.0	73.6	81.0	
pH		3.8	7.4	10.0	1.990
Mean depth	m	0.1	20.1	48.8	1.416
Richness		0.0	6.2	34.0	

[a]Species richness of tertiary watersheds.
[b]Year of lake inventory, i.e., from 1968 to 1981.

Table 5.5. A multiple regression model of fish species richness in Ontario's lakes ($N = 7,949$; $R = 0.696$; $F_{regr}{}^* = 1,243.8$).

Variable	b	Standard error of b[a]	$F_{b=0}$[a]
ln (area)	1.255	0.019	4,303.6
ln (elevation)	−2.720	0.123	492.5
ln (3 richness)	2.898	0.124	545.8
Survey year	0.145	0.093	242.4
ln (pH)	4.735	0.312	230.0
ln (mean depth)	0.349	0.045	59.2
Intercept	−14.297	1.260	128.8

[a]Significant $P = 0.001$.

richness is a measure of the pool of species in the region of each lake. Elevation is the second most important variable and points to access as a determinant of richness. Lakes at higher elevations in watersheds are more difficult for many fish species to gain access to through rivers and streams. Survey year reflects the steadily improving quality of the lake inventory process. Early inventories were incomplete, while in recent years the range of methods used to sample fish has been expanded and the sampling effort intensified. The equation can be standardized to a recent survey year, but obviously a portion of total richness will be unaccounted for. Mean depth is strongly related to lake area [ln (Area) : ln (Mean depth), $r = 0.324$], but for a given area richness increases with depth – further evidence of the role of habitat diversity in determining richness. The importance of the pH variable is related to the fact that different species have different pH tolerances; but that in general, as pH decreases, the number of fish species that can survive decreases (Eilers *et al.*, 1984).

To apply the richness model (*Table 5.5*), I used the means of the ln (elevation) and ln (tertiary richness) variables (*Table 5.4*) to control for their effect. I fixed the survey year at 1980. I considered the range of lake areas from 1 to 100,000 ha. Following the approach of Kent and Wong (1982), I used my data (Minns, 1984) on the size–frequency distribution of Ontario lakes to fit a function

$$P(> Ao) = 9.5/(9.5 + Ao^{1.1706})$$

$$(r^2 = 0.9997, F_{\text{regr}} = 6716, \text{signif. } P = 0.0001)$$

(5.5)

describing the percentage of lakes with an area greater than a value Ao. Given that the total number of lakes in Ontario south of latitude 52° between 1 and 100,000 ha is estimated at 137,415 (Jones *et al.*, 1989), I can integrate to calculate the number of lakes in any size interval Ao to $Ao + \delta A$. I divided the ln interval for 1–100,000 ha into 20 equal portions and determined the ln mean depth in each portion using the lake inventory database. I then applied the number-weighted pH frequencies generated from the regional impact model's deposition scenarios and integrated the richness model over the lake area intervals to obtain an estimate of the total number of species–populations (a population of a single species in a single lake) in Ontario's lakes.

The aggregate species richness (the set of all species–populations in all lakes) varies with the levels of acid deposition (*Table 5.6*). The model predicts there would be nearly 600,000 species–populations of fish spread over 137,000 lakes (or 4.35 species per lake) in Ontario in the absence of acid deposition. A 4.3% decline is predicted for the 1980 base deposition scenario, amounting to nearly 26,000 species–populations. Emission control programs are predicted to produce recoveries of 35.5% and 71.4%, respectively, for Canada only and Canada plus USA. A 10% increase over 1980 deposition is predicted to cause a further 0.8% resource loss of species–populations compared with the original condition, or an 18.0% increase over losses predicted given 1980 deposition levels.

Table 5.6. Predicted number of fish species–populations in Ontario's lakes in relation to different levels of acid deposition.

Scenario	Species–populations	% decrease[a]	% increase[b]
(a) Original	598,204	–	–
(b) 1980 baseline	572,532	4.3	–
(c) Canada only	581,654	2.8	35.5
(d) Canada + USA	590,873	1.3	71.4
(e) Worse 10%	567,911	5.1	−18.0

[a]Percent decrease from (a).
[b]Percent recovery of loss from (a) to (b).

The percentage changes in the fish species–population are less than the percentage of lakes predicted to have alkalinities less than zero. This is because lakes sensitive to acidification are expected to have lower species richnesses at the outset than the average lake. Lakes with a higher original pH are expected to decrease in pH less than those with a lower original pH for a given level of deposition, and the former lakes are expected to have a greater species richness as well.

5.4.2. Presence–absence

Declines in overall species richness should be reflected in the disappearance of particular fish species at different points on the pH gradient. Mills and Schindler (1986) showed that fish, such as fathead minnows (*Pimephales promelas*) and common shiner (*Notropis cornutus*), are rarely found in lakes with pH < 5.8; while others, such as pearl dace (*Semotilus margarita*) and northern redbelly dace (*Phoxinus eos*), are generally not found in lakes with pH < 5.3. As threshold indicators, these presence–absence relationships can be used to show qualitatively the impact of acidification.

There are, however, numerous problems with using such patterns to produce quantitative regional impact assessments. The presence–absence patterns are not determined solely by pH. As with species richness, there are regional and local factors influencing the likelihood that a species colonizes a lake and persists, or avoids extinction. In practice, this means we are not trying to distinguish between probabilities of 0 and 1 for presence, but rather between 0 and some value less than 1 – in many cases considerably less than 1 (designated an encounter probability).

Wales and Beggs (1986) have presented an analysis of encounter probabilities in relation to pH for a set of fish species found in Ontario lakes. In particular, they have examined six species that are widely distributed in lakes with pH < 6.4 and six species that are poorly distributed in lakes with pH < 6.4. They used a subset that has been cross-matched with recent water chemistry data. The encounter probabilities range as high as 0.80 for common species, such as yellow perch (*Perca flavescens*) and white sucker (*Catostomous commersoni*), while for other species they may peak in the 0.15–0.20 range, e.g., fathead minnow and blacknose shiner (*Notropis heterolepis*). Some of the large species, such as brook trout (*Salvelinus fontinalis*) and lake trout (*S. namaycush*), are over-represented because of a predilection to select inventory lakes containing top predators (Minns, 1986a).

Variation in encounter probability with pH for a species can be used to assess the change in resource status in the following way: The regional model of acidification predicts the steady state pH frequency distribution for all lakes for a given deposition scenario. This may be expressed as a probability distribution: $-p_{1i}$ = the proportion of lakes with a pH between i and $i + \delta$. The species' encounter probabilities p_{2i} can be derived along the same set of pH intervals, then the products of the two probabilities are summed and multiplied by the number of lakes to produce the number of lakes expected to contain the species:

$$N_j = N_T \cdot \Sigma p_{1i} \cdot p_{2i} \quad , \tag{5.6}$$

where i is the low pH to high pH by δpH; N_j is the number of lakes with species j; and N_T is the total number of lakes.

In some cases where the species is known to be over-sampled, the calculations can be done at a reference deposition level, i.e., 1980 base, and an expected value of N_j used to scale the results.

I determined encounter probabilities for the 12 species measured by Wales and Beggs (1986) but used the older, less reliable, but more abundant lake inventory pH values. I examined the pH range 4.0–10.0 by intervals of 0.5 units. The resulting probabilities are generally similar to those reported by Wales and Beggs (1986) (*Table 5.7*). Using the pH probabilities predicted by the regional model for the five deposition scenarios, I computed the expected numbers of lakes containing each species. For brook trout, lake trout, and walleye, I forced the 1980 lake numbers to equal those reported by Minns (1986a).

Some of the encounter probability distribution are uneven and, when combined with the pH distributions, can generate unexpected responses, e.g., for brook trout (*Table 5.7*). Overall, the results are similar to those obtained for species richness. Indeed, if all species were analyzed this way and the lake inventory data were a random sample for all Ontario lakes, the sum of species' responses would correspond to the species richness predictions. While the lake frequencies for three species were adjusted, those of lake whitefish and northern pike were not and so are likely too high. The smaller species, such as cyprinids, and common species, such as yellow perch and white sucker, are not selected for in the inventory process, and so their lake numbers are considered to be reasonable.

Application of this technique could be greatly improved through the use of better water chemistry data, correction for geographical distribution limitations, and perhaps regional stratification within distribution range to correct for uneven sample distribution.

Further extensions of the presence–absence evaluation of acidification effects involves making explicit allowance for the influence of other variables on encounter probability. Discriminant analysis has been widely used to develop predictive models (Johnson et al, 1977; Tonn and Magnuson, 1982; Beggs et al., 1985). More recently, others (Adler and Wilson, 1985; Reckhow et al., 1985) have suggested that logistic regression might produce more reliable presence–absence models.

5.4.3. Production/biomass/yield

As Kelso (1985) pointed out, there are a number of empirical models relating fish harvest and lake abiotic variables in lakes. In most cases, harvest has been used as a substitute for production, the variable of primary interest. The problem has been that there are few available measures of total lake fish production. Indeed, Kelso (1985) almost doubled the pool of data for the Canadian Shield with his report on lakes in the Turkey Lakes watershed, one of Canada's intensive study sites.

Much more fish production and biomass data are now emerging from a Department of Fisheries and Oceans project to characterize intraregional variability in catchments on the Shield containing up to 100–200 lakes. J.R.M. Kelso and M.G. Johnson of the Great Lakes Laboratory for Fisheries and Aquatic Sciences, Sault Ste. Marie and Owen Sound, Canada, have provided me with some

Table 5.7. Predicted damage to populations of 12 species of fish that are widely distributed in lakes with pH < 6.4 in Ontario.

Fish species		1980 base		% loss from original				% recovery from 1980		
		Prob.	No. of lakes	1980	Canada only	Canada + USA	Canada Worse 10%	Canada only	Canada + USA	Canada Worse 10%
Large distribution										
Brook trout	*Salvelinus fontinalis*	0.2112	2,110	1.89	-1.08	-0.02	1.62	157.2	101.0	14.6
White sucker	*Catastomous commersoni*	0.6178	84,890	4.47	2.95	0.80	5.58	34.0	82.0	-24.7
Northern redbelly dace	*Phoxinus eos*	0.2001	27,499	2.71	1.96	1.53	3.15	27.7	43.5	-16.2
Pumpkinseed	*Lepomis gibbosus*	0.3073	42,225	4.06	2.60	0.66	5.47	36.0	83.6	-34.9
Yellow perch	*Perca flavescens*	0.5586	76,761	6.82	5.04	2.06	8.42	26.1	69.7	-23.4
Fathead minnow	*Pimephales promelas*	0.1424	19,565	2.59	2.77	3.70	0.51	-6.7	-42.7	80.5
Limited distribution										
Lake trout	*Salvelinus namaycush*	0.1317	2,220	6.77	3.36	1.33	7.66	50.4	80.4	-13.1
Lake whitefish	*Coregonus clupeaformis*	0.1829	25,129	8.60	4.27	1.87	9.67	50.4	78.3	-12.4
Northern pike	*Esox lucius*	0.4129	56,737	6.55	4.39	1.51	7.88	33.0	76.9	-20.3
Common shiner	*Notropis cornutus*	0.1042	14,316	5.70	-0.07	-0.05	7.16	101.3	100.8	-25.7
Blacknose shiner	*N. heterolepis*	0.1723	23,672	6.43	6.51	3.04	7.91	-1.2	52.7	-23.0
Walleye	*Stizostedion vitreum vitreum*	0.2410	6,000	6.09	4.87	1.51	8.12	20.0	75.2	-33.3

preliminary fish production and biomass data for a series of Shield Lakes plus literature data from areas with similar water chemistry.

The best predictor of fish production (kg ha^{-1} yr^{-1}) in 16 lakes (areas 0.7 to 52.0 ha, mean depths 1.5 to 12.0 m, and alkalinities –9 to 358 μeq l^{-1}) was alkalinity:

$$\text{Production} = 7.126 + 0.07624 \cdot \text{Alkalinity}$$

$$(r = 0.556, F_{\text{regr}} = 6.27, \text{signif. } P = 0.025) \quad .$$

(5.7)

The best predictor of fish biomass was lake area. These results are not very robust but, until a larger database is developed, they will serve as an example to illustrate our approach to assessing regional impacts.

The fish production equation can be integrated over the area-weighted frequency distributions of alkalinity for each acid deposition scenario (*Table 5.8*). Original production was estimated to have been 222,000 metric tons per year. The 1980 level of acid deposition was predicted to have caused a 3.3% decrease. The Canada only emission controls are predicted to give a 20% recovery, while the combined Canada–US controls are predicted to restore 54% of the lost production. A worsening of deposition was predicted to increase the loss by a further 7%.

Table 5.8. Predicted fish production (tons per year) in Ontario's inland lakes relative to different levels of acid deposition (estimated total area of lakes = 6,014,719 ha).

Scenario	Fish production	% decrease[a]	% increase[b]
(a) Original	221,865.3	–	–
(b) 1980 baseline	214,560.4	3.29	–
(c) Canada only	216,004.8	2.64	19.8
(d) Canada + USA	218,481.1	1.53	53.5
(e) Worse 10%	214,037.6	3.53	–7.3

[a,b]See *Table 5.6*.

The level of changes in production is not startling. To place these results in perspective, recall that Minns and Kelso (1986) reported there were 23.4 million angler-days expended in Ontario's inland lakes in 1980. If we assume that the catch per angler-day was 1 kg of fish, not an unreasonable number, the annual yield would be 23,400 tons. Leach *et al.* (1987) estimated that potential or maximum sustained fish yield as a percentage of production averaged 21% (range 7–47%) in the Great Lakes. Minns *et al.* (1987) reported a range of 11–45% for the Bay of Quinte, Lake Ontario. So assuming yield is 20% of production, total potential yield in Ontario was 42,900 tons, and anglers removed 54.5%.

Overall, the indices of fishery response predict percentage responses less than those percentages associated with indicated levels of chemical sensitivity (alkalinity < 0 or pH < 5). The three fisheries indicators are based on a positive relationship with either alkalinity or pH, such that more sensitive lakes are presumed to have less richness or production to start with, compared with the average lake. Some highly valued resources, such as lake trout, are generally considered to be closely associated with sensitive conditions and thus prone to higher levels of damage. If we want to emphasize the importance of the fishery resource in acid-sensitive lakes, we will need to devise additional indicators.

It should also be remembered that the responses shown for Ontario are averaged for a number of secondary watersheds. As was shown by Minns (1986a), the level of damage in southern Ontario on the Shield, where high acid deposition levels prevail, will be much greater than in northern Ontario.

5.5. Discussion

To examine the modeling process and the results, we can consider the following questions:

- Why do we try to build models?
- How *good* are the models?
- What levels of damage or recovery can we expect?
- Who will use the models and results?

The extent to which we are able to answer these questions varies, but they are all important and need answers.

5.5.1. Why do we try to build models?

We build models because they present an explicit statement of how we expect systems to behave and because they allow us to extrapolate beyond the bounds of available measurements. The models are not a reproduction of the system, but a crude attempt to describe the processes we think are critical to understanding a system's behavior in response to certain stress.

In our case, we have chosen to represent few processes so we might consider more cases. Recently, similar types of models have been proposed by Small and Sutton (1986b) and Hornberger *et al.* (1986a). We have to extrapolate, as we are limited to measuring just a few attributes of a small proportion of the total resource. Eastern Canada likely has between 1 and 2 million lakes, so we cannot reasonably expect to sample more than a few percent of them.

5.5.2. How *good* are the models?

In this context, *good* is hard to define. By *good* do we mean *accurate, realistic* etc? The models consist of functional relationships and input data. If either is wrong, then is the model no good? If both are wrong, then obviously the model is no good. I favor a view that the essential responses of most systems can be well represented by relatively few equations and parameters, such as in the case of spruce budworm–forest dynamics (Ludwig *et al.*, 1978).

One of the most obvious weaknesses of these models is the lack of measures of statistical confidence. There are two main reasons for this omission: (1) our inability to secure an adequate representative sample of the resource at risk and (2) the extreme difficulties involved in measuring some of the critical parameters. This is not meant to understate the difficulty of determining the size of the resource in the first place, because there is much debate as to just how many lakes there are in Canada! Elements of the first main reason can probably be overcome with mesoscale surveys designed to assess heterogeneity in lake chemistry, fish presence–absence patterns, and other factors and to quantify the biases in existing pools of data. Uncertainties in key parameters may only be resolved in the long term as detailed watershed studies allow us to measure weathering rates and sulfate adsorption–desorption patterns. It may be that some parameters can only be measured after substantial changes in acid deposition have occurred.

Lake watershed model

The greatest uncertainties in the lake watershed model are centered on F_W and $[SO_4]_0$. Without knowledge of cation weathering rates, the cation exchange capacity reservoirs in watersheds, and how they respond to acid deposition, we can never be sure of the correct values for F_W and how they might vary. We have assumed that F_W is a simple function of $[Alk]_0$ and a surrogate for a supposed invariant sensitivity property of a lake and its watershed. If F_W were a sigmoid function of $[Alk]_0$, we might expect to see a sharp rise in impacts above an acid deposition threshold. Nearly all lakes with F_W less than 0.5 would be acidified. Given observed patterns of lake water chemistry (Jefferies *et al.*, 1986) and the predictions of models such as MAGIC (Cosby *et al.*, 1985c), F_W is probably not a sigmoid function, but other possible functions need to be explored. Schnoor and Stumm (1986) have reviewed the role of chemical weathering in acidification and presented a model with a similar F_W formulation to that presented here.

Another problem with the model presented here is the lack of a time dimension. The temporal response of lakes to changes in acid deposition was not modeled because of the extensive databases needed to parameterize elements that are strongly time-dependent. The Birkenes and MAGIC models (cf. Reuss *et al.*, 1986) consider the adsorption–desorption of sulfate in watershed soils. Cosby *et al.* (1986a) showed that the response times can be long when the watersheds have extensive soil accumulations. We have not been concerned with this omission in the regional impact model for two reasons. Much of the affected area on the Shield has relatively little soil, and soil sulfate is expected to reach steady state fairly quickly. Second, once watershed sulfate export matches deposition, predictions of lake chemistry should be the same for the empirical models of Henriksen and models such as MAGIC.

Fisheries models

Of the three fisheries models presented, the species richness one is best supported by available data. The analysis and prediction of assemblage size and structure have been the subject of considerable debate in recent years (cf. Strong *et al.*, 1984). More recently, Ricklefs (1986) provided a timely reminder that both local processes, such as competition and predation, and regional processes, such as speciation and dispersal, interact to determine the local richness of biological assemblages. For freshwater fishes in eastern Canada, both local and regional elements are evident in the regression model presented here. Area, depth, and pH stand for local processes, while elevation and tertiary watershed richness represent regional ones.

Much of the area of concern in Ontario and the rest of eastern Canada was glaciated until recently (10,000–20,000 years ago) (Briggs, 1986), and is likely still undergoing colonization by fish (Legendre and Legendre, 1983). The utility of such species–area models will increase with the development of additional explanatory models. In the case of acid deposition, they provide a powerful means of assessing resource impact.

5.5.3. What levels of damage or recovery can we expect?

The predicted levels of damage or recovery vary with the measurement indices and the emission strategy. The predictions presented here for Ontario suggest the lakes have already sustained appreciable damage: 7.1% of lakes with alkalinity less than zero, 4.3% of fish species–populations lost, 1.9% to 8.6% losses for particular species including important sportfishes, such as lake trout and walleye, and a 3.3% decrease in overall fish productivity.

Locally, these percentages can be higher where sensitive lakes are situated in areas of greatest acidic deposition. Muniz (1984) reported 15% loss of species–populations in southern Sweden, with species losses ranging from 4 to 43%; while in southern Norway fish stocks were affected in a region of 33,000 km^2 and virtually extinct in 13,000 km^2 of that area. Howells (1984) quoted reports showing 51% of Adirondacks lakes were fishless and 28% in the Sudbury region of Ontario. Schnoor *et al.* (1986) reported 0–19% acid lakes in parts of the Upper Midwest in the USA. We should be concerned with reduced productivity and species richness in moderately acid lakes (pH 5.0–6.0) as well as extinction of fish assemblages in extremely acid lakes (pH < 5.0).

5.5.4. Who will use the models and results?

There are many audiences for the results. Each audience defines the uses of the modeling. The regional impact models for aquatic and fisheries resources allow us to study a range of "What if . . . ?" questions. Sometimes on the local scale uncertainty in a model parameter may be thought to have considerable impact on the predictions, while on a regional scale that parameter's effect may be marginal. For example, the watershed area to lake area ratio r for individual lakes may be important in determining the degree of acidification where r has a low value, but analysis of regional distributions of r showed that high ratios predominate. Thus, emphasis on studying lakes with low r ratios might not be justified, given a total resource perspective.

The models can be used to evaluate and potentially to optimize management strategies. The impacts could be minimized by linking the emissions to deposition models with the aquatic fisheries models, as we have done already, and exploiting existing optimization procedures (Shaw, 1986). Equally, the models can be used to estimate the needs for liming in all lakes, if such a mitigation strategy were adopted. Jones *et al.* (1989) estimated with the regional impact model that, given 1980 levels of acidic deposition, the annual lime addition rate for all lakes in eastern Canada south of latitude 52° would be 150,000 to 1,500,000 tons. The range depends on the neutralization target of either liming to keep all lakes with an alkalinity greater than zero, or liming to restore all lakes to their original alkalinity. Needless to say, developing the logistical support and access to lakes, and identifying the target lakes would easily render lime mitigation prohibitively expensive, if not impossible, compared with the costs of emission controls.

Finally, the models are used to provide overall measures of impact, which are essential if resource protection and restoration are to be considered fairly, relative to the reported total costs for source emission controls. Natural renewable resources are notoriously difficult to value. The impact models provide at least the necessary multipliers, should the unit value be set or determined. In the meantime, total losses of species–populations, losses for individual species, and production losses will serve as surrogate for economic measures.

Acknowledgments

First, I want to thank my partners in this enterprise, Mike Jones and Dave Marmorek of ESSA Ltd. and Floyd Elder of DOE, with whom I have been engaged in this work for several years. Second, I extent my thanks to my colleagues in DFO who helped by providing information or time, especially John Kelso and Jim Moore.

Present and Future Acidification of Swedish Lakes: Model Calculations Based on an Extensive Survey

C. Bernes and E. Thörnelöf

6.1. Introduction

To combat acidification and its consequences, Sweden has in recent years been committing considerable resources to the liming of lakes and running water. Nearly US$20 million are now being spent annually on these efforts, and about 2,500 lakes had been limed at least once by 1985.

As part of the monitoring of acidification trends and effects of liming, Sweden's National Environmental Protection Board (SNV) and the Swedish county administrations made a nationwide survey of lakes during the winter of 1985. This survey was mainly intended to permit quantitative conclusions concerning the extent and degree of lake acidity, to reveal in some detail the regional acidification patterns, and to aid planning on national and regional levels of further liming operations. Present plans call for repeated surveys of this kind at five-year intervals, and in the future it will thus also be possible to follow long-term changes of the acidification status of Swedish lakes.

A fairly comprehensive account of the results obtained during the 1985 survey has been published by Bernes (1986), who has also developed and briefly discussed an acidification model that was applied to data from the survey. This model permits assessments of the degree to which air pollutants have changed the acidity of lake water in relation to preindustrial conditions. It also allows forecasts to be made of how the acidification status will change in response to future reductions of acid deposition. Many of the basic features of the model

J. Kämäri (ed.), Impact Models to Assess Regional Acidification, 83–107.

were originally developed by Henriksen (1980) and applied for prediction purposes by Wright and Henriksen (1983).

The present chapter is mainly devoted to a full account of this acidification model and thus gives the scientific basis for the assessments and predictions presented by Bernes (1986). Initially we also describe the lake survey. Some general results from the survey and from the model applications based on these data are presented as well.

6.2. The Survey

Compared with many other nations, Sweden is abundantly endowed with lakes. There are more than 85,000 lakes with an area in excess of 0.01 km^2. Thanks to a decentralized organization, with each of Sweden's 24 county administrations being responsible for the survey of lakes within its jurisdiction, more than 8% of the country's lakes could be visited in the winter of 1985. Samples were thus collected from 6,908 lakes larger than 0.01 km^2 (*Figure 6.1*).

Most of these lakes were selected randomly by SNV. The basis of this selection was a file of Swedish lakes created by the Swedish Meteorological and Hydrological Institute (1983). This file may be considered a complete inventory of all lakes larger than 0.01 km^2 except in the mountain areas of northwestern Sweden, where no lakes smaller than 0.1 km^2 have been registered. Every lake in the register is identified by a unique, 12-digit number based on the geographical coordinates (in a national grid) for the lake's outflow point.

The registered lakes are divided into five different classes according to size. These classes include lakes larger than 100 km^2 and lakes with areas 10–100 km^2, 1–10 km^2, 0.1–1 km^2, and 0.01–0.1 km^2. The first two classes comprise a total of 22 and 362 Swedish lakes, respectively, and all of these large lakes went into the selection to be sampled. The last three classes encompass several thousand lakes each. In these cases, SNV drew up a list of 40 "mandatory" lakes per class to be sampled in each county. The selection procedure included an element of geographical stratification: the lakes within a certain size class and county were first sorted in 24 × 25 km squares, and an appropriate number of lakes were then elected at random from each such square.

In addition to this selection of "mandatory" lakes, SNV made a random selection – without geographical stratification – of another 100 lakes per size and county. The county administrations were free to investigate as many of these additional lakes as the available time and resources would allow. The only provision was that the lakes be picked "from the top of the list and downwards," to avoid any nonrandom selection effects.

As a result, nearly 2,000 randomly selected lakes in each of the three smallest size classes were visited. Thus, 5,744 out of the 6,908 lakes sampled during the 1985 survey were selected at random. The remaining 1,164 lakes were selected by the county administrations themselves to satisfy local interests, e.g., to follow up individual research or liming projects. About 400 of the nearly 7,000 lakes investigated in 1985 had then been limed at least once.

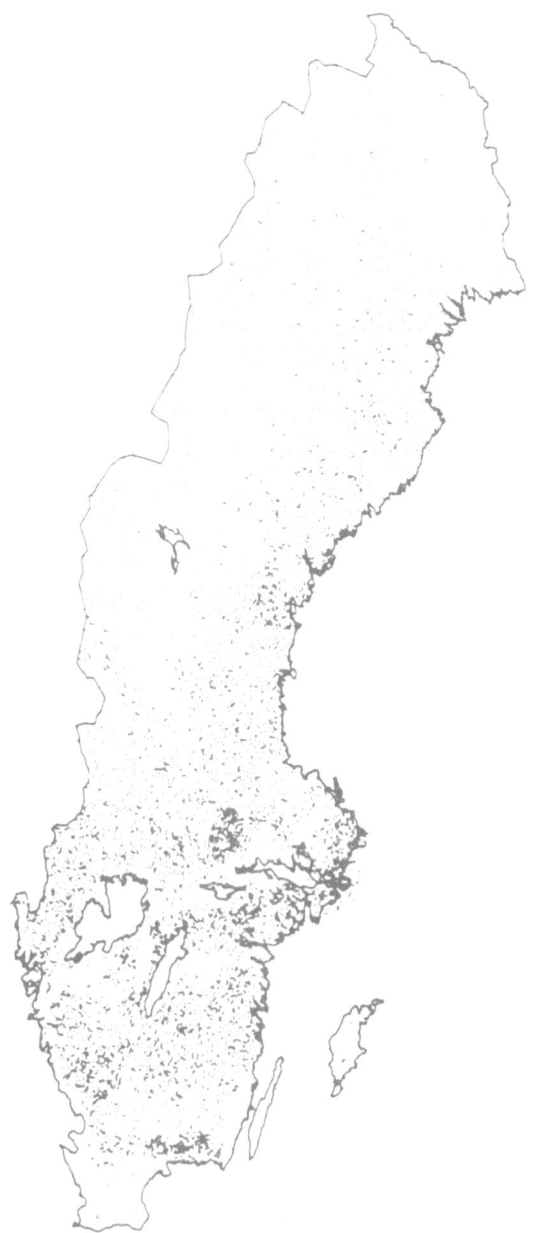

Figure 6.1. Lakes sampled in 1985. Of these lakes 5,744 were selected using objective random techniques, whereas the remaining 1,164 lakes were sampled to satisfy local interests.

The quantitative estimates of the total number of lakes in various acidification stages presented by Bernes (1986) and in this work are based on data from the randomly selected lakes only, whereas the maps in these two works are based on all available data.

Most lakes were visited in the late winter of 1985, before the snow started to melt. This meant that samplings were usually done during January–March in southern and central Sweden, but in April or even May in northern Sweden. The samples were brought up from a depth of 2 m through holes bored in the ice, as a rule far from shore.

The choice of winter to do sampling was partly governed by certain practical considerations. At least in northern Sweden, access to remote lakes is easier in winter than in summer. Another reason for preferring winter as the sampling period is that lakes then have low and fairly stable pH values. Even lower pH values may occur during the snowmelt and in connection with heavy autumn rains. However, since the acidity of small lakes can then change drastically from one week to the next, comparison between spring or autumn data from different lakes and different years would be difficult.

Summertime conditions in the lakes are also relatively stable, but then the pH values are usually higher owing to primary production, thus inviting the risk of portraying an acidification picture that would be unrealistically positive. After all, the main determinants of survival capabilities for sensitive species are a lake's pH "lows."

It is risky nevertheless to draw far-reaching conclusions about the acidification status on the basis of a single survey, even if made in winter, since temporary changes in climate can cause very noticeable year-to-year variations in lake water acidity. Therefore, to assess such variations, SNV and the county administrations have selected about 170 nonlimed reference lakes for investigation three to four times a year. Although the monitoring of these lakes did not start until 1983, the data collected so far suggest that conditions in the lakes were fairly normal during the winter of 1985. This conclusion is supported by studies of groundwater-level changes over the past few years (Aastrup and Johnson, 1986).

For the 1985 survey, SNV chose an analytical program that included determinations of pH, alkalinity, hardness (calcium plus magnesium concentrations), color, and conductivity – all analyses following Swedish standard procedures (Swedish Standards Institution, 1981). In the case of hardness, about half the laboratories used atomic absorption spectroscopy to determine calcium and magnesium concentrations separately, while the other half used a titrimetrical method to measure total hardness (and, in some cases, calcium as well).

This rather limited program was thought adequate, bearing in mind the possibilities for comparisons afforded by the considerably more comprehensive analyses routinely made by SNV of reference-lake samples.

The analyses were made by each county administration itself or by laboratories commissioned by them. For practical and economical reasons, this was preferable to a centralized procedure; but at the same time it entailed a risk of uneven quality and systematic differences in the results from different counties. However, reference-lake data provide us with the opportunity to check some of the data: 90 lakes went into the selection sampled during the national survey and were thus visited independently by SNV and local sampling teams in the winter of 1985.

Most of the evidence indicates that the majority of analyses made by regional laboratories adhered to a high standard of quality. We found a difference of 0 to 37 μeq l^{-1} between SNV and regional alkalinity data from the 90 reference lakes. The standard deviation quoted above corresponds to a standard error of 4 μeq l^{-1}. Two regional laboratories had problems with their hardness measurements, though. In these cases, we made certain corrections based on comparison with SNV data from the counties in question. After the corrections were made, the mean difference between SNV and regional hardness analyses of reference-lake samples turned out to be -6 (standard deviation 61 μeq l^{-1}; standard error 6 μeq l^{-1}).

Apart from analytical data, the county administrations provided SNV with some general information on the sampled lakes, such as lake area; height above sea level; and whether the lakes were limed, influenced by liming further upstream, or not influenced by liming at all.

6.3. A Quantitative Estimate of Lake Acidity in 1985

Because the properties of randomly selected lakes within a certain county and size class may be considered as representative of all lakes within that county and class, the data collected from this selection of lakes can be used as a basis for various quantitative conclusions regarding the whole population of Swedish inland waters.

Calculations for all counties and size classes show that approximately 33,500 – i.e., more than 40% – of Sweden's nearly 85,000 lakes larger than 0.01 km^2 had pH values below 6 in the winter of 1985. In about 4,600 of these lakes the pH was lower than 5. However, the majority of the acid lakes is small. Thus, almost no lakes larger than 10 km^2 had pH values below 6.

In the smaller size classes, every randomly selected lake often represents several hundred similar lakes, especially in the large counties of northern Sweden. In these cases, obviously, estimates of the total number of lakes within different categories will be relatively uncertain. Nevertheless, we consider national totals of the kind specified above to be accurate within a few percent.

6.4. Definition of Acidification

Because the main environmental concern is not with lake acidity itself but rather with its consequences for living organisms, pH data of the kind presented above are of relatively limited value. Actually, as long as a lake has been sampled and analyzed only once or just a few times, alkalinity often turns out to be more useful than the less stable pH values as a measure of the possibilities of survival for various species.

The experience gained by SNV over a number of years indicates that lakes with a winter alkalinity of at least 50 μeq l^{-1} are usually able to support a diverse and rich flora and fauna. There the pH value normally stays above 6 all year. Winter alkalinities lower than 50 μeq l^{-1} signify a considerable risk for acid surges in connection with snowmelt, autumn rains, or both. Such surges may wipe out the more sensitive species, such as roach and crayfish, or at least disturb their reproduction (e.g., Almer *et al.*, 1978). Current regulations in Sweden thus entitle liming projects to governmental subsidies if they deal with waters having an alkalinity lower than 50 μeq l^{-1}.

In lakes that have zero alkalinity in winter, all but the hardiest species are gone, especially if the winter pH value falls below 5.0. Such lakes are usually almost or entirely devoid of fish (Almer *et al.*, 1978).

Building on this experience, we have designed a classification of life conditions in Swedish lakes that is based on winter measurements of alkalinity and pH (cf. *Figure 6.2*). According to this classification, lakes with alkalinity greater than 50 μeq l^{-1} are in phase 1; whereas lakes with alkalinity less than 50 μeq l^{-1} have reached phase 2; or if the alkalinity is 0 and the pH is below 5.0, phase 3. A similar classification, also mainly based on alkalinity, is used by Henriksen (1980).

Data from the 1985 survey indicate that some 61,500 Swedish lakes were then in phase 1, whereas about 17,000 had reached phase 2, and 4,500 were in phase 3. Thus, about 21,500 lakes – or 26% of the entire population – were acidic enough to make survival difficult or impossible for sensitive species. Together these lakes cover some 3,000 km^2, or about 8% of Sweden's total lake area.

It would be misleading, though, to classify all these lakes as acidified, at least if that term is meant to imply a change of acidity caused by anthropogenic emissions. There is no denying that every Swedish lake receives more sulfur compounds and other acidifying air pollutants now than during preindustrial times, but in many lakes the pH decrease has probably been rather insignificant. Some lakes are remote from emission sources and have received only small amounts of air pollutants; others are surrounded by soils with a high capability of neutralizing acid fallout. Today, most such lakes still have a diverse flora and fauna.

Other lakes, again, were fairly acid even before humans began to exert their impact. Some of the lakes are humic and, hence, acidic because of high concentrations of carbon dioxide (cf. Section 6.7) and humic acids; whereas certain waters are surrounded by very poor soils and therefore have naturally low alkalinity, with concomitant risk of temporary pH drops. In other words, owing to natural supplies of acid, these lakes had reached phase 2 – and, in a few instances, phase 3 as well – long before the air was contaminated. We may describe the lakes in question as naturally influenced by acids.

The acidification caused by air pollutants cannot be considered serious unless it has impaired or threatens to impair the conditions under which plants, animals, or human beings live. Such impairments will have incurred if a body of water, in consequence of increased acid supply, has passed over from one acidification phase to another.

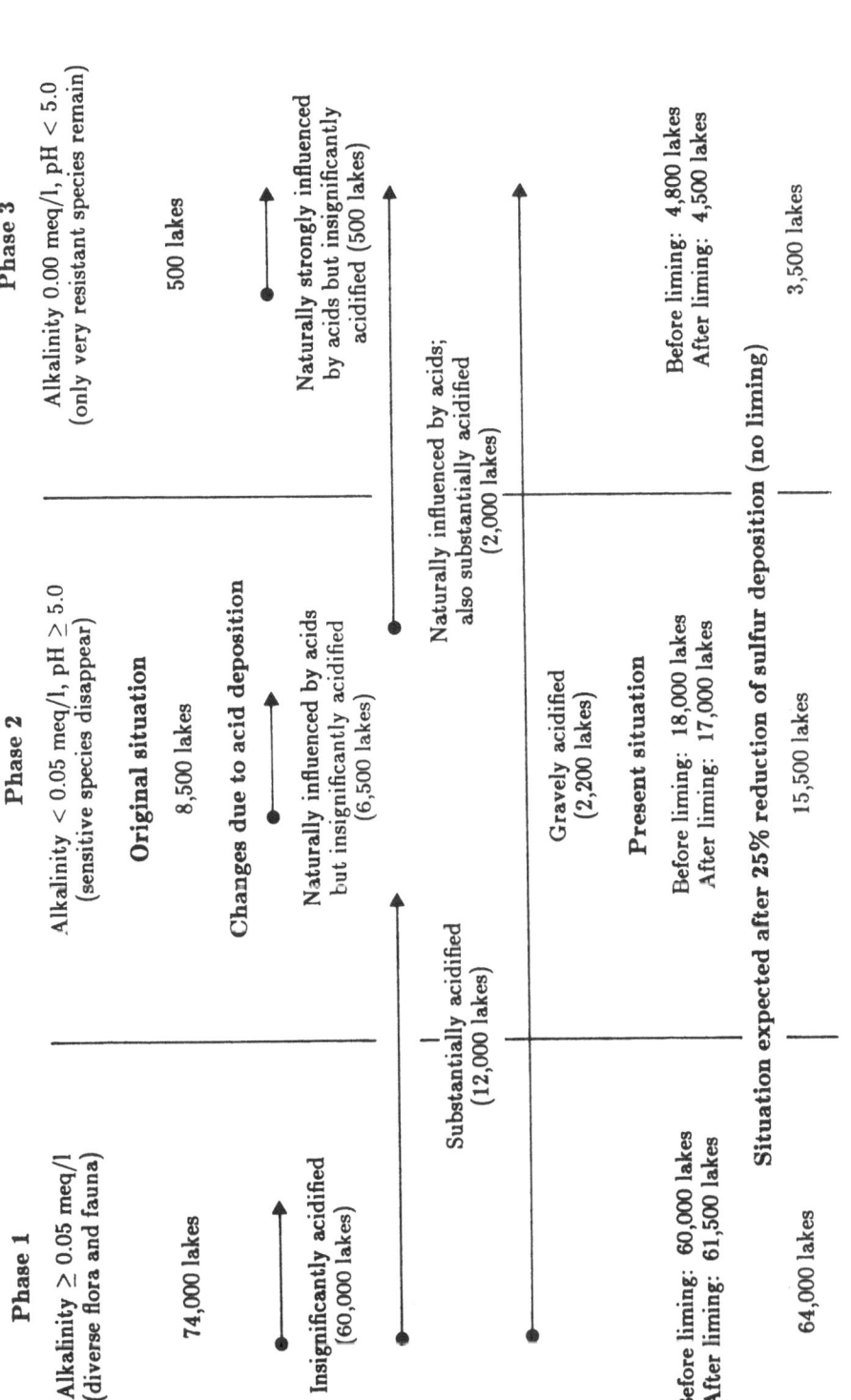

Figure 6.2. A summary of the definitions of phases and acidification used in this chapter. Also shown are the calculated number of lakes in various phases and acidification stages under preindustrial, present, and future conditions (cf. Sections 6.8 and 6.9). The arrows indicate changes from preindustrial to present conditions (dots and arrowheads, respectively) caused by acid deposition.

Here we elect to define a lake as being substantially acidified if, because of air pollution, it has passed over to phase 2 after having originally been in phase 1, or if it has passed over to phase 3 after first having been in phase 2. If air pollution has pushed the lake all the way from phase 1 to phase 3, we call it gravely acidified. If a lake has not changed phase in spite of increased acid supply, we consider it insignificantly acidified.

The main difficulty with these definitions of acidification and natural acidity is that they require knowledge of the lakes' preindustrial properties. Actually, of course, we have no measurements from that period of time. Instead, the original properties of the lakes must be assessed on the basis of contemporary data.

The following five sections deal with this central problem and with the related problem of how the lakes will respond to future changes of the acid deposition. We first derive a generalized version of Henriksen's (1980) "acidification equation," which is a simple relation between the concentrations of certain key substances in lake water. In Section 6.6 we use this equation to calculate present-day sulfate concentrations in all of the lakes investigated in 1985.

We then proceed in Section 6.7 with a general theory of how pH and alkalinity in lakes change in response to changes of the sulfur deposition. This theory is mainly based on the acidification equation and on simple assumptions concerning the relations between sulfate concentrations in precipitation and lake water and between sulfate and hardness in the lakes. It allows us to estimate lake alkalinities and pH values both in the complete absence of anthropogenic sulfur deposition (cf. Section 6.8), i.e., under preindustrial conditions, and after future reductions of the present-day level of deposition (cf. Section 6.9).

6.5. The Acidification Equation

Ionic relationships in oligotrophic lakes have been studied extensively by Henriksen (1980). His approach is based on the fact that soil weathering releases proportional and nearly equivalent amounts of calcium and magnesium, on the one hand, and bicarbonate [HCO_3], on the other hand, to the lakes. When acids are supplied to the lakes, the result resembles a large-scale titration: the bicarbonate neutralizes the acids and is replaced by their anions.

Sulfuric acid usually dominates completely among the acids received by Swedish clearwater lakes. We may therefore expect that the concentration of calcium plus magnesium is balanced in such lakes by an approximately equal concentration of bicarbonate plus sulfate.

Analyses of samples taken in the winter of 1985 from 71 reference lakes with highly transparent water and nonzero alkalinity (cf. *Figure 6.9*) thus indicate that

$$[HCO_3] + [SO_4^*] = 0.95 \, [Ca^* + Mg^*] \tag{6.1}$$

if we make the reasonable assumption that the alkalinity in clearwater lakes is practically identical to the concentration of bicarbonate. The asterisks denote

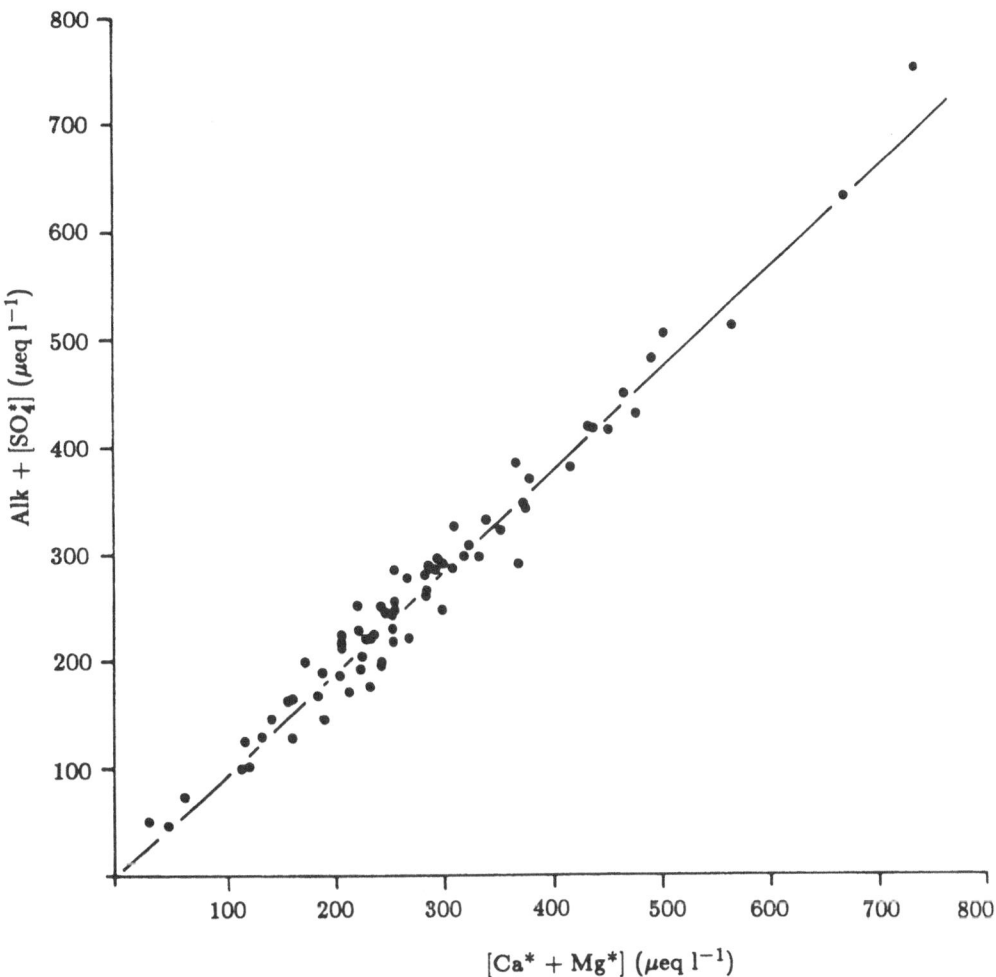

Figure 6.3. The relation between concentrations of nonmarine calcium and magnesium and alkalinity plus concentration of nonmarine sulfate in 71 reference lakes with clearwater and nonzero alkalinity. The regression line has a slope of 0.95.

that the marine components of the sulfate, calcium, and magnesium concentrations have been deducted.

In addition to sulfuric acid, considerable amounts of nitric acid are also deposited over Sweden. However, nearly all nitric acid falling over land areas is trapped in soil and vegetation, and the amounts that reach surface waters are therefore still relatively small. On the average, nitrate concentrations in the reference lakes thus amount to no more than 5% of the sulfate concentrations. This means that the contribution of nitric acid to the acidification of Swedish inland waters has so far been rather insignificant.

Therefore, in the calculations of lake pH and alkalinity described below, we have only taken into account changes of the deposition of sulfuric acid. The present-day concentrations of nitrate in the lakes are assumed to remain constant even if the sulfate concentrations are changed. Nevertheless, it may be noted that the constant in equation (6.1) would have been even closer to unity if the concentration of nitrate had been included on the left-hand side.

It may further be noted that equation (6.1) implies that the alkalinity in clearwater lakes would equal on the average 0.95 $[Ca^* + Mg^*]$ if no sulfuric acid was supplied. An almost identical relation was derived by Henriksen (1980), who used data on oligotrophic waters in parts of Scandinavia and North America that were relatively unaffected by acid deposition at the time of sampling.

In areas where the long-term acid deposition exceeds the release of bicarbonate through weathering, essentially all bicarbonate is consumed by neutralization processes. Excessive amounts of acid either are not neutralized or are buffered by aluminum compounds present in soil or water. Under such circumstances, we may expect a balance in the lakes between the sulfate concentration, on the one hand, and the sum of the concentrations of calcium, magnesium, hydrogen, and aluminum ions, on the other hand – still assuming that sulfuric acid is the dominating acid. We may then write a more complete version of equation (6.1) as follows:

$$[HCO_3] + [SO_4^*] = 0.95 \, [Ca^* + Mg^*] + [H^+ + Al^{n+}] \tag{6.2}$$

where $[HCO_3]$ will be very small if $[SO_4^*] > 0.95 \, [Ca^* + Mg^*]$, whereas $[H^+ + Al^{n+}]$ will be small if $[SO_4^*] < 0.95 \, [Ca^* + Mg^*]$.

No analyses of aluminum ions $[Al^{n+}]$ were performed during the 1985 lake survey. Experience gained from lake liming indicates, however, that owing to the presence of aluminum in strongly acidified lakes, the amounts of lime needed to restore them are usually roughly twice as large as would have been expected from the concentration of hydrogen ions (Almer *et al.*, 1978). We have, therefore, used the following simple relation in our calculations:

$$[H^+ + Al^{n+}] = 2 \, [H^+] \quad . \tag{6.3}$$

A more elaborate relation used by Wright and Henriksen (1983) gives similar results over a large part of the relevant pH range.

Equation (6.2) is essentially identical to the acidification equation formulated by Henriksen (1983). In this form, however, it is only applicable to clearwater lakes. In the general case natural organic acids must also be accounted for. Considerable amounts of such acids are present in humic brown-water lakes, which are very common in several parts of Sweden.

Humic acids enter the balance in equation (6.2) in the same way as sulfuric acid, and we thus obtain:

$$[HCO_3] + [SO_4^*] + [A^-] = 0.95 \, [Ca^* + Mg^*] + [H^+ + Al^{n+}] \qquad (6.4)$$

where A^- is a general expression for the anions of humic acids, which themselves may be denoted as HA. Because most humic acids are relatively weak, they may not be completely dissociated in lakes with low pH values. This means that the presence of such acids also affects the alkalinity, which in effect is a measure of the total concentration of buffering substances in the water. Bicarbonate is the most important of these substances, but humic acids also contribute to the buffering against acidity increases, since their anions bind an increasing amount of hydrogen ions when the pH value decreases.

We can thus no longer assume equivalence between alkalinity and bicarbonate concentration. The Swedish standard procedure for measuring alkalinity (Swedish Standards Institution, 1981) involves a determination of the amount of acid required for lowering the pH value of a sample to 5.4. We should therefore be able to express the alkalinity (Alk) of lake water as follows:

$$Alk = [HCO_3] + [A^-] - [A^-]_{5.4} \qquad (6.5)$$

where $[A^-]_{5.4}$ is the concentration of dissociated humic acids at pH 5.4.

Combining equations (6.4) and (6.5), we can finally derive the following generalized acidification equation:

$$Alk + [SO_4^*] + [A^-]_{5.4} = 0.95 \, [Ca^* + Mg^*] + [H^+ + Al^{n+}] \quad . \qquad (6.6)$$

6.6. Calculations of Humic Acid and Sulfate Concentrations

As pointed out in Section 6.2, the set of data collected during the 1985 survey was quite limited. Since neither chloride nor sodium was measured, the subtraction of the marine component of the calcium and magnesium concentrations thus had to be based on regional averages of the chloride concentration as derived from measurements in the 170 reference lakes. Within about 30 km from the west coast of Sweden, this correction may exceed 100 μeq l^{-1} but it is smaller than 50 μeq l^{-1} in at least 90% of the Swedish lakes and smaller than 10 μeq l^{-1} in about half of them.

A more serious shortcoming is that sulfate was not included in the analytical program. In retrospect, that decision must be judged as unfortunate in view of the dominating role played by this substance in present-day lake chemistry. Moreover, no analyses of humic substances have been made either during the 1985 survey or of samples from the reference lakes.

It thus turns out that even with calcium and magnesium concentrations having been corrected for sea salts as outlined above, we still have data on only three out of the six quantities involved in the generalized acidification equation.

However, Dickson (1985) has suggested that data on the water color in lakes can be used as simple and very approximate measures of the total concentration of humic acids, dissociated or not, i.e., of $[HA + A^-]$. As described below, we have adopted this suggestion and used color data from the 1985 survey as a basis for rough calculations of $[A^-]_{5.4}$ in each lake investigated. The acidification equation then enabled us to calculate sulfate concentrations as well in these lakes. Had sulfate actually been measured during the survey, we would instead have used the acidification equation for a more reliable calculation of the concentrations of humic acids.

Dickson draws upon data from Andersson and Borg (unpublished, SNV), who found an approximate relation between water color (in mg Pt l^{-1}) and the concentration of organic carbon (in mg l^{-1}) using data from 301 Swedish lakes:

$$\text{color} = 10.6 \, [\text{organic C}] \quad . \tag{6.7}$$

Furthermore, Henriksen and Seip (1980) have found the following average relation between the concentrations of organic acids (in mmol l^{-1}) and organic carbon in Norwegian and Scottish waters:

$$[HA + A^-] = 0.0055 \, [\text{organic C}] \quad . \tag{6.8}$$

Combining these results, Dickson derived the following approximate result:

$$[HA + A^-] = 5 \cdot 10^{-4} \, \text{color} \quad . \tag{6.9}$$

When applying this formula to data from a small number of Swedish lakes, he found a quite satisfactory correspondence between the anion deficit (i.e., the difference between the sum concentrations of inorganic cations and anions) and the predicted concentration of organic acids.

To separate $[A^-]$ – or, more specifically, $[A^-]_{5.4}$ – from $[HA + A^-]$, one still has to know the average strength of the humic acids, i.e., their average degree of dissociation at various pH levels.

As discussed by Perdue (1985), simple carboxylic acids of the kinds that might be present in natural waters have widely different strengths. Their acid dissociation constants (pK_a values) range roughly from 1 to 9, but the pK_a distribution has a rather well-defined peak near 4.5. This result is in good agreement with analyses performed by Lee (1980) of a few brown-water lakes in Sweden. She established the presence of at least one organic acid, probably of the carboxylic type, with a pK_a value in the range 4.4–4.7.

We adopted an average pK_a value of 4.7. The condition of acid-base equilibrium then gave us the following relation between $[A^-]$ and $[HA]$:

$$\log \, [A^-] \, / \, [HA] = pH - 4.7 \quad . \tag{6.10}$$

Using equation (6.10) in combination with equations (6.3), (6.6), and (6.9), we calculated the $[A^-]_{5.4}$ and sulfate concentrations in each of the 6,908 lakes surveyed in 1985. The individual sulfate results usually agreed reasonably well with the regional average sulfate concentration as derived from reference-lake data. A certain amount of scatter was observed, however. Local variations in, e.g., the degree of evapotranspiration will thus lead to different concentrations of sulfate even in lakes situated close to one another.

In approximately a tenth of these cases, however, the difference between the individually calculated sulfate concentration in a lake and the regional average was unreasonably large. If the deviation exceeded a factor of 2, the calculated concentration of humic acids was assumed to be in error. If a calculated sulfate concentration exceeded the "upper limit" (twice the regional average), it was thus adjusted down to that limit, whereas the concentration of humic acids was increased by the same amount to satisfy equation (6.6). Similarly, a calculated sulfate concentration below the "lower limit" (half the regional average) was increased to that limit and the concentration of humic acids correspondingly decreased.

Changes leading to negative concentrations of humic acids were not allowed, however. This means that obviously erroneous and even negative sulfate values remained, after the adjustment described above, in those cases where, according to the analyses, alkalinity exceeded total hardness. Here, the source of the error was most probably in the analyses of alkalinity or hardness or both. Such errors fell mainly upon high-alkalinity lakes that were not significantly acidified and thus of limited interest for this evaluation. No attempt was made to adjust the data further.

How, then, can the overall reliability of the individually calculated (and, in some cases, adjusted) sulfate concentrations be assessed? One way is to compare the sulfate concentrations measured by SNV in those 90 lakes that were sampled both during the national survey and within the reference-lake program.

The average difference between individual calculated and measured concentrations amounts to 12 ± 71 μeq l^{-1}. We may conclude that the average result of the calculations is quite satisfactory, whereas the individual variation is large. However, much of the variation is due not to the method of calculation itself but rather to errors in the hardness analyses – to some extent also the alkalinity analyses – performed by the regional laboratories. Part of the variation may also be attributed to the fact that calculated and measured sulfate concentrations do not refer to the same samples.

This becomes evident if the sulfate concentrations in the 90 lakes are calculated using SNV's alkalinity and hardness analyses of the ordinary reference-lake samples rather than analyses made by the regional laboratories of the national survey samples from these lakes. The average difference between calculated and measured sulfate concentrations then becomes 6 ± 36 μeq l^{-1} – both the difference itself and its standard deviation are thus halved.

A similar comparison shows the importance of including humic acids in the calculations. It thus turns out that the average difference of 12 ± 71 μeq l^{-1} between calculated and measured sulfate concentrations in the reference lakes would instead have been 39 ± 82 μeq l^{-1}, had we ignored humic acids and attributed the difference between hardness and alkalinity to the presence of sulfuric acid.

By taking humic acids into account, we obviously improved the sulfate estimates considerably, notwithstanding the very approximate relation between humic acids and water color on which the calculations in question were based.

6.7. The Response of Lakes to Deposition Changes

The relation between atmospheric sulfur deposition and sulfate concentrations in freshwater lakes plays a key role in any prediction of how deposition changes affect the acidification status of the lakes. One way of establishing this relation would be to compare present-day deposition and concentration values in regions under different acid loading.

However, we cannot expect to find a general relation valid for all lakes. In southeastern Sweden, where evapotranspiration is fairly large compared with the precipitation amounts and where dry sulfur deposition is relatively important, the mean ratio between nonmarine sulfate concentrations in lake water and precipitation is thus quite large (3–5). In northwestern Sweden, where both evapotranspiration and dry deposition are less important but precipitation amounts are larger, this ratio is much smaller (1.5 or even less). As pointed out in Section 6.6, moreover, the ratio may be different even for lakes within the same region.

Instead we simply assume that the long-term ratio between sulfate concentrations in lake water and precipitation remains constant for any individual lake, regardless of the level of sulfur deposition. A change of the nonmarine sulfur deposition is thus expected eventually to cause a proportional change of the concentrations of nonmarine sulfate in the lakes. Implicit in this hypothesis is the further assumption that the local relation between dry and wet deposition of sulfur remains unchanged everywhere.

Equation (6.6) may seem to imply that a certain change of the sulfate concentration in a lake will be reflected by an equivalent – but opposite – change of the alkalinity. It must be realized, however, that the rate of neutralizing reactions in the soil will probably change as well, if the deposition of acids is altered.

The neutralization rate is especially likely to change in calcareous soils, since the weathering of calcite (and, hence, the release of calcium, magnesium, and bicarbonate from such soils) is highly dependent of the acidity. Weathering of silicates – the dominating class of minerals in most parts of Sweden – seems, on the other hand, to proceed at a more or less constant rate within a considerable pH range (e.g., Helgeson *et al.*, 1984; Fölster, 1985). However, the rate of another neutralizing process – ion exchanges in the uppermost soil layers – may at least temporarily change in siliceous as well as calcareous soils in response to alterations of the acid load.

Increased weathering and increased ion exchange rates both lead to enhanced leaching of calcium and magnesium from the soil – and vice versa. This means that a temporary increase of the deposition of sulfuric acid will not only cause an increase of the sulfate concentration in lakes but also a certain increase of the hardness and – according to equation (6.6) – a correspondingly smaller decrease of the alkalinity than might have been expected if the rate of neutralization had remained unchanged. The effects of deposition changes on the acidification status in lakes are thus to some extent counteracted by changes of the neutralization rates.

Henriksen (1983; 1984) has introduced the ratio F between hardness change and change of sulfate concentration in a lake (following a change of the sulfur deposition) as a measure of the importance of neutralization rate changes in the soil, i.e.,

$$F = \Delta \, [Ca^* + Mg^*] \, / \, \Delta \, [SO_4^*] \quad . \tag{6.11}$$

This ratio depends on local soil characteristics and will thus be different for different lakes. It is also likely to change with time, one reason being that depletion (or enrichment) of exchangeable ions may alter the rate of ion exchanges even under a constant load of acids.

Nevertheless, Henriksen (1982b; 1984) has tried to calculate average F values by comparing historical "pre-acidification" data with present-day data on lakes and rivers, by studying lakes over a gradient in acid loading, and by using data on strongly acid lakes. He concluded that F factors in the range 0–0.4 would probably yield reasonable results for prediction purposes. Wright and Henriksen (1983) thus used an F value of 0.2 in their forecast of lake restoration following a reduction in sulfur deposition.

We have adopted the same F value (0.2) for the majority of Swedish lakes sampled in the 1985 survey. However, silicates are likely to dominate completely in soil layers surrounding strongly acid lakes, and there the weathering rate may thus be more or less independent of the acid deposition. Moreover, recent analyses of soil from areas in southern Sweden, where such lakes are common, show that exchangeable cations are strongly depleted (Hallbäcken and Tamm, 1986; Falkengren-Grerup et al., 1987), which means that ion exchange processes must be relatively insignificant there today. It thus turns out that the concentration of nonmarine calcium plus magnesium in many of Sweden's strongly acid lakes is quite low (usually < 200 μeq l^{-1}). This indicates that 0.2 would probably be an overestimate of the F value for these lakes, and we have therefore adopted $F = 0$ for strongly acid waters or, to be more specific, for lakes where $[SO_4^*]$ exceeds $[Ca^* + Mg^*]$.

The above assumptions can be summarized as follows: An $X\%$ change of the sulfur deposition will lead to an $X\%$ change of the sulfate concentrations in lakes. If the latter change amounts to Y μeq l^{-1} in a certain lake, it will be accompanied by a hardness change amounting to 0.2 Y μeq l^{-1} in that lake (unless the lake is strongly acidified).

These assumptions, together with equations (6.3) and (6.6) and the additional assumption that the concentration of humic acids is independent of the deposition of sulfuric acid, would in principle enable us to calculate the effects of a deposition change on the alkalinity of a lake. However, since the concentration of hydrogen ions is involved as well in equations (6.3) and (6.6), we must simultaneously calculate how the pH value changes in response to the deposition change. This requires careful consideration.

The relations between pH values and alkalinities reported during the 1985 survey thus indicated that the pH values in most cases were significantly lowered by carbon dioxide (CO_2) oversaturation. Such oversaturation is due to CO_2 being released during biological decomposition of organic substances in the water (cf. Norton and Henriksen, 1983). It is particularly evident in winter, when degassing of CO_2 from the water is almost completely prevented by the ice cover.

Moreover, we have found a significant relation between the amount of CO_2 oversaturation and the concentration of humic substances as reflected by the water color. The pH values in brown-water lakes sampled in the winter of 1985 were thus usually 1–2 units lower than might have been expected from the alkalinity if the amount of CO_2 in the water had been in equilibrium with CO_2 in the atmosphere (cf. *Figure 6.4*).

To calculate pH values consistent with those measured during the 1985 survey, we assume that the wintertime CO_2 concentration in a lake is constant and thus independent of the sulfate concentration and acidity of the water. The carbonate equilibrium equations then require that

$$[H^+]\,[HCO_3] = \text{constant} \tag{6.12}$$

in each lake. The value of the constant was derived individually from pH values and alkalinities found in 1985 in each of the sampled lakes, using equation (6.5) to convert from alkalinity to HCO_3 concentration. In those cases where the 1985 alkalinity was zero, however, we used instead the average relation between water color and wintertime CO_2 (or H^+ and HCO_3) concentrations found during the survey:

$$[H^+]\,[HCO_3] = 1.41 \cdot 10^{-5}\ \text{color}\ 0.42 \quad . \tag{6.13}$$

By solving equation (6.12) or (6.13) together with equations (6.3), (6.5), (6.6), (6.9), and (6.10), we were finally able to calculate how both pH and alkalinity in each lake investigated in 1985 are likely to change in response to changes of the sulfur concentrations in precipitation and lake water.

6.8. Preindustrial Acidity and Present-Day Acidification

Our first choice was to study the preindustrial case, i.e., the situation that prevailed when no anthropogenic sulfur was deposited at all. This is not equivalent to assuming that $[SO_4{}^*] = 0$, since a small proportion of the sulfuric acid deposited today surely have a natural origin.

Several studies (e.g., Rodhe, 1982) indicate that the concentration of non-marine sulfate in precipitation over northern Europe probably is about ten times greater now than before industrialization. The average present-day $SO_4{}^*$ concentrations in precipitation over Sweden range from 20 to 80 $\mu eq\ l^{-1}$, and we

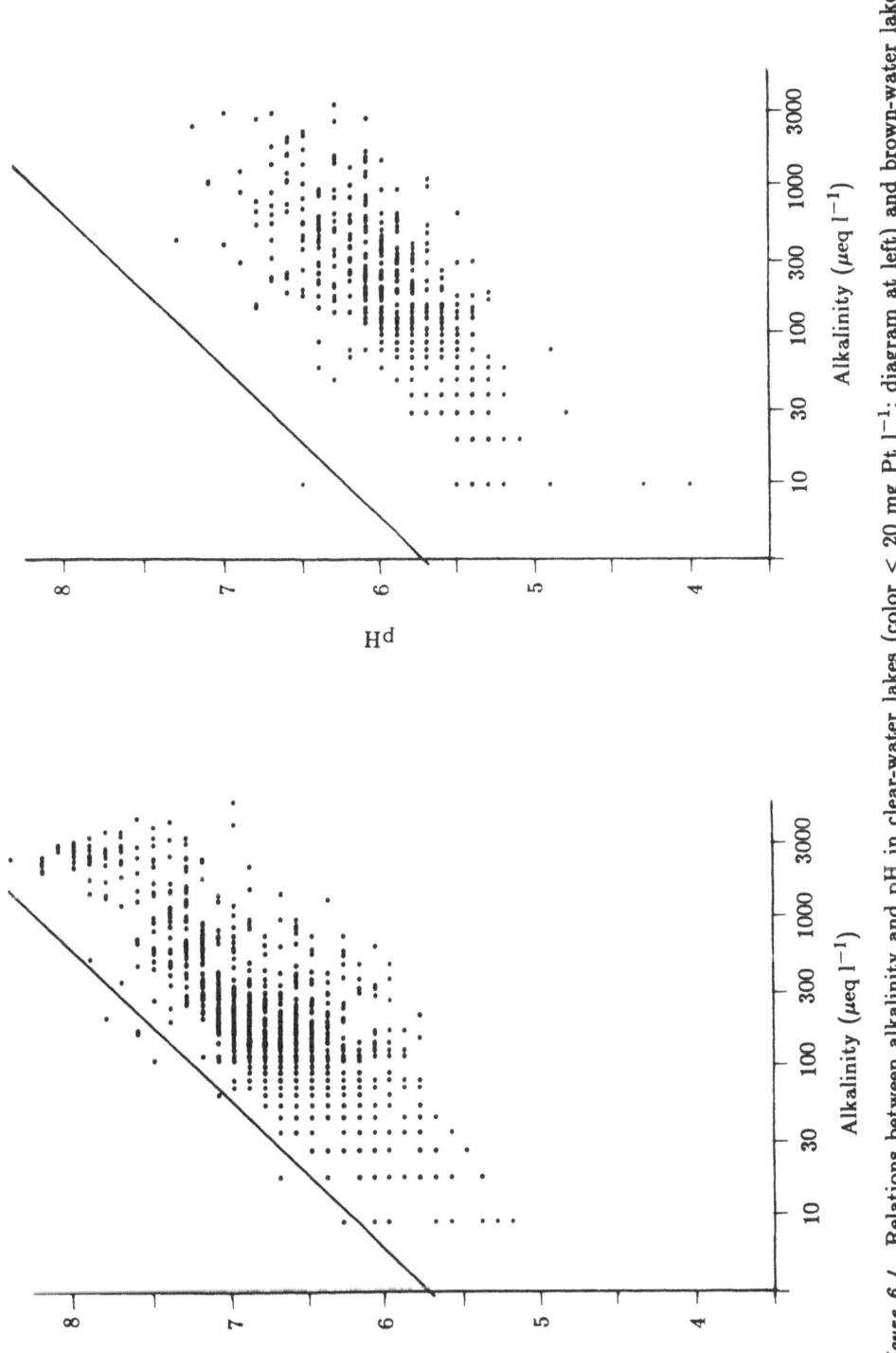

Figure 6.4. Relations between alkalinity and pH in clear-water lakes (color < 20 mg Pt l^{-1}; diagram at left) and brown-water lakes (color > 200 mg Pt l^{-1}; diagram at right); data from the winter survey in 1985. The diagonal lines show the theoretical relation between alkalinity and pH in water at equilibrium with CO_2 in the atmosphere.

therefore found it reasonable to assume that the preindustrial SO_4* concentration was 5 μeq l^{-1} all over the country.

Using the SO_4* concentration field in precipitation over Sweden as mapped by Granat (1986) in 1983–1985, we calculated the ratio between present-day and preindustrial SO_4* concentrations in precipitation at the location of every lake surveyed in 1985. The assumption of proportionality between SO_4* concentrations in precipitation and lake water then directly gave us the preindustrial SO_4* concentrations in the lakes, which allowed us to estimate preindustrial alkalinities and pH values.

These estimates enabled us to decide whether a certain lake was in phase 1, 2, or 3 before it was affected by air pollution. This was done individually for each lake investigated in 1985. The results indicate that a total of about 74,000 Swedish lakes were originally in phase 1, whereas about 8,500 were in phase 2, and 500 in phase 3.

This means that some 9,000 lakes may be considered as naturally influenced by acids – they were so acidic even in their original state that sensitive species could not live there. Most of these lakes are situated in parts of northern Sweden where the soil's capacity to neutralize acids has always been very limited (cf. lake D in *Table 6.1*). The pH value of the lake water probably stayed around 6 during most of the year, but the alkalinity was so low that snowmelt and heavy rainfalls could cause acid surges even though the pH in precipitation was then considerably higher than it is today.

By checking if a lake was still in the same phase in 1985 as under preindustrial conditions, we could decide whether it should be classified as acidified or not (cf. *Figure 6.5*). For those lakes that by 1985 had been limed at least once, we compared preindustrial data with data collected just before the first liming operation, which in most cases took place in the early 1980s.

It turned out that about 14,000 Swedish lakes had been substantially acidified by 1985, whereas about 2,200 lakes had been gravely acidified. These lakes have aggregate areas of about 3,500 km^2 and 320 km^2, respectively. The 14,000 substantially acidified lakes include some 2,000 lakes that are also naturally influenced by acids. These lakes were thus originally in phase 2, but air pollutants have since brought them to phase 3.

Most of the acidified lakes can be found in southwestern Sweden, where the acid deposition is high owing to the proximity to large foreign and domestic sources and also to the fairly large amounts of precipitation. In some counties in this part of Sweden, more than half the lakes are substantially or gravely acidified. Even in northernmost Sweden, however, roughly one out of ten lakes is substantially acidified. Only in those relatively small parts of the country where soils, bedrock, or both are calcareous is lake acidification altogether insignificant.

The situation may be briefly summarized as follows: Natural processes keep about 8% of Sweden's lakes so acidic that they affect the ability of many organisms to live there. In another nearly 20% of the lakes, the flora and fauna have suffered injury during our century because air pollution has acidified the water.

Table 6.1. Measured and calculated data of five sample lakes.

	A	B	C	D	E
Measured data (1985)					
pH	5.6	4.2	6.3	6.2	4.5
Alkalinity (μeq l^{-1})	20.0	0.0	10.0	20.0	0.0
Ca + Mg (μeq l^{-1})	490.0	220.0	80.0	60.0	380.0
Color (mg Pt l^{-1})	170.0	25.0	5.0	1.0	800.0
Calculated data (1985)					
Ca^* + Mg^* (μeq l^{-1})	400.0	130.0	70.0	50.0	340.0
SO_4 (μeq l^{-1})	290.0	240.0	60.0	30.0	230.0
A^- (humic anions) (μeq l^{-1})	80.0	0.0	0.0	0.0	150.0
Estimates of preindustrial conditions					
pH	6.8	6.3	7.0	6.5	5.0
Alkalinity (μeq l^{-1})	240.0	100.0	50.0	40.0	0.0
Estimates of conditions after 25% deposition decrease					
pH	6.2	4.5	6.6	6.3	4.6
Alkalinity (μeq l^{-1})	70.0	0.0	20.0	30.0	0.0
Phase and acidification classification					
Preindustrial phase	1	1	1	2	2
Present-day phase	2	3	2	2	3
Present-day acidification	subst.	grave	subst.	insignif.	subst.
Phase after 25% deposition decrease	1	3	2	2	3
Acidification after 25% deposition decrease	insignif.	grave	subst.	insignif.	subst.

A. Svarta sjö (S Sweden). High S deposition but also high neutralizing capacity in surrounding soils (reflected by high SO_4^* concentration and hardness, respectively). Acidified today, but even a limited deposition decrease should lead to recovery.

B. Tjärnesjön (SW Sweden). High S deposition; low neutralizing capacity. F value assumed to be 0. Today gravely acidified; needs considerable deposition decrease to recover.

C. Bredåsjön (NW Sweden). Low S deposition and neutralizing capacity. Acidified today; needs considerable deposition decrease to recover.

D. Njuongerjaure (N Sweden). Low S deposition and neutralizing capacity. Low alkalinity under preindustrial conditions; thus, naturally influenced by acids. Further acidification caused by air pollutants not significant.

E. Nålgöl (SE Sweden). Brown-water lake with high content of humic acids; thus, naturally influenced by acids. Fairly high S deposition has caused substantial further acidification. Needs considerable deposition decrease to recover.

However, about 1,250 lakes, which a few years earlier were substantially or gravely acidified, had been successfully restored to their original phase through liming by 1985. Together these lakes cover more than 1,000 km^2.

6.9. Consequences of a Future 25% Deposition Decrease

Recent model calculations of the long-range transport of air pollutants (EMEP/MSC-W, 1986) indicate that about 17% of the nonmarine sulfur

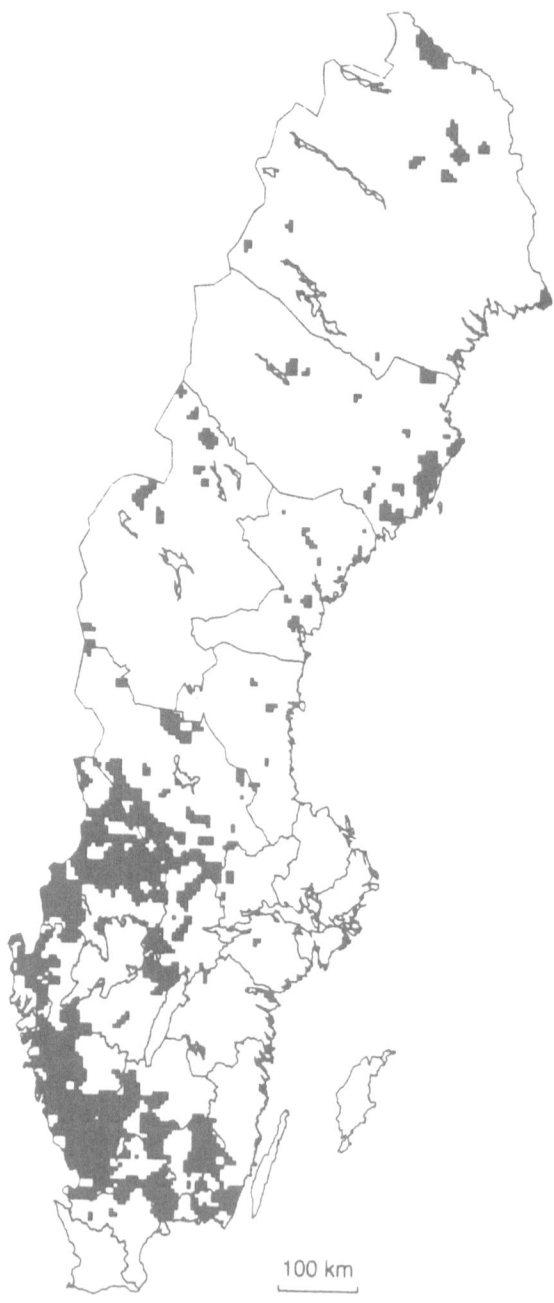

Figure 6.5. Areas (black) where a majority of the lakes sampled in 1985 were substantially or gravely acidified at that time (if unlimed) or before the first liming (if limed in 1985).

deposited over Sweden in 1983–1984 originated from domestic sources, while 40% came from emissions in seven other European countries (the German Democratic Republic, Poland, the Federal Republic of Germany, the United Kingdom, the Soviet Union, Czechoslovakia, and Denmark). Another 5% could be attributed to minor contributions from other countries in Europe, whereas the remaining 38% were ascribed to unidentified emissions that are probably mainly anthropogenic but may include certain natural emissions as well.

Many European countries have agreed to reduce their sulfur emissions by at least 30% from 1980 to 1993. Even larger reductions are planned in Sweden and some of its neighboring countries, while the United Kingdom and Poland have not yet agreed to a 30% reduction. A comparison between available emission estimates for the mid-1980s and the goals set up for the mid-1990s indicates that the total deposition of sulfur from identifiable European sources over Sweden can be expected to decrease by about 25% over the next ten years, assuming that the EMEP model mentioned above holds true generally. Let us also assume that the contributions from unidentified sources will decrease by the same amount.

What, then, would the consequences be for the acidification status of Swedish lakes if the anthropogenic deposition of sulfur were reduced by 25%? The acidification model described in the previous sections allows us to check if the alkalinity in a lake that is now in phase 2 or 3 and is substantially or gravely acidified would increase enough after this deposition reduction to bring the lake back to phase 1 and enable us to classify it as restored.

The results indicate that about 4,400 acidified lakes with an aggregate area close to 1,500 km^2 would be restored if the sulfur deposition were reduced by one-fourth (cf. *Figure 6.6*). We have then disregarded all liming efforts, both those already made and those to be made in future years.

Most of the lakes likely to be restored are located in the southernmost parts of Sweden, where the neutralizing capacity of the soil layers is fairly high, but where many lakes nevertheless are acidified today due to the heavy acid deposition (cf. lake A in *Table 6.1*). However, greater emission reductions than 25% will generally be needed to restore acidified lakes surrounded by this and strongly leached soils. Many such lakes can be found in western and northern Sweden (cf. lakes B and C in *Table 6.1*).

According to the above results, the number of lakes that would be restored through a 25% reduction of the anthropogenic sulfur deposition is more than three times higher than the number of lakes that up until 1985 had been restored by liming. However, this does not mean that present liming efforts could just as well be reduced or stopped. The objects so far limed are a careful selection of valuable and relatively large lakes, which in many cases would need a deposition decrease considerably greater than 25% to recover without liming, whereas most of those 4,400 lakes expected to recover without such treatment after a 25% deposition decrease are fairly small.

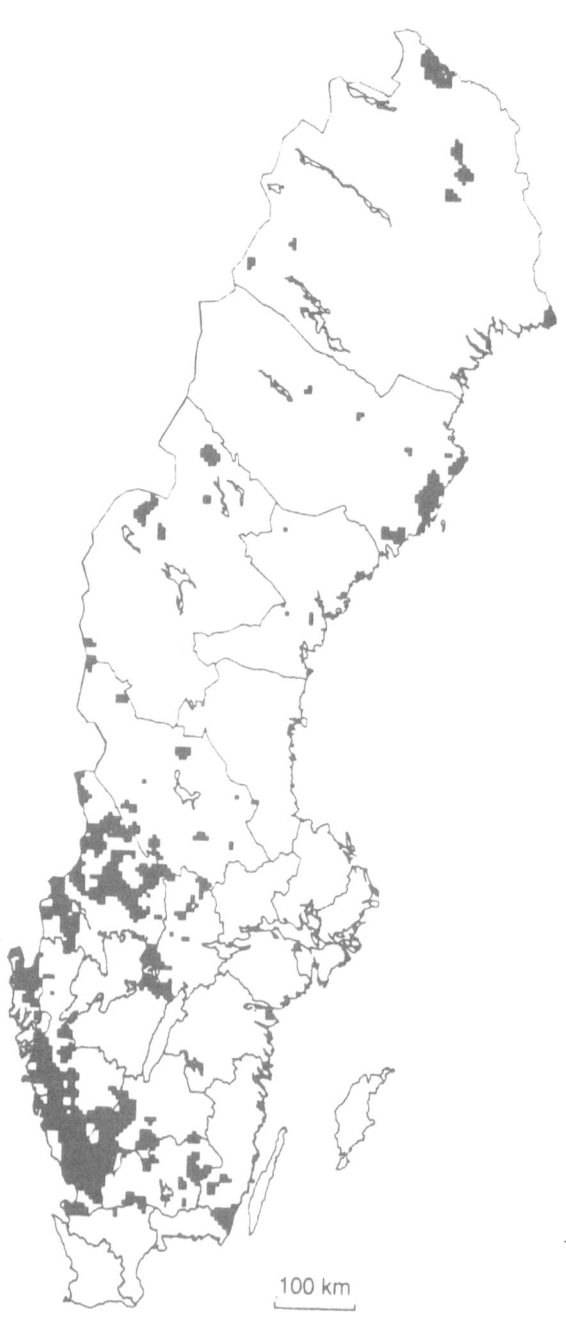

Figure 6.6. Areas (black) where a majority of the lakes sampled in 1985 will probably remain substantially or gravely acidified even after a 25% reduction of sulfur deposition (liming is disregarded).

6.10. When Will the Recovery Take Place?

The acidification model we use does not predict how soon the expected changes will occur in the lakes. It may well be that lakes react with a considerable delay to changes of the sulfur deposition, since some of the deposited sulfur may be retained in the soil for several years before it reaches lakes and waterways. Substances stored in the lake sediments could act to delay the response to deposition changes in a similar way. Furthermore, it must be remembered that the water volume itself may have a retention time of several years or even decades in large lakes.

Nevertheless, a significant decrease – typically 10%–20%, but in some cases as large as 30% – of the sulfate concentration has been observed since the mid-1970s in many rivers and lakes in southern and central Sweden (Ahl, 1986). It is tempting to see this change as a response to the 10%–20% decrease of the average sulfate concentration in precipitation that has occurred in this area during the same period (Granat, 1986).

On the other hand, no general increase of alkalinities or pH values in lakes and rivers has been observed during the past decade. Instead, the sulfate decrease has been accompanied by an almost equivalent hardness decrease. If interpreted as responses to the current deposition decrease, these changes would indicate an F value close to 1.0, i.e., that a decrease of the acid deposition leads to a decreased rate of soil acidification rather than an improvement of the acidification status of surface waters.

As pointed out by Ahl (1986), however, one must be careful with any such interpretation. The mid-1970s marked not only a turning point in the sulfur-deposition trend, but also the end a six-year period of relatively small precipitation amounts and low groundwater levels.

The rise of the groundwater level that has taken place since then has probably caused a decreased rate of sulfur oxidation in the soil and, thus, a slow-down of the sulfate leaching to lakes and waterways. Such a rise would also be expected to cause a decrease of the surface-water hardness, since the groundwater supplied to surface waters is then more superficial and thus less affected by weathering processes than usual. Moreover, all substances would tend to appear in smaller concentrations than usual owing to the simple fact that they are diluted in larger-than-average volumes of water (cf. *Table 6.2*).

The concentration changes observed in lakes and rivers during the past decade could mainly be a net effect of these climate-related factors. The deposition decrease may also have had certain effects, but these could well have been masked or counteracted by the climatic effects. However, if this interpretation is correct, we may expect considerable improvement of the acidification situation when the precipitation amounts decrease again. The short-term climate-related effects should then combine with the long-term deposition decrease to increase pH and alkalinity in the lakes quite significantly.

We may expect effects of the deposition decrease to appear first in small lakes and brooks located in areas where the soil's capacity to retain sulfur and other substances is very limited. The observations by Forsberg *et al.* (1985) of a recent pH increase in a few small and strongly acidified lakes in southwestern

Table 6.2. Changes in sulfate concentrations, hardness, and alkalinity in lake water that may be expected as a consequence of (A) an increase in precipitation and (B) a decrease of sulfur deposition.

	Sulfate	*Hardness*	*Alkalinity*

A. Precipitation increase
Neutralization caused by
   weathering less efficient
   due to more superficial
   groundwater flow
		Decrease	Decrease

Decrease of sulfur oxidation in soil	Decrease	Decrease	Increase
All substances "diluted"	Decrease	Decrease	Decrease
Net result:	Decrease	Decrease	Relatively small change
B. Deposition decrease	Decrease	Decrease	Increase

Sweden may thus be seen as a first sign of a development that should become noticeable in other waters in the relatively near future.

6.11. Discussion and Conclusions

Applying data from an extensive lake survey to a simple acidification model, we have assessed the preindustrial, present, and future acidification status of Swedish lakes. Since a majority of the sampled lakes had been selected at random, we were able to make quantitative estimates of the total number of lakes in various acidification stages under different acid loadings.

The analyses of the selected lakes were restricted to a very limited set of parameters. Certain important variables – notably, the concentrations of sulfate and humic acids – thus had to be calculated. This means that the results obtained for an individual lake must be regarded as uncertain.

The main sources of uncertainty in our treatment are our assumptions that changes of the sulfur deposition eventually lead to proportional changes of the sulfur concentrations in lakes and to more or less equivalent (but opposite) changes of lake alkalinity. These assumptions may seem reasonable, but they are at present almost impossible to verify. This is because data on lake chemistry from the period when sulfur deposition was increasing (until the early 1970s in the case of Sweden) are rather incomplete, and because there is still no actual experience of the extent to which acidified lakes are able to recover in response to a decreased deposition of acids.

More specifically, our results are quite sensitive to the choice of F values. The calculated number of currently acidified lakes and the number of lakes expected to recover after a deposition reduction are both roughly proportional to $(1 - F)$. Had we adopted $F = 0.4$ instead of $F = 0.2$ as our main hypothesis, our estimates of the number of acidified and recovering lakes would thus have been approximately 25% lower than those quoted in Sections 6.8 and 6.9, respectively.

We are aware that F values considerably higher than 0.2 are likely for many lakes surrounded by rich soils, where weathering rates may be strongly dependent on the acid supply. However, few of these lakes are significantly acidified today. The majority of those lakes that are substantially or gravely acidified, according to our definitions and calculations, is surrounded by thin and poor soil layers. There, an F value of 0.2 or less may actually be a good approximation, since weathering rates are probably not very pH-dependent and since exchangeable cations are more or less depleted.

Nevertheless, it may be worthwhile to develop this model further, taking into account, e.g., sulfur retention in the soil and variations of the base saturation. That would allow predictions of when an acidified lake is likely to recover in response to a specific reduction of the acid deposition. In its present form, the model simply predicts the acidification status that may be expected once all soil processes have reached a steady state.

Acknowledgments

The authors gratefully acknowledge the dedicated efforts of field, laboratory, and administrative personnel at the county administrations in collecting and analyzing samples and data during the great lake survey in 1985. We also thank Georg Moberg for generous help with the data handling and William Dickson for many fruitful discussions. This work was funded by a governmental grant for measures against acidification.

CHAPTER 7

A Regional Model of Surface-Water Acidification in Southern Norway: Calibration and Validation Using Survey Data

B.J. Cosby, G.M. Hornberger, and R.F. Wright

7.1. Introduction

The acidification of freshwaters and the disappearance of natural fish populations from hundreds of lakes and streams in southern Norway during the past 50 years has been viewed with increasing concern. Effects on inland fisheries became apparent during the 1940s and 1950s, when brown trout disappeared from an increasing number of lakes. Acidification of these freshwaters was not related to catchment land-use changes, such as draining of bogs and peatlands, and these waters received no significant industrial, municipal, or agricultural effluents. The cause is thus clearly acid precipitation, owing to the emission, oxidation, and long-range transport of sulfur and nitrogen oxides from highly industrialized and densely populated regions of Europe (Overrein *et al.*, 1981).

We undertake here the reconstruction of the changes that have occurred in surface-water chemistry in the lakes of southern Norway in response to increased atmospheric acidic deposition during the last 140 years. We use a mathematical model, MAGIC (Model of Acidification of Groundwater In Catchments) (Cosby *et al.*, 1985a, 1985b) to reconstruct these changes.

Rather than explicitly model hundreds of catchment lake systems, we use a regionalization of the model to simulate the observed statistical distributions of

J. Kämäri (ed.), Impact Models to Assess Regional Acidification, 109–130.
© 1990 *International Institute for Applied Systems Analysis.*

water chemistry variables. The regionalization procedure relies on Monte Carlo simulation techniques to select joint distributions of input parameters for the mechanistic model of catchment hydrogeochemistry. The procedure produces an ensemble of model simulations whose output variables have statistical properties that match those of measured chemical variables obtained from a synoptic lake survey conducted in 1974 (Wright and Snekvik, 1978). Using this ensemble of simulations, past changes in the distributions of water quality variables can be estimated, and the critical model parameters necessary to match the 1974 distribution can be deduced. The calibrated regional model can also be used to forecast changes in the distributions of water quality variables that might be expected in response to different scenarios of future increases or decreases in deposition.

We have previously applied our regionalization procedure to Shenandoah National Park, Virginia (Hornberger and Cosby, 1985b; Hornberger *et al.*, 1986a, b) and, in a preliminary way, to southern Norway (Hornberger *et al.*, 1987). The regionalization procedure is similar in many respects to modeling protocols that have been applied to other studies in earth sciences and ecology (particularly the modeling of forest succession; cf. Shugart and West, 1981). In essence, the procedure substitutes space for time as the dimension along which data are collected to provide constraints in calibrating the model. That is, many data points in space are taken to reflect the different stages in the temporal evolution of discrete, individual systems with different inputs, parameters, etc. An ensemble of model simulations is collected whose outputs can be constrained to be consistent with the observed spatial distributions of water quality variables, resulting in a "calibration" of the model to the region rather to any individual system. In collecting the ensemble of model simulations, no attempt is made to preserve a one-to-one correspondence between any particular simulation and any particular lake in the survey. Rather, the regionalization procedure attempts to match the bulk statistical properties of the simulation ensemble to those of the survey data set.

The appropriate questions to address using the regional model concern the relative proportion of surveyed lakes that exhibit a certain behavior (e.g., how many lakes might become acidic in response to a given deposition scenario). Questions concerning the behavior of any individual lakes are inappropriate given the relative paucity of data for any single lake in the survey data set. The regional model is intended for use with a different type of data (spatially extensive, temporally sparse) than site-specific models (which traditionally use temporally extensive, spatially sparse data).

In this chapter we describe the conceptual model and the survey region in southern Norway; present the regionalization protocol and use a trial and error procedure to calibrate the model input parameter distributions for the survey region; describe a validation procedure that uses forecasts and hindcasts from the calibrated mode with historical fisheries and resurvey data; discuss the robustness of and possible biases in the calibration–validation procedures; and suggest an alternative calibration methodology. In Chapter 13 we implement the alternative calibration methodology; develop a resampling scheme to evaluate uncertainties in calibration, forecasts, and hindcasts; provide a brief sensitivity analysis of the input parameters; and present forecasts for the region using a different (longer-term) scenario of future deposition.

7.2. The Mathematical Model

Acidic deposition causes decreased pH and alkalinity and increased base cation and aluminum concentrations in sensitive surface waters. We postulate that a relatively small number of important soil processes aggregated over whole catchments produce these responses. We have incorporated these processes into MAGIC, a dynamic model of catchment response to acidic deposition (Cosby *et al.*, 1985a, b, c).

MAGIC consists of (1) a section in which the concentrations of major ions are assumed to be governed by simultaneous reactions involving sulfate adsorption, cation exchange, dissolution/precipitation of aluminum and dissolution/speciation of inorganic carbon and (2) a mass balance section in which the flux of major ions to and from the soil is assumed to be controlled by atmospheric inputs, chemical weathering inputs, net uptake in biomass, and losses to runoff (*Figure 7.1*). At the heart of MAGIC is the size of the pool of exchangeable base cations in the soil. As the fluxes to and from this pool change over time, owing to changes in atmospheric deposition, the chemical equilibria between soil and soil solution shift to give changes in surface-water chemistry. The degree and rate of change in surface-water acidity thus depend both on flux factors and the inherent characteristics of the impacted soils.

Sulfate adsorption is treated in the model by a Langmuir isotherm. Aluminum dissolution/precipitation is assumed to be controlled by equilibrium with a solid phase of $Al(OH)_3$. Speciation of inorganic carbon is computed from known equilibrium equations. Cation exchange is also treated using equilibrium (Gaines–Thomas) equations. Weathering rates are assumed to be constant with time in the model. Finally, a set of mass balance equations for the base cations and strong acid anions is included. MAGIC consists of 32 simultaneous equations. Given a description of the historical deposition at a site, the model equations are solved numerically to give long-term reconstructions of surface-water chemistry (Cosby *et al.*, 1985a, b, c).

Application of MAGIC requires data for (1) soil physical and chemical characteristics, (2) rainfall/runoff characteristics, (3) precipitation chemistry covering the period from the onset of acidic deposition to the present, and (4) estimates of base cation weathering rates. For this regional application of the model, the distributions of these inputs for southern Norway are needed. Unfortunately, data do not exist that adequately describe *a priori* the distribution of all these inputs. The regional application of MAGIC must be accomplished by "calibrating" some of the input distributions for the model – particularly the weathering rates and certain soil physical/chemical properties. This procedure is described below.

The distributions of average annual rainfall/runoff and soil and surface water temperatures for southern Norway are derived from climate atlases for the region (Holtan, 1986). Historical precipitation chemistry for the region is

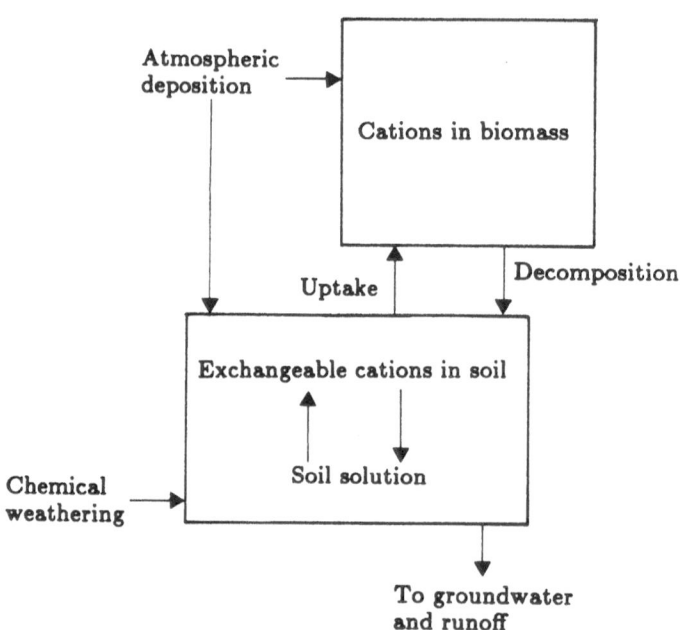

Figure 7.1. Schematic view of MAGIC. The model is essentially composed of two parts: (1) a series of equilibrium chemical reactions between exchangeable cations and soil solutions and (2) fluxes to and from the soil controlled by deposition, chemical weathering, biomass utilization, and runoff.

determined by scaling currently observed precipitation chemistry to a reconstruction of emissions for northern Europe (Bettelheim and Littler, 1979). This scaling procedure has been described elsewhere for regions in North America (Cosby *et al.*, 1985b). In this regional application, MAGIC is implemented using yearly time steps and a single aggregated soil layer. Average soil and surface-water partial pressures of CO_2 are set to 0.02 atm and 0.002 atm, respectively (R.F. Wright, unpublished data). Simulations are run for 130 years (1844–1974) to reconstruct the historical acidification of lakes in the region.

7.3. Region of Application

Southern Norway is an area that has been severely impacted by atmospheric acidic deposition. Surface waters are dilute and inherently sensitive to acid deposition. In 1974–1975, water samples were obtained for 715 lakes in the region to determine the major ion chemistry of these surface waters (Wright and Snekvik, 1978). The 715 lakes are of medium size, with the longest dimension between 1–10 km. Lake areas thus vary between about 0.1–30 km^2. Lakes on major rivers (i.e., those with very large watersheds) were generally excluded. One sample from each lake was collected from the surface (0.5 m) or from the lake outlet during the ice-free season (May–November).

Chemical analyses proceeded according to routine methods used at the Directorate for Wildlife and Freshwater Fish (pH, specific conductance) and at the Norwegian Institute for Water Research (Na, K, Ca, Mg, Cl, SO_4, NO_3). The pH and specific conductance were determined potentiometrically; Na, K, Ca, and Mg were determined by atomic-absorption spectrophotometry; Cl, SO_4, and NO_3 by automated colorimetry; and alkalinity estimated by difference from the ionic balance (Wright and Snekvik, 1978). Forty-one low-elevation lakes (which lie below the limit of seawater incursion during the last interglacial period) were deleted from the data set for this work, leaving a total of 674 lakes to be considered.

The region is underlain by granites and felsic gneisses. Soils are young, podzolic, and generally thin and patchy. Information on soil depths, bulk densities, cation exchange capacities, and base saturations are derived from several studies in southern Norway (Lotse and Otabbong, 1985; Seip et al., 1979a, b; Stuanes and Sveistrup, 1979; Frank, 1980; Abrahamsen et al., 1976, Wright et al., 1977). These soil studies were not necessarily conducted in the same catchments as the survey lakes, but the soils described are similar to the soils in the catchments of the surveyed lakes.

The chemistry of precipitation in southern Norway can be depicted as a mixture of marine aerosols (Cl, Na, Mg, and to a lesser extent K, Ca, and SO_4) and long-range transported pollutants (H^+, SO_4, NO_3, and NH_4) (Wright and Dovland, 1978). These components are also deposited by dry deposition and impaction. The concentrations of sea salt components in precipitation decrease sharply with distance from the coast, reflecting the rapid deposition of the relatively large marine particulates. The concentration of sulfate in precipitation displays a gentler gradient, reflecting lower deposition rates for gaseous and small particulate long-distance transported components. We chose to divide the set of lakes into two classes corresponding to low sea salt inland lakes (Cl < 100 μeq l^{-1}) and high sea salt coastal lakes (Cl > 100 μeq l^{-1}). The regionalized model is applied to each class of lake as described below. The results for the two classes are combined to produce the final calibrated model for southern Norway.

7.4. Monte Carlo Methods

Fourteen parameters are varied in the Monte Carlo simulations. The first 11 parameters are related to soil hydrochemical properties. These are: the annual runoff (Qp); the cation exchange capacity (CEC, intended here as the total areal cation exchange capacity of a system – a product of depth, bulk density, and measured exchange capacity per unit mass); the logarithm of the solubility constant of the aluminum hydroxide solid phase (K_{Al}); the weathering rates of the four base cations (WE_{Ca}, WE_{Mg}, WE_{Na}, and WE_K); and the logarithms of the selectivity coefficients that specify the cation exchange affinity of soils for aluminum and the four base cations (S_{AlCa}, S_{AlMg}, S_{AlNa}, and S_{AlK}).

The remaining three parameters (Cl_{atm}, $SO_4{}^*$, and $NO_3{}^*$) describe the atmospheric deposition for each simulation. Deposition is determined by selecting a value for Cl_{atm}, using sea salt ratios to calculate deposition of based cations and background sulfate based on that Cl_{atm} value, and then incorporating additional sulfate (excess sulfate, SO_4^*) and nitrate (excess nitrate, NO_3^*) to represent anthropogenic deposition of acidic compounds.

We assume that the soils were in steady state relative to inputs from weathering and the atmosphere in 1844, the initial year of simulation. The input parameters listed above all exercise important control on cation dynamics of soil and surface water. Because the soils of southern Norway are young, thin and have low contents of sesquioxides, sulfate absorption is minor. Sulfate concentrations in soil solution and in surface water should respond rapidly to changes in sulfate deposition. Thus, although we include sulfate adsorption parameters, they are fixed at a value consistent with a sulfate steady state assumption in 1974 and are not varied independently in the Monte Carlo runs.

The particular joint distribution of the 11 soil input parameters that is "correct" for this region of southern Norway cannot be decided *a priori*. Our approach is to define broad ranges for acceptable values of the individual soil parameters based on available soil data from this and other similar regions. All soil parameter values are selected from distributions contained within these broad ranges (*Table 7.1*). The input distributions for the soil parameters are assumed to be independent.

Table 7.1. Ranges for the 11 soil hydrochemical parameters used in the Monte Carlo simulations for southern Norway. The parameters are defined in the text. The ranges are the same for each Cl class.

Parameter	Units	Minimum	Maximum
Q_P	cm	50.0	300.0
CEC	eq m^{-2}	10.0	500.0
WE_{Ca}	meq m^{-2} yr^{-1}	0.0	100.0
WE_{Mg}	meq m^{-2} yr^{-1}	0.0	30.0
WE_{Na}	meq m^{-2} yr^{-1}	0.0	30.0
WE_K	meq m^{-2} yr^{-1}	0.0	10.0
S_{AlCa}	logarithm (base 10)	−1.0	5.0
S_{AlMg}	logarithm (base 10)	−1.0	5.0
S_{AlNa}	logarithm (base 10)	−3.0	3.0
S_{AlK}	logarithm (base 10)	−3.0	3.0
K_{Al}	logarithm (base 10)	8.0	11.0

The input distributions of the three deposition parameters are determined from the survey data set. Observed distributions of atmospheric chloride, excess sulfate, and excess nitrate concentrations in the surveyed lakes (as approximated by triangular distributions; see *Table 7.2*) are corrected for rainfall and runoff effects (a water yield of 80% is assumed for all simulations) to provide input deposition distributions for these parameters. The distributions of these three input parameters are not adjusted in the calibration procedure described below because these distributions are considered "fixed" by the data. Both sea salt and anthropogenic inputs of sulfates and nitrates increase toward the coast in southern Norway. These input parameters are expected to be highly correlated.

When sampling from the Cl_{atm} and $SO_4{}^*$ input distributions, we implemented a scheme to reproduce the linear correlation between these inputs that was observed in the survey data (*Table 7.2*). No strong correlation with excess nitrate was found in the data. The distribution of this input parameter was considered independent of other factors within each Cl class, although the upper limit of the excess nitrate distribution was decreased in the low-Cl class.

Table 7.2. Input distributions for the three deposition parameters. The parameters are described in the text. Mode, maximum, and minimum are given for triangular distributions. Units are $\mu eq\ l^{-1}$.

Parameter	High-Cl class			Low-Cl class		
	Mode	*Maximum*	*Minimum*	*Mode*	*Maximum*	*Minimum*
Cl_{atm}	120	350	100	20	100	10
$SO_4{}^*$	45	140	30	0	80	0
$NO_3{}^*$	3	20	0	3	15	0

Relationship between $SO_4{}^*$ and Cl_{atm}	High-Cl class		Low-Cl class	
	Observed	*Simulated*	*Observed*	*Simulated*
Slope	0.22	0.21	0.57	0.56
Intercept	54.50	51.40	19.60	24.40
r^2	0.21	0.21	0.32	0.27

Given the data-derived distributions of deposition parameters for each Cl class, the regionalized model is calibrated by an *ad hoc* trial and adjustment procedure. Simple distributions within the prespecified ranges for each Cl class are postulated for each input soil parameter. Rectangular distributions are used in the first iteration of the calibration procedure; subsequent refinements of these input distributions use simple triangular distributions specified by a maximum, minimum, and mode for each parameter. Soil parameter values are selected at random from these assumed distributions. Simulations are run from 1844 to 1974. The simulated 1974 results are compared with the range of concentrations for each major ion observed in the 1974–1975 survey (*Table 7.3*). If the simulated concentrations fall within the observed "windows" (defined by the maximum and minimum of the observed data), the simulation is accepted; if they do not, the simulation is considered unacceptable as a representation of a catchment in the region.

This sampling–simulation procedure is repeated many times for the assumed input distributions. When a large number of "acceptable" simulations are accumulated for each Cl class, the distributions of the 1974 output variables of this ensemble of simulations are compared with the observed distribution of water quality variables from the survey data set. If the match is not satisfactory, the *a posteriori* distributions of input parameters for the acceptable simulations are compared with the assumed *a priori* distributions of input parameters as an aid in determining how the latter distributions should be modified. The "old" input distributions of soil parameters are adjusted, and the sampling–simulation

Table 7.3. Maximum and minimum values of lake chemistry variables observed in the lake survey of southern Norway. The survey data are divided into two classes based on Cl concentrations. These maximum/minimum values define "acceptable" simulations for each Cl class (see text). All are in μeq l^{-1}, except soil base saturation (BS %). The high-Cl class contains 342 lakes; the low-Cl class contains 332 lakes.

Variable	High-Cl class		Low-Cl class	
	Maximum	Minimum	Maximum	Minimum
Ca	140	5	100	5
Mg	120	20	55	5
Na	315	80	125	20
K	25	2	25	1
SO_4	210	45	180	10
Cl	350	100	100	10
NO_3	30	0	20	0
H^+	50	0	50	0
CALK[a]	80	-70	80	-70
SBC	600	115	300	45
SAA	560	165	240	25
BS %	15	1	15	1

[a] Alkalinity is calculated from ionic balance, CALK = SBC – SAA, where SBC is the sum of base cation concentrations (Ca + Mg + Na + K) and SAA is the sum of acid anion concentrations (SO_4 + NO_3 + Cl).

Table 7.4. Final "calibrated" input distributions for the 11 soil hydrochemical parameters. The parameters are described in the text. Units are given in *Table 7.1.* Mode, maximum, and minimum are given for triangular distributions.

Parameter	High-Cl class			Low-Cl class		
	Mode	Maximum	Minimum	Mode	Maximum	Minimum
Q_p	150	300	50	150	300	50
CEC	10	500	10	10	500	10
WE_{Ca}	20	100	0	0	60	0
WE_{Mg}	0	10	0	0	10	0
WE_{Na}	0	10	0	0	30	0
WE_K	0	10	0	0	10	0
S_{AlCa}	2	5	-1	2	5	-1
S_{AlMg}	2	5	-1	2	5	-1
S_{AlNa}	0	3	-3	0	3	-3
S_{AlK}	0	3	-3	0	3	-3
K_{Al}	10	11	8	10	11	8

procedure repeated until a new comparison with the observed variable distributions can be made.

In this manner a set of soil parameter input distributions is finally derived (*Table 7.4*), which reproduces (approximately) the observed variable distributions for each Cl class. In this work, two iterations of the procedure are employed: the initial pass with rectangular distributions, followed by a single pass with triangular distributions.

After the two Cl classes have each been "calibrated," the simulation ensembles are combined to produce the final model for the region. On combining the final simulation ensembles, the number of surveyed lakes in each Cl class (*Table 7.3*) is used to weight the simulation ensembles for that class.

7.5. Results

The initial Monte Carlo procedure consisted of 3,000 simulations for each class, drawing soil parameter values from independent rectangular distributions with the limits given in *Table 7.1*. Exactly 994 simulations satisfied all constraints (i.e., were within the measured ranges for all constituents of the 1974 measured water chemistry and base saturation; see *Table 7.3*) for the high-Cl class; and 1,127 simulations satisfied all constraints for the low-Cl class (approximately 65% unacceptable runs). By comparing the distributions of parameters for the acceptable simulations with the distributions of the same parameters for the unacceptable simulations, we were able to derive modified input distributions that increased the proportion of acceptable runs.

For example, WE_{Ca}, S_{AlCa}, S_{AlMg}, and K_{Al} for the high-Cl class all show very different distributions when the "window" constraints were met as opposed to when the constraints were not met (*Figure 7.2*). To improve the number of acceptable runs in the next Monte Carlo iteration, we modified the input distributions of these parameters to make them more like the posterior distributions of acceptable runs from the first iteration (*Figure 7.2*). In this manner a new set of triangular distributions was specified for the 11 soil parameters (*Table 7.4*). Unless the analysis strongly suggested otherwise (as was the case only for WE_{Ca} and WE_{Na}), the modified distributions were assumed to be the same for each Cl class (i.e., soil properties were assumed to be similarly distributed in coastal and inland regions).

The independent triangular distributions were used as inputs, and 3,000 additional simulations were performed for each Cl class. The new distributions produced 1,922 and 1,693 acceptable simulations for the high- and low-Cl classes, respectively (approximately 40% unacceptable runs). While the proportion of acceptable runs increased, the posterior distributions of acceptable versus unacceptable parameter values were much more similar after this iteration (*Figure 7.3*) than after the first (*Figure 7.2*). The procedure could be applied again to achieve a further refinement of the input distributions, but the marked similarities between the posterior univariate distributions of the input parameters do not provide much information as to how the input distributions might be modified. In other words, the triangular input distributions are as likely to produce unacceptable as acceptable simulations (viewed in the univariate sense; see *Figure 7.3*), and any further significant improvement in the number of acceptable simulations is not likely to be achieved by changing the triangular shape of any individual parameter distribution. We infer from this that the remaining information concerning what constitutes acceptable simulations lies predominantly in the covariance structure of the input parameters.

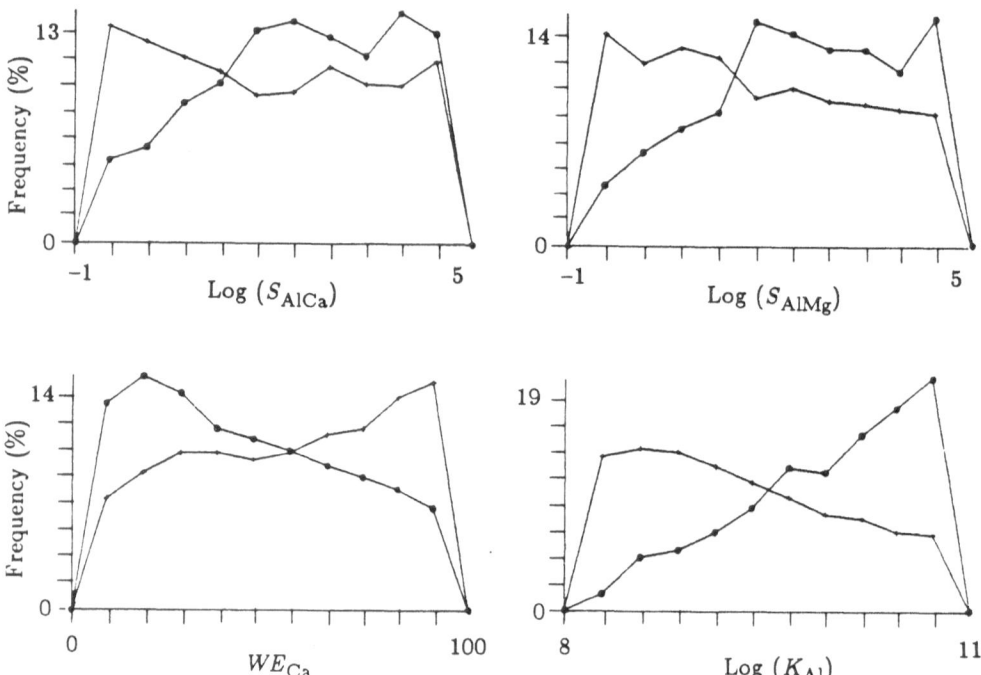

Figure 7.2. A comparison of the distributions of four input parameters for acceptable [window (•)] versus unacceptable [not window (+)] simulations derived from rectangular input distributions for the high-Cl class. The dissimilarities between distributions for each parameter suggest that more acceptable runs may be achieved if the input distributions are modified. The shapes of the distributions for acceptable simulations suggest that the input distributions should be approximately triangular.

There is no information available concerning the covariance of soil hydro-chemical properties in the Norwegian catchments. We chose, therefore, not to attempt to specify a covariance structure for the *a priori* input distributions of soil parameters and perform another iteration of the procedure. In that the current ensembles of acceptable simulations for each class are entirely defined by the observed data, we infer that these ensembles necessarily contain a covariance structure consistent with the data. Having used the Monte Carlo procedure to specify independent univariate input distributions that result in a large number of acceptable simulations, we use the observed data "windows" as a "filter" to reject a subset of the simulations generated from those input distributions. The rejection of this subset will result in a *dependent* joint distribution of the parame-ters of the remaining acceptable simulations with an "induced" covariance struc-ture. The acceptable simulations from the second iteration of the Monte Carlo procedure, along with the induced correlations among parameters (*Table 7.5*), therefore define the empirically "calibrated" regional model for each Cl class.

The statistical properties of the water quality variables simulated for each Cl class by the "calibrated" regional model agree well with those of the survey data set. The maximum difference between the mean of a simulated variable

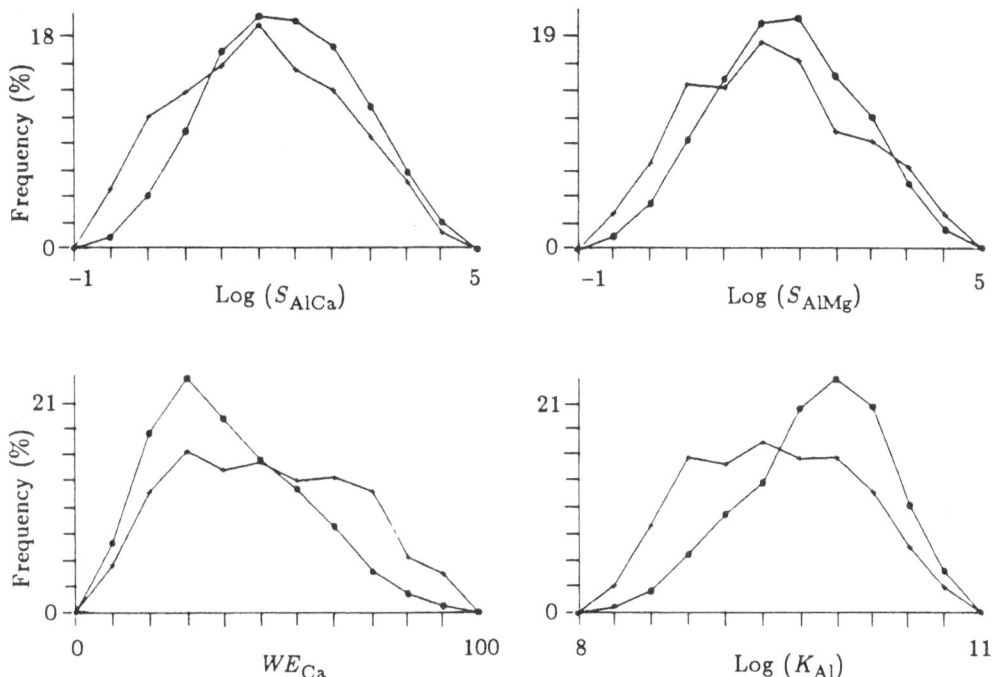

Figure 7.3. A comparison of the distributions of four input parameters for acceptable [window (•) versus unacceptable [not window (+)] simulations derived from triangular input distributions for the high-Cl class. The similarities between distributions for each parameter suggest that a larger proportion of acceptable runs is not likely to be achieved by further modification of the univariate input distributions (see text).

and the mean of that observed variable for either Cl class is 8 μeq l^{-1} (*Table 7.6*). The means of all but one variable (high-Cl calcium) differ by 5 μeq l^{-1} or less. The maximum difference between simulated and observed standard deviations for any variable in either Cl class is 19 μeq l^{-1} (*Table 7.6*). The standard deviations of all but two variables (high-Cl and low-Cl sum of base cations) differ by 10 μeq l^{-1}) or less. If the differences between simulated and observed means or standard deviations given in *Table 7.6* are expressed as percentage of the ranges of each variable given in *Table 7.3*, we find that all differences in means or standard deviations (for either Cl class) are less than 10% of the range of interest for all variables.

This is not to say that the differences between the statistical measures of the simulated and observed distributions are not significant. Given the large sample sizes of the survey data and model simulations in each Cl class, even small differences in means or variances will prove to be statistically significant in appropriate tests. The purpose of the comparison in *Table 7.6* is to demonstrate that the differences between the first two moments of the simulated and observed distributions are "operationally" small (even though they may be statistically significant) and are therefore unimportant.

Table 7.5(a). Correlation matrix (r) for the ensemble of soil parameters comprising the acceptable simulations for the high-Cl class (second Monte Carlo iteration, triangular distributions).

Parameter	Q_p	CEC	WE_{Ca}	WE_{Mg}	WE_{Na}	WE_K
Q_p	1.00	.09	.15	.03	−.01	.03
CEC	−	1.00	−.07	.01	−.01	−.03
WE_{Ca}	−	−	1.00	.01	−.01	−.06
WE_{Mg}	−	−	−	1.00	−.02	.02
WE_{Na}	−	−	−	−	1.00	−.03
WE_K	−	−	−	−	−	1.00

Parameter	S_{AlCa}	S_{AlMg}	S_{AlNa}	S_{AlK}	K_{Al}
Q_p	−.08	−.02	−.03	.02	−.11
CEC	.02	.04	.03	−.04	.04
WE_{Ca}	.10	.14	.05	.01	.14
WE_{Mg}	−.03	.06	.01	.00	.03
WE_{Na}	−.01	.02	−.02	−.05	.00
WE_K	.01	.05	−.01	−.01	−.03
S_{AlCa}	1.00	−.07	−.06	.01	−.13
S_{AlMg}	−	1.00	−.09	.02	−.14
S_{AlNa}	−	−	1.00	−.02	−.02
S_{AlK}	−	−	−	1.00	−.03
K_{Al}	−	−	−	−	1.00

The final product of the calibration exercise is a specified joint distribution of input parameters, which will (when sampled in a Monte Carlo framework) produce distributions of water quality variables that can (1) be compared with measured distributions, (2) be used to infer historical distributions, or (3) be used to forecast future responses. Having achieved an operationally defined good fit to the distributions of water quality variables in each Cl class, the acceptable simulation ensembles for each class are combined to produce the final regional model for the lakes of southern Norway. In that the regional model is intended to simulate distributions of variables, the total number of model simulations included in the combined ensemble is arbitrary. The only constraint here is that the combined ensemble consist of weighted contributions from each Cl class that correctly represent the proportion of survey lakes in each class (see *Table 7.3*). Accordingly, all 1,693 acceptable simulations from the low-Cl class were combined with 1,744 (randomly chosen) acceptable simulations from the high-Cl class.

The univariate distributions of the resultant input parameter set [*Figure 7.4(a)*, (*b*), (*c*)] are similar to the *a priori* triangular distributions used for each Cl class. The process of "filtering" the parameters using the observed data windows has resulted in some modification of the theoretical inputs, but the general shapes of the univariate distributions have been preserved. These empirical parameter distributions [*Figure 7.4(a)*, (*b*), (*c*)] with their induced covariance structure (see above) form the "calibrated" inputs to the regional model for southern Norway.

Table 7.5(b). Correlation matrix (r) for the ensemble of soil parameters comprising the acceptable simulations for the low-Cl class (second Monte Carlo iteration, triangular distributions).

Parameter	Q_p	CEC	WE_{Ca}	WE_{Mg}	WE_{Na}	WE_K
Q_p	1.00	.02	.17	.02	.01	.06
CEC	–	1.00	−.06	−.02	−.03	−.01
WE_{Ca}	–	–	1.00	.05	−.05	−.02
WE_{Mg}	–	–	–	1.00	−.03	−.01
WE_{Na}	–	–	–	–	1.00	−.01
WE_K	–	–	–	–	–	1.00

Parameter	S_{AlCa}	S_{AlMg}	S_{AlNa}	S_{AlK}	K_{Al}
Q_p	−.12	−.04	−.04	−.02	−.10
CEC	.03	.03	.01	−.02	−.02
WE_{Ca}	.11	.05	.05	.01	.13
WE_{Mg}	.00	.01	.04	−.02	.04
WE_{Na}	.04	−.01	.01	−.02	.04
WE_K	.02	−.03	.02	−.01	.01
S_{AlCa}	1.00	−.12	−.10	.06	−.16
S_{AlMg}	–	1.00	−.10	−.04	−.06
S_{AlNa}	–	–	1.00	−.02	−.04
S_{AlK}	–	–	–	1.00	.00
K_{Al}	–	–	–	–	1.00

The distributions of simulated water quality variables produced by the "calibrated" regional model closely match those measured in the lake survey for cations [*Figure 7.5(a)*], anions [*Figure 7.5(b)*], and alkalinity and pH (*Figure 7.5(c)*). The simulated distribution of soil base saturation (*Figure 7.5(a)*] is remarkably similar to the distribution derived from soil survey in southern Norway (see site description). As mentioned above, the discrepancies between simulated and observed distributions would prove statistically significant if appropriate tests were performed. The operational discrepancies between simulated and observed distributions are, however, not large.

7.6. Discussion

This regional model of lake chemistry in southern Norway provides a means for predicting the long-term changes in water quality distribution resulting from changes in acid deposition. It is difficult to conduct a test of such a model for two reasons: (1) the model outputs are statistical distributions and (2) time scales are involved. These two problems are considered separately.

A true test of the model (in the classical sense of model validation) would require an independent data set with which the model simulations could be compared. One method that is traditionally used to provide an independent data set is to divide the available data into two subsets: one for calibration of the mode, and the other for validation of the model. When the observed data and model

Table 7.6. Means and standard deviations (μeq l^{-1}) of water quality variables observed in the lake survey compared with those simulated by the "calibrated" regional model.

Variable	Sim mean	Obs mean	Sim S.D.	Obs S.D.
	High-Cl class			
Ca	51	59	25	35
Mg	56	53	19	18
Na	172	168	50	52
K	7	9	3	5
SO$_4$	107	112	29	30
Cl	185	181	54	57
NO$_3$	8	8	4	5
H$^+$	15	19	10	11
CALK	-15	-12	29	34
SBC	285	290	75	93
SAA	300	301	75	80
	Low-Cl class			
Ca	30	29	17	26
Mg	18	19	8	9
Na	54	51	20	22
K	4	4	2	3
SO$_4$	55	51	22	25
Cl	48	46	20	23
NO$_3$	6	6	3	4
H$^+$	10	14	8	9
CALK	-3	0	25	30
SBC	106	103	33	52
SAA	108	103	37	44

outputs are statistical distributions, however, this subsampling scheme is not applicable. In that the original survey data provide estimates of the parent distributions of water quality variables in the region, any subsample of the survey data would be expected to provide equivalent estimates of those distributions (subject, of course, to the constraint that the subsamples are randomly chosen from the original data). In the limit as the calibration and validation subsets become large (i.e., many lakes have been surveyed), the two subsets will have identical distributions; and a good model fit to the calibration subset must necessarily result in a good model fit to the validation subset. Stated differently, the distributions of random subsets of a survey data set are not independent. If non-random subsets of the original survey data are selected, the calibration and validation distributions are by definition different and a model calibrated to one subset is not expected to agree with the other subset. Subsets of a survey taken at one point in time do not permit a valid test of the model.

The alternative approach to obtaining an independent validation data set is to use data from a survey taken at a different time. This approach will, in theory, permit valid model tests, but the time scales of acidification introduce practical problems in implementation. Changes in water quality in response to acidic deposition typically occur over decades rather than years. Validation surveys must be sufficiently separated in time to allow statistically detectable

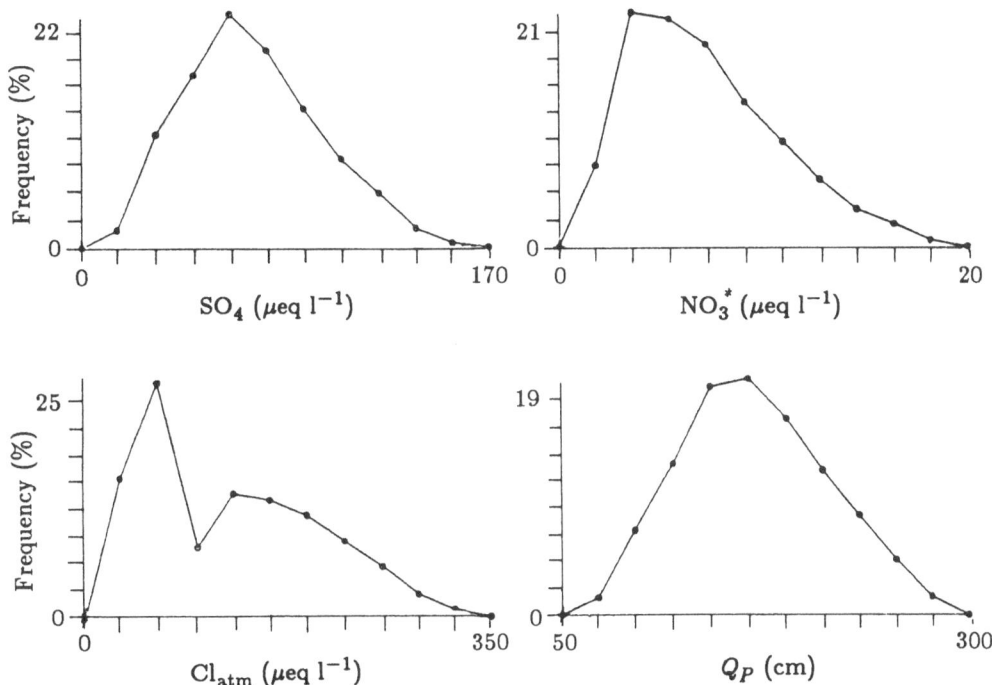

Figure 7.4(a). "Calibrated" input distribution of deposition and runoff parameters for the surveyed lakes.

changes in the observed variable distributions to occur, if a robust test of the model is to be achieved. These practical problems notwithstanding, this approach provides the only feasible method of generating tests of the model. We apply this method to test the regional model of southern Norway using both hindcasts and forecasts.

The "calibrated" regional model was used to reconstruct the distribution of lake alkalinity that existed in southern Norway before the onset of acidic deposition (*Figure 7.6*). Compared with the 1974 distribution of alkalinity in the region [*Figures 5(c)* and *7.6*], we conclude that better than half of the surveyed lakes have lost their alkalinity as a result of acidic deposition. The model was also used to forecast the changes in alkalinity distribution that occurred between 1974 and 1986. Deposition monitoring data over that period suggest that sulfur deposition in the region has declined only slightly – 7% (R.F. Wright, unpublished data) during the 12-year period. Using this observed decline as an input, the model forecasts essentially no change in the distribution of alkalinity for the region by 1986 (*Figure 7.6*).

There are two means by which the model can be evaluated using these hindcasts and forecasts. First, the hindcast alkalinity distribution in 1844 (*Figure 7.6*) can be compared with historical water quality data. Historical water quality data, unfortunately, do not exist, but historical fisheries data for the lakes are available. The reconstructed chemistry can be used to reconstruct the

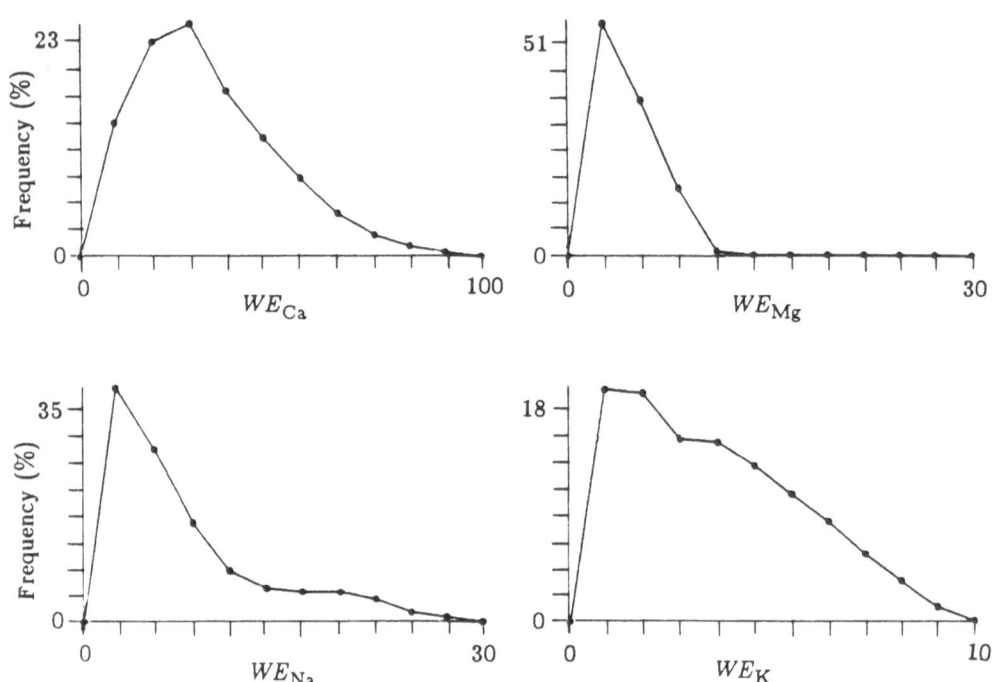

Figure 7.4(b). "Calibrated" input distribution of base cation weathering parameters for the surveyed lakes.

observed changes in fisheries status by virtue of their empirical relationship (Wright and Henriksen, 1983). Using this relationship, the number of fishless lakes expected in the region can be reconstructed for any year. A cumulative plot of the percentage of fishless lakes as a function of time (*Figure 7.7*) suggests that, by 1974, approximately 60% of the surveyed lakes should have lost their fish populations. Wright and Snekvik (1978) report that, by 1974, 80% of the surveyed lakes either had lost their fish populations or had sparse populations (40% in each category). The correspondence between these observations on fisheries status and our reconstructions of alkalinity are encouraging. Further work along these lines is under way to compare not only the cumulative fish losses by 1974, but also the losses that occurred in each decade. Such comparisons should provide additional constraints on dynamic model responses.

The second approach to model evaluation uses the forecast alkalinity distribution based on observed declines in deposition between 1974 and 1986 (*Figure 7.6*). These results suggest that the Norwegian systems are near steady state and, given the small decrease in deposition, little change is expected in the alkalinity distribution by 1986. A large number of the lakes included in the 1974 survey were resurveyed in 1986 (R.F. Wright, personal communication). This resurvey data will provide an independent data set with which this forecast can be compared. During the preparation of this chapter, the resurvey data were not available, so this evaluation cannot yet be completed. Data sets consisting of

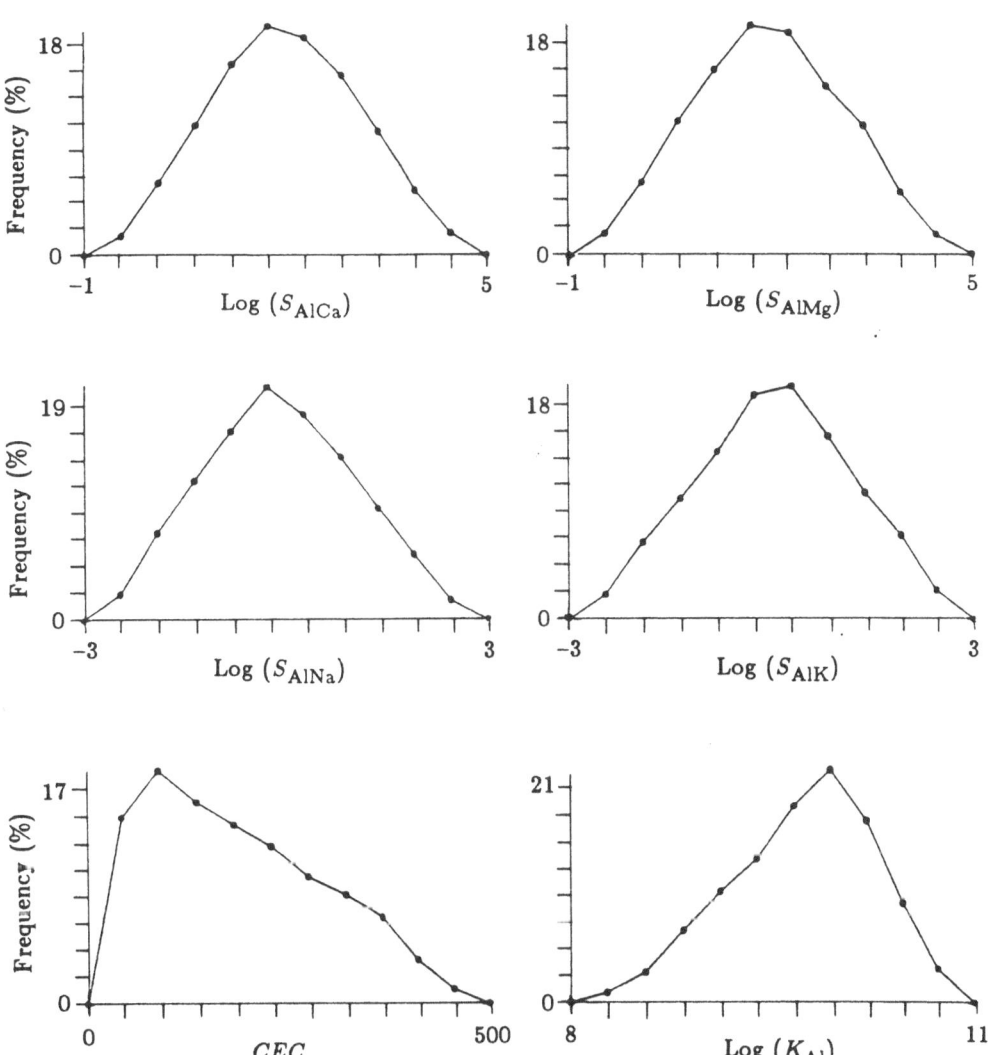

Figure 7.4(c). "Calibrated" input distribution of cation exchange parameters for the surveyed lakes.

successive surveys of a large number of lakes are uncommon, and this work presents a unique opportunity to evaluate the regional dynamic model. Unfortunately, both the observed deposition change and the forecast water quality response are small, thereby producing a less than robust test of the model.

Setting aside considerations of model validation, the calibration procedure illustrated here needs to be examined in two respects: is this the "best" calibration of the model for the region (a question of uniqueness), and just how "good" is the fit of the calibrated model (a question of uncertainty).

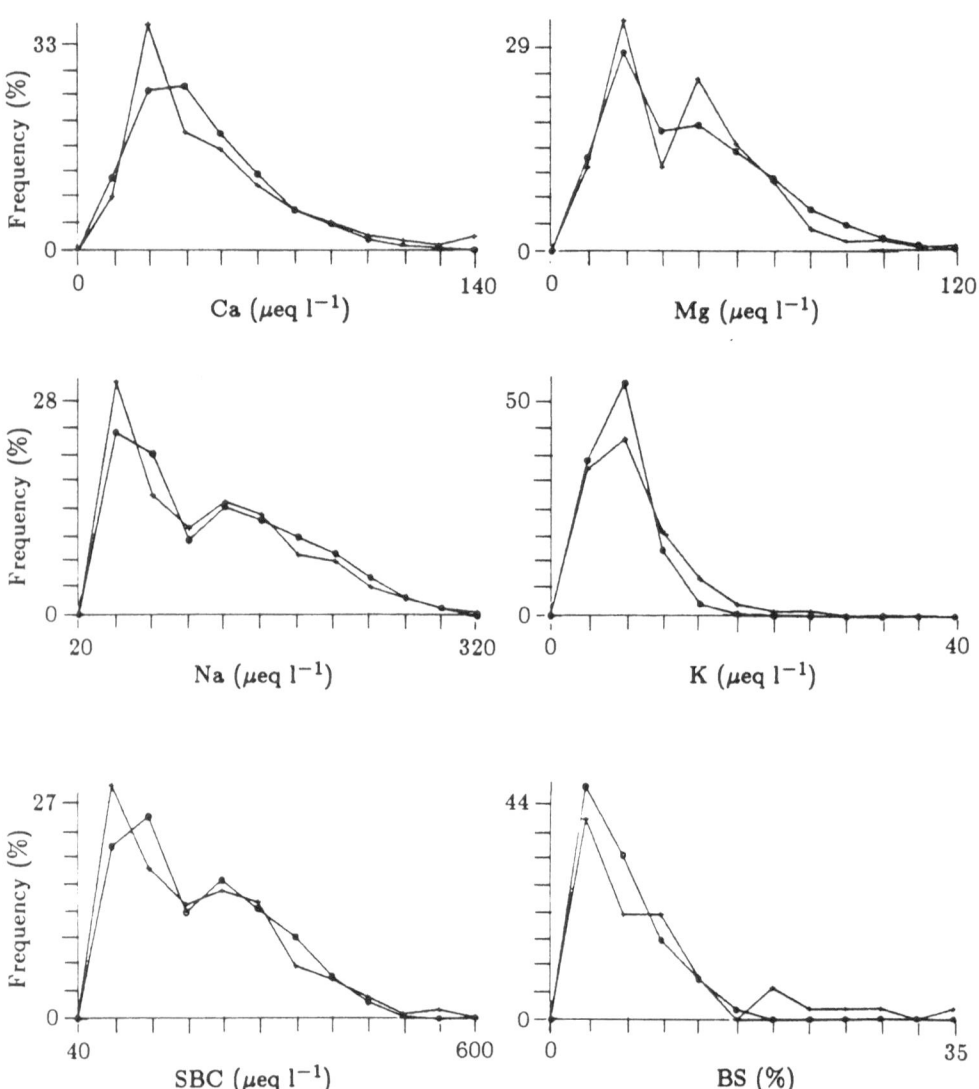

Figure 7.5(a). Simulated (•) and observed (+) 1974 distributions of base cation concentrations (upper four panels), sum of base cation concentrations (bottom left panel), and soil base saturation (bottom right panel).

First, the calibration procedure as presented is essentially a "seat of the pants" procedure. Different input distributions are arbitrarily selected, model output is evaluated, and the input distributions are modified in a manner calculated to improve model fit. It is the modification of the input distributions that lacks objective rigor. It is not possible to imagine all possible input shapes that might produce good fits. As a result, only simple shapes are tried. The distributions of soil characteristics in the Norwegian catchments are undoubtedly

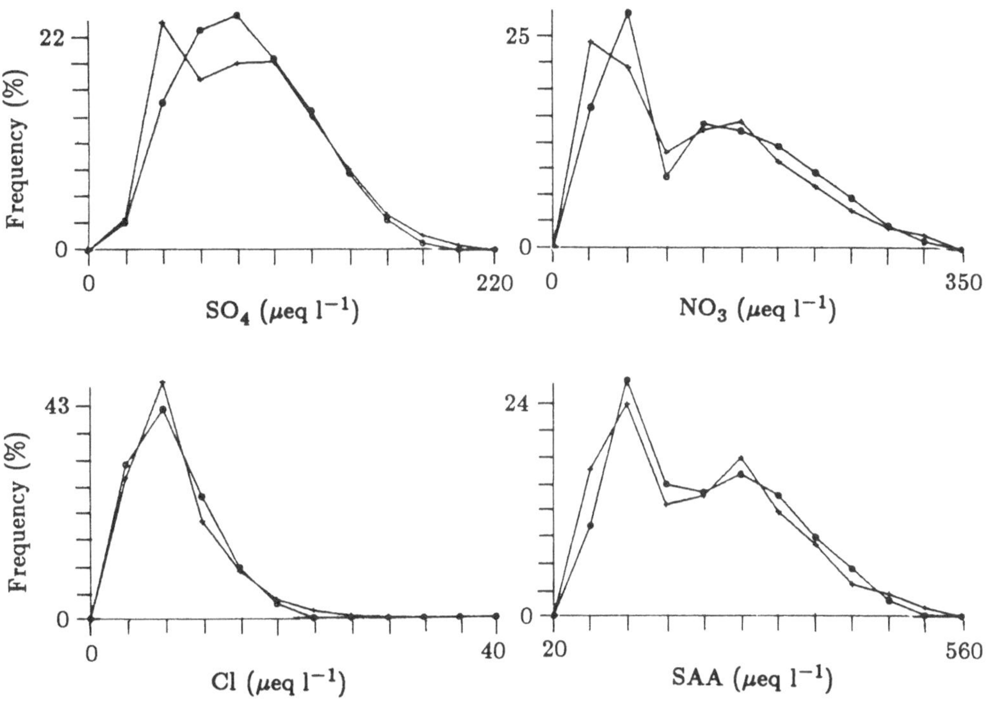

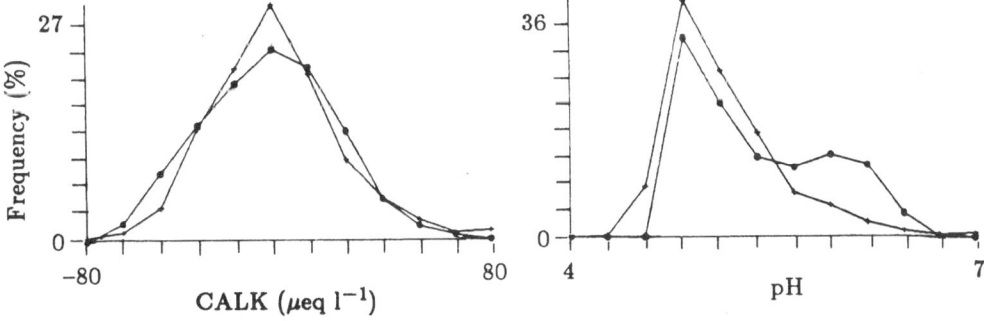

Figure 7.5(b). Simulated (•) and observed (+) 1974 distributions of acid anion concentrations and sum of acid anion concentrations (lower right panel).

Figure 7.5(c). Simulated (•) and observed (+) 1974 distributions of pH and alkalinity.

complex, and describing those distributions in the model with flat or unimodal shapes may severely limit the ability of the model to reproduce the observed water quality variable distributions. Furthermore, covariance among parameters is ignored. These difficulties are overcome to a certain extent by "filtering" the simulations using the ranges of observed variable values (only those simulations deemed acceptable by this criterion were retained in the final calibrated model). The "filtered" input distributions, however, retain much of the character of the arbitrarily assumed *a priori* shape. This suggests that the calibration procedure

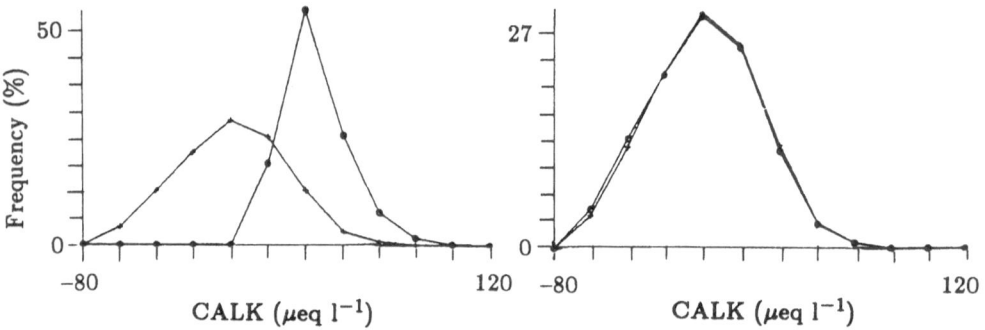

Figure 7.6. Comparison of hindcast [simulated 1844 (•)] and current [1974 (+)] simulated alkalinity (left panel). Comparison of forecast (1986) and current alkalinity, assuming a 7% decrease in sulfur deposition between 1974 (+) and 1986 (•) (right panel).

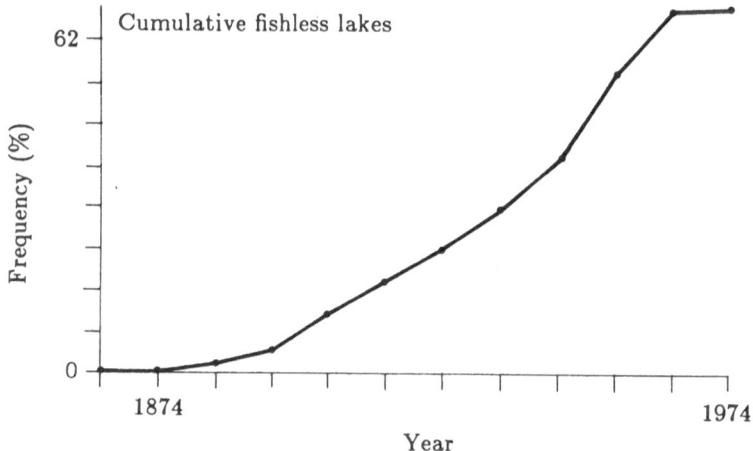

Figure 7.7. Reconstructed fisheries history for southern Norway. The plot shows the total percentage of lakes that are not expected to have viable fish populations (based on reconstructed water quality) in each decade from 1874 to 1974.

should have ended after the first Monte Carlo iteration, which used rectangular input distributions. That is, equal probabilities should be allowed for any parameter value, with the final distribution of parameter inputs determined solely by the "filtering" process based on observed data. Given little knowledge of the "true" shapes of the input distributions, we perhaps unintentionally introduce biases by using input distributions that result in *a priori* unequal probabilities for different parameter values. This aspect of calibration uniqueness and uncertainty is examined in more detail in Chapter 13.

Second, even though the distributions of water quality variables simulated by the regional model are markedly similar to the observed distributions [*Figure 7.5(a), (b), (c)*], differences exist. Given the large sample sizes for both the lake survey and the simulation ensemble, even much smaller differences are likely to be statistically significant. The objective of the calibration procedure should therefore be to minimize these differences. If they cannot be rendered insignificant, they should at least be reduced to the level where they are operationally unimportant.

The calibration procedure as described in this chapter used only the ranges of the observed variables to define data "windows" that were used to constrain the model. That is, only those runs that produced simulated values for all variables within these "windows" were retained. Each simulation was judged on its own merits, and an ensemble of acceptable simulations was accumulated. No constraint was placed on the shape of the accumulated distributions within the windows. In general, one would not necessarily expect that fitting "rectangular windows" would result in simulated frequency distributions that would match observed distributions. This calibration procedure attempted to improve the match between observed and simulated variable distributions within each window by modifying the distributions of the input parameters The relatively good fit that was obtained in this manner argues for the robustness of the model. As mentioned above, however, arbitrarily selecting input distributions may result in a biased model.

An alternative procedure to obtain good fits to variable distributions within the windows can be found (Kämäri et al., 1986). The observed distributions of the survey variables can be used along with the observed ranges to provide additional constraints on the model during calibration. Once an ensemble of acceptable simulations has been identified using the window constraints, the simulated distributions within each window can be "trimmed and shaped" by discarding simulations in certain class intervals until (relatively) enough have been accumulated in others to match the observed distributions. This amounts to applying a second data-based filter to the simulation. Following this procedure (and given enough simulations from which to choose), an exact match to the observed frequency distribution can be achieved. Rather than manipulate the input distributions in an arbitrary manner to achieve an approximate fit to the "filtered" simulation ensemble, the simulations are "twice-filtered," using data-defined criteria to achieve an exact match to histographs that approximate the observed joint distribution of variables.

This "bin filling" procedure, however, is no less *ad hoc* than a trial and error adjustment of input distributions. The arbitrary nature of the "bin filling" procedure becomes apparent when one is faced with the necessity of discarding simulations within some class interval of the output distributions. Which simulations should be discarded, and which should be retained? In the absence of an objective criterion for this decision, the possibility exists of introducing bias in the calibration if the "wrong" simulations are discarded. This alternate calibration procedure and its associated difficulties are also discussed in Chapter 13.

Finally, every model calibration should be accompanied by a quantitative analysis of the uncertainty in the simulation of the calibration data set and uncertainty in forecasts and hindcasts made with the calibrated model. *Ad hoc* calibration procedures make such analyses difficult. Alternative (and equally *ad hoc*) calibrations can be found against which the "final" calibration can be compared. There is no objective way, however, to decide on the likelihood of one calibration *versus* another. In other words, using "trial and error" or "bin filling" calibration procedures, we cannot be sure that the differences between simulated and observed distributions in our final calibration are the smallest possible. Such determinations can only be made through repeated calibrations using random or directed search algorithms to obtain the best fit. If such algorithms are implemented, a quantitative analysis of calibration uncertainty can be achieved by comparing the "best" fit with other fits that are slightly "less good" (cf. Bard, 1974). If several *ad hoc* calibrations result in equally good fits to observed data, we are faced with the additional difficulty that forecasts or hindcasts made with each calibration might differ. In these cases of non-unique calibrations, an analysis of the degree to which forecasts and hindcasts diverge for each "equally good" calibration is essential. In Chapter 13 Hornberger *et al.* perform such an analysis for this regional model of southern Norway.

Despite these difficulties, the regional model presented here performs remarkably well. The simulated distributions of water quality variables are operationally very consistent with observed data from lakes in southern Norway [see *Figure 7.5(a)*, *(b)*, *(c)*]. Even though confidence limits are not presented here, the calibrated model can be used in simulation experiments to examine the likely consequences of long-term changes in deposition on the distribution of water quality in southern Norway. Combined with the uncertainty analysis in the companion analysis (Chapter 13), the model provides a useful tool for examining the implications of proposed policy decisions regarding reductions of acidic deposition

Acknowledgments

This research was made possible by generous grants from the Norwegian Ministry of the Environment and the Norwegian Institute for Water Research. The research was funded in part by the EPA Direct/Delayed Response Program. It has not been subjected to EPA's required peer and policy review and therefore does not necessarily reflect the views of the Agency, and no official endorsement should be inferred.

CHAPTER 8

Regional Modeling of Acidity in the Galloway Region in Southwest Scotland

T.J. Musgrove, P.G. Whitehead, and B.J. Cosby

8.1. Introduction

The southwest region of Scotland (*Figure 8.1*) is believed to be particularly vulnerable to the effects of acidic deposition (Wright and Henriksen, 1979). Reports of long-term changes in acidic status of lakes and streams combine with reports of a decline in fish populations to present a picture of concern to freshwater fisheries and environmentalists alike (Battarbee and Flower, 1986; Burns *et al.*, 1984).

Action to reverse the observed trends can be effectively prescribed only once the processes within the catchment soils that transform acidic precipitation into stream water pollution are understood. Practical experimentation in the laboratory and in the field can help to explain part of the story. The picture for soils, however, is obscured in the field by their heterogeneous nature, and in the laboratory by the inability to reproduce field conditions exactly. Mathematical models are very useful as an aid to understanding the intrinsic processes in complex systems, such as the soil–water chemical interaction through time and space. By using models, hypotheses may be tested by simulating processes and the results compared with their measured counterparts. Useful information may often be deduced, irrespective of the outcome of the comparison.

In this chapter we use a well-documented and extensively used model of groundwater in catchments (the MAGIC model) to perform a regional analysis of lakes and streams in southwest Scotland. Instead of modeling each lake and stream separately, we pool them all and model the overall probability density

J. Kämäri (ed.), Impact Models to Assess Regional Acidification, 131–144.
© 1990 *International Institute for Applied Systems Analysis.*

Figure 8.1. Map of the British Isles showing location of the Galloway area.

functions of water chemistry variables. We use the Monte Carlo technique to run the model 880 times, taking for each run a different set of parameters randomly sampled from specified distributions. The runs that enable us to match the observed distributions are investigated to determine the sensitivity of the model results to the individual parameters. The changes that have occurred in the region as a result of increased deposition during the past 140 years are simulated, and a possible future scenario of deposition is investigated.

8.2. Analysis of Chemical Data for the Galloway Region

The Galloway Region in southwest Scotland contains many lochs and streams that drain moorland, forest, and pasture catchments. The bedrock consists mainly of lower Paleozoic rocks of Ordovician and Silurian systems with a few intrusions of granite of the Old Red Sandstone era (Harriman et al., 1987). These catchments have only a thin covering of unconsolidated glacial till, often closely related to the nature of the underlying rocks (Greig, 1971). Many catchments in the region are covered by a blanket layer of peat of between 50 and 100 cm thickness. The southern part of the region is reported to have a higher calcite availability than the northern part (Welsh et al., 1986; Edmunds and Kinniburgh, 1986b). The mean yearly rainfall ranges from 0.8 to 2.4 m with a mean of 2.0 m.

The concentrations of the major ions indicate three important sources: terrestrial input of Ca, Mg, Al, HCO_3 as products of weathering; atmospheric pollution input of H, NH_4, NO_3, and SO_4; and atmospheric input of seawater salts Na, Cl, Mg, and SO_4 (Wright et al., 1980a). The distribution of these ions varies spatially with a regional character. Concentration of all the major ions is higher nearer the coast to the south. Pollutant concentrations are higher as a result of the location of the sulfur emission areas to the south and the southeast of Galloway. Weathering is highest in the more calcareous Silurian and Ordovician rocks to the south rather than in the slowly weathering, harder granitic intrusions to the north. Sea salts are more evident nearer the coast. The pH of precipitation is 4.1 to 4.4 and contains a large excess sulfate proportion (Wright and Henriksen, 1979). The region as a whole is believe to be highly sensitive to acidic deposition and is therefore a useful area for scientific study.

Data sets from four separate investigations in Galloway were used in the study. First, Edmunds and Kinniburgh (1986b) sampled sites during the wet summer of 1985 as part of a larger survey of groundwater throughout the United Kingdom. They took spot samples from springs, shallow wells, boreholes, and river baseflow from a 20 by 25 km area extending from Wigtown in the south to Loch Macaterick in the north and from the River Luce in the east to the Waters of Ken in the west. The study encountered the full range of Galloway bedrock, but sampled few coastal sites. Second, Wright and Henriksen (1979) took spot samples from 72 lochs and 39 streams during the wet period of 19–26 April 1979. The data were compared with an area of similar acidic susceptibility in Norway. The area of study was 70 km by 50 km and covered the entire Galloway region. These samples thus enveloped the complete spectrum of Galloway chemistry. Third, Flower et al. (1985) looked at the water quality and diatom content of 34 Galloway lochs in November 1984 and again in July 1985. The study area once again encompassed the whole of Galloway. Finally, Harriman et al. (1987) sampled 22 lochs and 27 streams, looking at their chemistry and their fish populations. The region of study was limited mainly to the granitic area between Loch Dee and Loch Doon in a 20-km by 15-km area.

A summary of each data set is presented in Table 8.1. The highest mean pH levels were recorded by Flower et al. (1985) in their July sample and by Edmunds and Kinniburgh (1986b), for whom the shallow well and borehole samples show the more alkaline groundwater chemistry. The latter study also shows the highest base cation concentrations. The data from Harriman *et al.* (1987) show low base concentrations indicative of the slowly weathering, acidic granite region away from the coast and the emission sources. Of the major anions, SO_4 levels are more or less similar at approximately 158 meq m^{-3}; but Cl is more variable, ranging from 148 to 385 meq m^{-3}.

Table 8.1. A summary of data from each study (minimum, mean, standard deviation, maximum).[a]

Data set		pH	Al	Na	K	Ca	Mg	Cl	SO$_4$	NO$_3$
Edmunds &	Min	4.20	.01	96	0.0	19.0	16.5	28.2	37.5	0.0
Kinniburgh	Mean	5.80	.21	277	26.7	329.0	158.0	253.0	159.0	49.5
(1986)	Std	.85	.21	153	65.3	38.7	171.0	192.0	112.0	125.5
	Max	7.30	.80	1,095	644.0	2,136.0	1,242.0	1,291.0	1.041.0	728.0
Wright &	Min	4.30	.01	101	1.0	6.0	21.0	37.0	47.0	0.0
Henriksen	Mean	5.40	.19	210	19.8	174.5	107.0	233.0	163.8	35.5
(1980)	Std	.92	.18	113	16.1	300.0	95.0	143.0	77.0	65.0
	Max	7.60	1.10	761	176.0	1,811.0	403.0	930.0	478.0	460.0
Harriman	Min	4.20	.01	100	3.0	1.0	34.0	100.0	88.0	1.0
et al.	Mean	5.20	.17	143	9.0	139.0	78.0	148.6	150.5	11.3
(1986)	Std	.60	.10	29	2.3	92.0	31.0	32.0	36.0	6.6
	Max	8.00	.62	257	13.0	1,325.0	429.0	232.0	279.0	45.0
Flower	Min	4.40	.10	1,748	4.6	14.0	38.0	90.0	54.0	.7
et al.	Mean	5.50	.21	286	13.4	134.0	121.0	385.0	161.5	17.0
(1985)	Std	.90	.11	138	12.0	114.0	83.0	179.0	85.0	31.0
	Max	7.00	.43	765	56.3	400.0	340.0	935.0	358.0	171.0
Flower	Min	4.50	.07	135	3.1	23.0	37.0	87.0	70.8	.7
et al.	Mean	6.00	.27	276	14.7	166.3	116.0	298.0	139.0	7.5
(1985)	Std	1.30	.40	164	14.6	142.0	88.0	163.0	67.0	7.2
	Max	7.20	.59	717	64.0	518.0	356.0	899.0	320.6	31.4
Overall	Mean	5.65	.21	248	21.5	248.6	132.0	250.4	158.0	37.0

[a]All values are shown in meq m^{-3} except for those for Al, which are mg l^{-1}.

The probability density distributions of corresponding chemicals within data sets are roughly triangular with a tail of varying length. The tail values affect the mean results in *Table 8.1* by shifting them from their respective median values. This effect is greatest for the three data sets with a low number of samples. The distribution of pH has three peaks – at 4.5, 6.1, and 7.1 – for each of the studies except for that of Harriman *et al.*, who sampled the most acidic area and hence only show the lower two.

The data were pooled together to yield a large data set whose samples covered a variety of conditions produced by both spatial and temporal variation within the region. The samples were all taken within a five-year period, and it was assumed that the pooled set could represent the mean state of the region during those five years. The time scale of the model is longer, but the errors introduced by such an approach were considered to be outweighed by the benefit of utilizing all of the data to get a more accurate representation of region as a whole during that time. The overall mean results are presented in *Table 8.1*.

Probability density functions showing the three-peaked pH distribution and the long-tailed "triangular" distribution for calcium, but otherwise typical of each of the major ions, are shown in *Figures 8.2* and *8.3*.

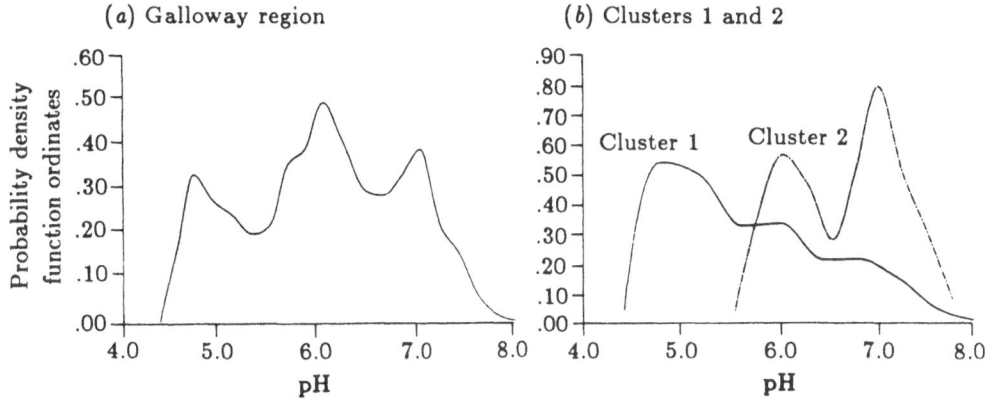

Figure 8.2. pH distributions in (a) Galloway region and (b) clusters 1 and 2.

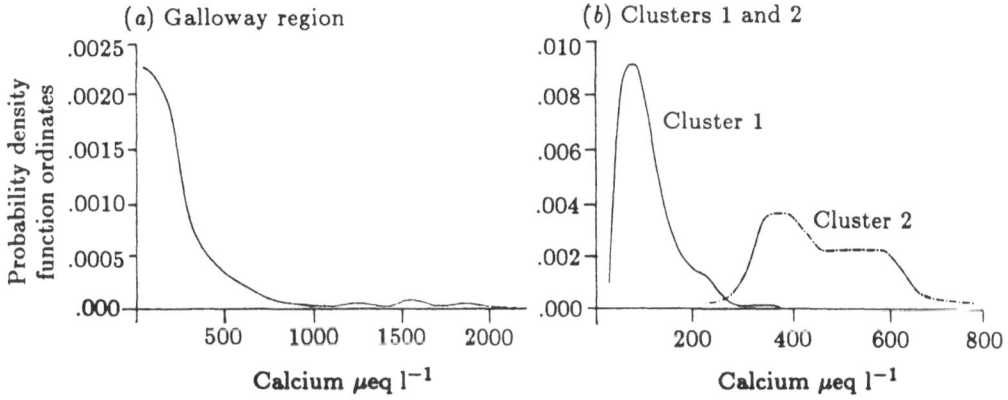

Figure 8.3. Calcium distributions in (a) Galloway region and (b) clusters 1 and 2.

A cluster analysis was performed in an attempt to categorize groups of data linked by a common chemical makeup. The minimum variance method of Ward (1963) was used to cluster the data, and the algorithms used were those available in the Statistical Analysis Systems computer package (*SAS*). The clearest division of the data was obtained by forming six clusters. Two main clusters contained 322 and 74 of the total sample of 453 sets of measurements, and four minor clusters effectively accounted for the long tails of the distributions for the appropriate chemicals.

Table 8.2 summarizes the two major clusters. Cluster 1 has a mean pH of 5.3 and includes all the low-pH sites. Concentration of all major ions is low, e.g., mean Ca is 83 meq m^{-3} and mean SO_4 is 120 meq m^{-3}. This contrasts with cluster 2, which has high values for ions (Ca is 412 meq m^{-3} and SO_4 is 216 meq

Table 8.2. Summary of the two major clusters.

Variable	Number	Mean	Standard deviation	Minimum	Maximum
Cluster 1					
Al	322	.312	.16	.01	1.05
Ca	322	82.90	52.90	6.00	339.30
Cl	322	189.80	80.10	28.20	541.50
SO_4	322	120.30	42.70	37.40	341.50
pH	322	5.34	.79	4.18	7.38
NO_3	322	8.75	13.30	3.57	89.00
Mg	322	69.00	31.70	16.50	310.20
Na	322	184.70	60.20	1.00	404.30
K	322	8.60	11.40	2.60	130.50
Cluster 2					
Al	74	.172	.11	.01	.42
Ca	74	412.10	120.00	196.10	748.50
Cl	74	332.20	135.50	87.40	631.70
SO_4	74	216.50	80.90	99.90	449.70
pH	74	6.40	.56	5.38	7.44
NO_3	74	54.80	57.80	3.57	298.40
Mg	74	204.90	73.10	90.50	477.20
Na	74	361.30	110.20	169.60	776.10
K	74	31.60	30.40	2.50	181.60

m^{-3}; see *Figure 8.2*), with a mean pH of 6.4. The differences may be interpreted by assuming that cluster 1 contains the more acidic sites on and around the granitic regions to the north of the Galloway region, and that cluster 2 represents the area with Silurian and Ordovician bedrock near the coast. Cluster 2 chemistry thus reveals the prevalence of higher deposition rate of sea salt and of atmospheric pollution and also the higher source of weathered minerals. The main interest focuses upon cluster 1, representing the most acid-sensitive subdivision. Therefore, only the simulation for cluster 1 is presented here.

8.3. Description of Modeling Techniques

The region was modeled using the Model of Acidification In Catchments (MAGIC), developed by Cosby *et al.* (1985a; 1985b). This is a lumped parameter, long-term simulation model of soil and surface-water quality. In the MAGIC model, five soil processes are identified as holding the key to understanding the acidification processes in soils:

(1) Formation of alkalinity in the soil from dissolution of carbon dioxide held at high partial pressure (Reuss and Johnson, 1985).
(2) Mineral weathering of base cations as a source within the catchment.
(3) Anion retention by the soil, e.g., sulfate adsorption.
(4) Aluminum mobilization and dissolution of aluminum minerals.
(5) Cation exchange by the soils.

Dynamic variation in the catchment soil condition is included in the model by forming a mass balance for each of the major anions and cations (presented by Cosby *et al.*, 1985b). Equilibrium is assumed for each time step, with changes between time steps monitored by the ion flux budgets. The sulfate subsection of the model is investigated in detail by Cosby *et al.* (1986a).

Initial conditions for the model are calculated by assuming an equilibrium with background chemistry prior to industrialization in 1844. This background chemistry is calculated using the present sea salt rainfall contribution as the 1844 precipitation concentration, and by taking the weathering rates to be constant throughout the time period considered by the model. Deposition of SO_4, Cl, and NO_3 are specified as an input to the model. The trend of sulfate deposition during the past 130 years is taken from findings of the Warren Springs Laboratory (1983). The effect of catchment flora is modeled as sink terms for the appropriate ions. Neal *et al.* (1986) and Whitehead *et al.* (1987) use this facility to simulate the effect of forest growth on stream acidity. The model has a simple component for the additional effect of organic chemistry, which has been applied by Lepistö *et al.* (1988) to simulate an organically rich catchment in eastern Finland.

An analysis of the uncertainties inherent in the model is proffered by Hornberger *et al.* (1986a). Verification of the model has been attempted by comparing model simulations to paleoecological reconstructions of pH for sites in Norway, Scotland, Sweden, and America (Wright *et al.*, 1986; Musgrove and Whitehead, 1988). The use of the model for regional analysis is illustrated by the study of Hornberger *et al.* (1987), who simulate the regional characteristics of the chemistry of 208 lakes in Norway.

For this study the model was set up for Monte Carlo analysis. In this technique, several arbitrary parameters are randomly selected from a prescribed parent distribution prior to running the model. This is repeated many times to build up a set of runs, each with a different set of parameter values. The statistical properties of the collection of runs are then analyzed. Examples of this technique applied to water quality problems are presented in Whitehead and Young (1979) and Spear and Hornberger (1980; 1983). The 12 independent model parameters that were varied are the mean annual precipitation, Q_p; the sulfate deposition rate for 1982, SO_4; the cation exchange capacity, CEC; the weathering rates of the base cations WE_{Ca}, WE_{Na}, WE_{Mg}, and WE_K; the $\log_{10}$ selectivity coefficients S_{AlCa}, S_{AlMg}, S_{AlNa}, and S_{AlK}; and the thermodynamic equilibrium constant for aluminum hydroxide dissolution, K_{Al}.

The results of the present analysis were investigated using the generalized sensitivity approach of Spear and Hornberger (1980). This method compares the simulations that accurately reproduced observed features of chemistry (the "behavior criterion") with those that did not. It then investigates the merits of the hypothesis in which the set of parameters giving rise to a successful simulation and the set of parameters failing to do so have the same parent distribution. This hypothesis is tested for each parameter individually, assuming them to be independent. The force with which the hypothesis is rejected is strongest for those parameters to which the achievement of a successful result is most sensitive. A pattern recognition technique based on the Fukunaga–Koontz transformation (1970) is employed to determine any linear combinations of parameters to which the model response is sensitive. This technique has been previously applied to MAGIC by Hornberger *et al.* (1986) and Cosby *et al.* (1987).

8.4. Monte Carlo Results

Cosby *et al.* (1986b) set up the MAGIC model to simulate one of the Loch Dee catchments. This catchment is located in the middle of the region described by cluster 1. For the present analysis, we initially set up the model in the same way to simulate cluster 1. *Table 8.2* indicates that for cluster 1 the effect of NO_3 is small. This is possibly due to a near equilibrium between the input sources to the catchments and the plant uptake sinks within the catchment. The uptake rate was set more or less to balance the source for this simulation.

The ranges from which the parameters were drawn in the Monte Carlo analysis are shown in *Table 8.3*. The mode indicates the peak for the triangular distributions (a negative mode indicates that the distribution was rectangular). The specified ranges are wide to encompass the entire spread of parameter combinations that are able to reproduce cluster 1 chemistry. The extent of the range is limited, however, to those values that may be experienced in the field. A behavior constraint was chosen such that a run was considered successful if its simulated chemistry for 1982 fell within the range of values between the appropriate 5 and 95 percentiles of the cluster 1 chemistry.

Table 8.3. Parameter variation for the Monte Carlo simulation.[a]

Parameter	Range for Monte Carlo variation			Successful simulations			
	Min	Max	Mode	Mean	Std	Min	Max
Q_p	1.2	3.0	2.0	1.7	.50	1.2	3.0
SO_4	50.0	220.0	150.0	121.50	32.40	52.2	184.1
CEC	40.0	250.0	120.0	178.50	119.90	40.0	229.0
WE_{Ca}	10.0	250.0	85.0	69.40	45.10	10.1	162.2
WE_{Mg}	0.0	50.0	25.0	13.40	8.90	6.5	39.4
WE_{Na}	0.0	50.0	25.0	10.30	6.93	.1	29.0
WE_K	0.0	30.0	10.0	3.70	2.53	.1	10.5
S_{AlCa}	−1.0	5.0	3.0	2.08	1.28	−.7	4.8
S_{AlMg}	−1.0	5.0	3.0	1.87	1.24	−.9	4.7
S_{AlNa}	−3.0	3.0	0.0	.08	1.22	−2.9	2.9
S_{AlK}	−3.0	3.0	0.0	.03	1.28	−2.8	2.9
K_{Al}	8.0	11.0	9.5	9.65	.6	8.1	10.9

[a] WE_{Ca}, WE_{Mg}, WE_{Na}, WE_K, and SO_4 have units of meq m^{-2} year^{-1}; Q_p is in meters depth over the catchment; CEC in meq/kg.

Of 880 runs, 530 were successful in reproducing the features of cluster 1 regional chemistry, and these are summarized in *Table 8.3*. *Table 8.4* shows the behavior constraint "windows" and a summary of the simulated chemistry.

Figure 8.4 shows the observed and simulated probability density functions of the cluster for Cl, SO_4, and Ca. Good fits about the mean value are seen for

Table 8.4. Summary of behavior criterion windows and the simulated chemistry (concentrations in meq m^{-3}) that satisfied the behavior criterion for cluster 1.

| Variable | Behavior criterion windows | | Successful runs | | |
	Min	Max	Mean	Min	Max
pH	4.4	7.6	5.4	4.3	6.5
Ca	20.0	245.0	91.5	21.5	224.0
Mg	18.0	190.0	27.0	19.0	85.0
Na	10.0	330.0	196.0	11.0	372.0
K	3.0	35.0	12.0	3.0	34.0
SO$_4$	40.0	245.0	115.0	39.0	242.0
NO$_3$	9.0	25.0	16.0	9.0	22.0
Cl	79.0	360.0	203.0	125.0	401.0

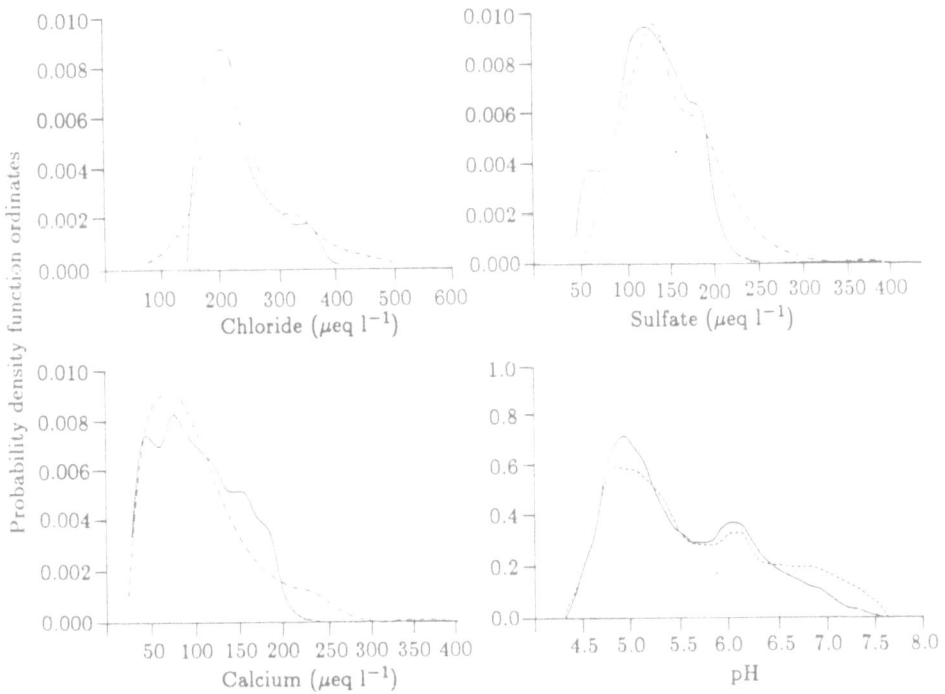

Figure 8.4. Cluster 1 distributions: simulated (——) and observed (-----).

Ca, Cl, and SO_4; and the shapes of the simulated distributions for these three variables are also good. The fits are worst at the tail ends, resulting either from specifying a too narrow range of parameters for the Monte Carlo analysis or from failure to pick up the extremes in sampling the triangular distributions.

The simulated distribution for pH is compared with the observed distribution in *Figure 8.4*. The simulation enabled the complete range of pH value, present in the region defined by cluster 1, to be reproduced. The bias toward reproducing the lower pH chemistry possibly lies with the model being set up to reproduce a poorly buffered, highly acidic catchment response. While this is adequate for the majority of cluster 1 sites, it may be less good for the remainder. Alternatively, a better match might also have been achieved by selecting the parameters from more sophisticated distributions than the rather crude triangular ones that were used. By matching more closely the uppermost part of the pH distribution, the upper tail end of the other distributions would also have been better matched.

8.5. Regional Chemistry Changes over Time

Historical changes in the distributions of values of the simulated chemicals were ascertained by saving the output for 1844 and 1982. Future deposition was modeled as a linear decrease to 30% of the 1982 level by the year 2001. This is merely one of many possible scenarios and is used only to illustrate the potential use of the model for regional prediction. The regional levels of chemical concentrations in 2062, under this deposition sequence, were also looked at. The results are shown in *Figure 8.5*.

For cluster 1, the MAGIC model shows a large drop in both pH and alkalinity over the past 140 years. The pH level falls by 0.8 pH unit and alkalinity falls by approximately 70 meq m^{-3}. This is in accord with the findings of Battarbee and Flower (1986), who report changes in pH level of up to 1.0 pH unit during the same period, for those lochs in the granitic region of Galloway. A small recovery is seen during the future scenario. The small size of this recovery reflects the depletion in the soil of base cations (to between 2% and 5% in 1982). The low rate of soil weathering in the region enables only a slow recovery rate.

The sulfate distributions clearly show the shift in regional concentrations simulated as the result of industrialization. The preindustrialized levels are between 0 and 38 meq m^{-3}, whereas the 1982 levels are much higher with values between 42 and 190 meq m^{-3}. The 2062 levels are reduced by about 30%, closely matching the decline in sulfate deposition. This confirms the status of the soil as having low sulfate adsorption capability. Aluminum is seen to increase in extent over the years, up to 36 meq m^{-3} for the most acidic lakes. The majority of sites, however, shows only a small increase. Of the base cations, calcium levels are seen to rise only slightly, whereas magnesium levels show a much more substantial rise. The future scenario shows small decreases in these cations, as a result of the partial recovery of the soil.

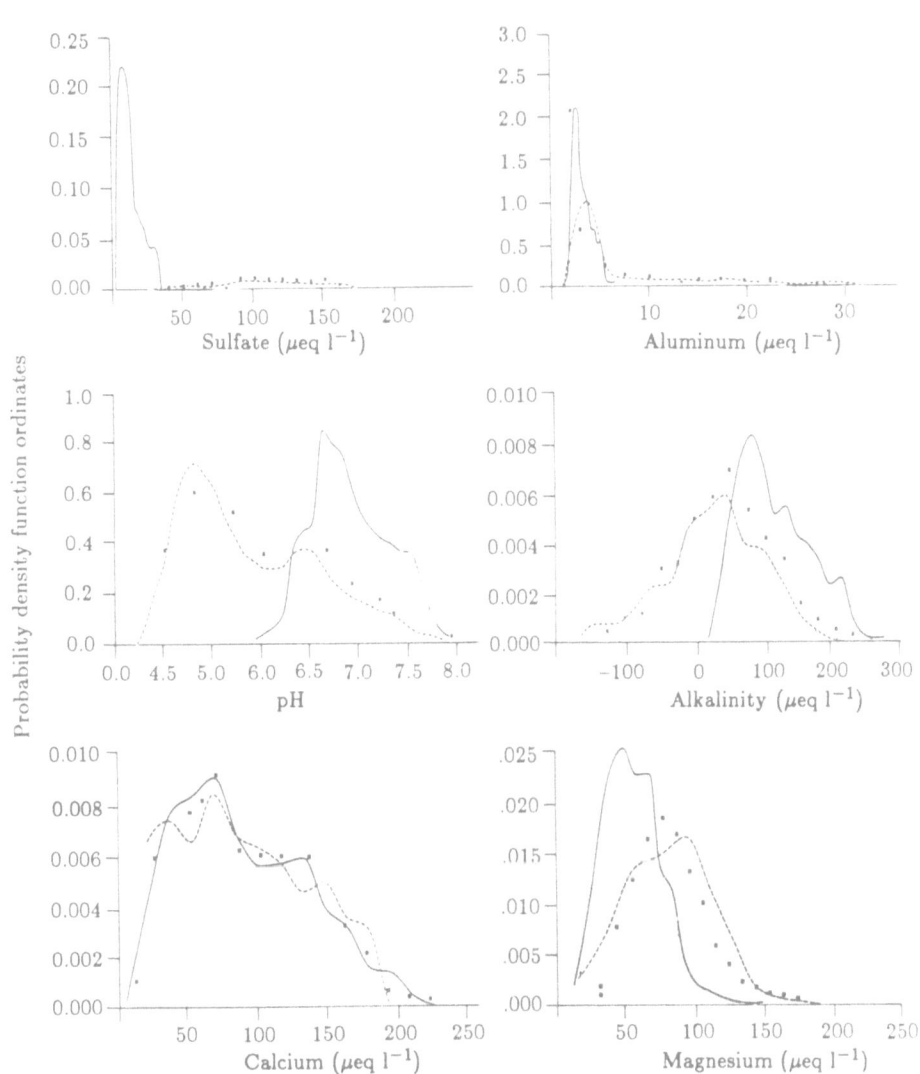

Figure 8.5. Cluster 1: simulated change through time (∗∗∗ 2062 level; ----- 1982 level; —— 1844 level).

8.6. Sensitivity Analysis

The results of the generalized sensitivity analysis are presented in *Table 8.5*, which describes the difference between the distributions of parameters that give rise to a "behavior" and those that do not. The Mann–Whitney statistic (M–W) indicates whether the distributions are significantly separated along the parameter axis, and the Kolmogorov–Smirnov statistic (K–S) gives the greatest separation of the two corresponding cumulative distribution functions. In both cases, a large number indicates a high significance and sensitivity of simulation to that parameter.

Table 8.5. Results of the generalized sensitivity analysis.

Parameter	Comparison of behaviors with non-behaviors	
	Mann–Whitney z statistic	*Kolmogorov–Smirnov d statistic*
Q_p	3.19	.139
SO_4	2.19	.204
CEC	1.59	.105
WE_{Ca}	.76	.219
WE_{Mg}	.51	.045
WE_{Na}	.26	.052
WE_K	.52	.065
S_{AlCa}	.26	.079
S_{AlMg}	1.21	.096
S_{AlNa}	.24	.042
S_{AlK}	1.08	.092
K_{Al}	.01	.054

The greatest sensitivity, as indicated by these two statistics, is seen for Q_p (M–W of 3.19) and for WE_{Ca} (K–S of 0.219), respectively. Parameter SO_4, the 1982 deposition level, is high in both cases. WE_{Ca} has a high value for the K–S statistic and a low value for M–W statistic. This indicates that one of the two distributions (behaviors or non-behaviors) is centrally placed within the domain of the other. The central distribution for WE_{Ca} is the one for behaviors, showing a peaked, triangular distribution with a well-defined band of acceptance. Thus, while the WE_{Ca} parameter was outside the range 10 to 162 meq m^{-2}yr^{-1}, there was no chance of a successful simulation.

A method based on the Fukunaga–Koontz pattern recognition technique was used to look at the sensitivity with regard to linear combinations of the parameters. This analysis reveals two eigenvectors of high sensitivity. The first was more sensitive than any individual parameter, with M–W of 3.63 and K–W of 0.214. The weight of each component of the normalized eigenvector indicates the relative importance of that component in explaining the variance associated with it. In this way the most important components were SO_4, explaining 63% of the variance, and WE_{Ca}, explaining 22%. The results indicate that the success of the model in reproducing the regional characteristics for areas of high acid susceptibility is strongly dependent on the selection of values for the WE_{Ca}, SO_4, and Q_p parameters (the weathering rate of calcium, the sulfate deposition rate in 1982, and the mean annual rainfall during the duration of the simulation).

8.7. Discussion

This analysis tends to confirm both the view that the Galloway region is suscep-
tible to acidic deposition and the view that a significant change in soil conditions
has occurred during the past 140 years. Falls in pH level of as much as 1.0 pH
unit have been discovered for the Galloway by Battarbee and Flower (1986),
using the method of reconstruction from study of diatom assemblage in the sediment of lochs. The simulation results show a drop in mean level of 0.8 pH unit.
This is increased by the inability to match the highest portion of the 1982 pH
distribution at the same time as matching its low end.

This inability also spills into the simulated distribution for alkalinity, by
cutting off the highest levels in the simulation. A fair view of the changes caused
by the acidification process can still be gleaned, however, by noting that the
lakes with higher alkalinity in 1982 are also those with the higher alkalinity in
1844. Hence, the simulated results are valid for those lower alkalinity lakes that
it represents well. The recovery under the scenario of a 30% deposition reduction is seen to be slight, owing to the depletion of base cation resources at the
available cation exchange sites on the soil. In particular, alkalinity levels are
seen to recover very little in the most highly acidified lakes.

Monte Carlo simulations are an excellent means of encountering the complete range of model response to uncertainty in parameters. Overall, the simulation was successful using triangular distributions for the parameters in the
Monte Carlo analysis. Better reproduction of the present-day distributions could
have been achieved using more complicated distributions to describe the parameters, noting that certain results in the behavior criterion are more in tune to
particular parameters. For example, calcium levels are strongly affected by its
weathering rate and selectivity coefficient.

The lack of available data for verification of the model calibration, however, limits the usefulness of such a study. Without verification, fine points
raised by such a simulation might be mere quirks of the model. The simulations
obtained work only under the assumption that the model structure is valid and
that its lumped representation of parameters is valid on a regional basis. This is
a difficult assumption to prove; one can only look at the simulations and see
whether the model appears to represent the region well.

Given the validity assumptions, many uncertainties within the model itself
need to be borne in mind when interpreting the results or when using the model
to project assumed deposition scenarios into the future. Uncertainties arise in
measurement of both precipitation and stream chemistry as a result of intrinsic
imprecision both in the measuring technique and in the representative nature of
the sampling plan used. In this study the results of five surveys are used, four of
which are single-sample surveys. The findings of each survey have been pooled
to produce an observed chemistry that may average out, to a degree, any
unrepresentative feature of any one survey.

Each submodel within MAGIC also has an uncertainty associated with it.
The true shape of the sulfate deposition trend curve, used to define the timing of
precipitation pollutants, has only been estimated, and errors may seriously affect
the timing of the onset of acidification. This aspect of the model has been

investigated by Hornberger *et al.* (1986a) who looked at the effect of uncertainty on a single-site prediction using MAGIC. The aluminum submodel contains a cubic equilibrium between aluminum and hydrogen activities. Assuming an equilibrium model to be valid, there is uncertainty in the order of the relationship. A fit to observations at one point in time may be found for many different orders, and the future predictions and past reconstructions may be very different for each one.

8.8. Conclusions

The MAGIC lumped-parameter model of groundwater in catchments can be coupled with Monte Carlo techniques to provide a useful method of simulating long-term regional response to acidic deposition. The sensitivity of simulation to parameters in the model is seen to be dependent on the weathering rate of calcium, the level of deposition of sulfate, and the mean annual rainfall for a region with low weathering rates and low sulfate adsorption.

   The response of the region to a future deposition scenario is also gleaned from the model. After a prolonged period of deposition, the ability of the region to recover is seen to decrease throughout the lifetime of the deposition sequence. Major future reductions in deposition are required to produce a significant recovery in the Galloway region of southwest Scotland.

Acknowledgments

The authors are particularly grateful to the Solway River Purification Board for providing data from the Loch Dee study and to the British–Scandinavian SWAP Committee and the Commission of the European Communities for providing research funding.

CHAPTER 9

Modeling Regional Acidification
of Finnish Lakes

M. Posch and J. Kämäri

9.1. Introduction

Conceptualizations of the acidification phenomenon have been based on recent findings in soil and water chemistry. Scientists have, however, not always chosen the same description for the key mechanisms that are thought to determine the systemic behavior. Moreover, different types of models have been developed with different modeling objectives.

First of all, simplified equilibrium models have been developed, which allow the estimation of the future steady state chemical composition of lakes resulting from changes in the load of strong acids. These models are either based on observed ionic relationships in present conditions (e.g., Henriksen, 1980) or on the assumption of steady state chemical weathering (e.g., Schnoor *et al.*, 1984). Second, models have been developed for simulating daily variations of water quality in streams, caused by variations in deposition, as well as in catchment hydrology and meteorology (e.g., Christophersen *et al.*, 1982). To date, the models have been mainly applied to providing information on the importance of different processes in the dynamics of the studied catchments. The third type of model utilizes mechanistic, process-oriented descriptions for hydrology and soil chemistry, as well as for stream and lake water quality, to link the evolution of acidic deposition with long-term surface-water acidification (e.g., Chen *et al.*, 1983; Cosby *et al.*, 1985). It has been shown that the observed surface-water chemistry can be reproduced by models that largely retain the simplicity of equilibrium models, but that also incorporate some mechanistic, process-oriented explanations in their structure (Cosby *et al.*, 1985; Kämäri *et al.*, 1985).

J. Kämäri (ed.), Impact Models to Assess Regional Acidification, 145–166.
© 1990 *International Institute for Applied Systems Analysis.*

To date, mechanistic models have been applied mostly to single catchments. Descriptions of quantitative consequences of alternative scenarios can assist in formulating policies for emission control. From a decision maker's point of view, however, the behavior of a single catchment is not very interesting. The assessment should investigate broad-scale aspects of alternative policy formulations, and thus analyze the behavior of a representative selection of catchments. As an output, the model should produce well-defined illustrative information that can easily be related to the effectiveness of the scenario being evaluated.

Two basic methods for regionalizing model structure have been introduced. Both methods attempt to modify the rather uncertain *a priori* parameter distributions such that a satisfactory fit with observed model output distributions is obtained. The first method draws on the use of Monte Carlo techniques. Uncertain model parameters and variables are assigned frequency distributions. In the regional application, not only the uncertain parameters, but also parameters that exhibit spatial variability are randomized within feasible *a priori* ranges. The criteria for accepting or rejecting model "postdictions," and thus parameter vectors used, are based on synoptic surveys of lake chemistry. The Monte Carlo runs accepted produce a set of theoretical lakes that are assumed to be representative for that region (Kämäri and Posch, 1987). Regional lake acidification scenarios are then constructed by running the model for the desired deposition scenario with each accepted parameter set. This technique for selecting parameter vectors is referred to as the Monte Carlo Filter Procedure (MCFP).

The second technique uses *a priori* criteria to select a satisfactory set of model parameters to create future distributions. This approach intends to modify the most important parameter distributions, as identified by a sensitivity analysis of the model, and to produce reliable parameter distribution estimates for most critical parameters (Chapter 11).

This chapter continues on the first path and applies to Finland a lake acidification model developed at IIASA as part of the integrated Regional Acidification INformation and Simulation (RAINS) model (Alcamo *et al.*, 1987). The lake acidification model is called the RAINS Lake Module (RLM). In this application two new features of the regional model structure are presented for the first time. First, a long calibration period from 1900 to 1980 (as opposed to 1960–1980 in an earlier version of the model; see Kämäri and Posch, 1987) is introduced to allow the theoretical lakes to experience a realistic acidification history. One would expect the accepted parameter ensembles to differ when the calibration period is varied. The second new feature is the possibility of creating a joint pH–alkalinity target distribution (compared with marginal distributions in the earlier version).

9.2. Model Structure

9.2.1. Catchment model

Lake acidification involves mechanisms in both the terrestrial and the aquatic catchment. The model provides a simple representation of those processes believed to be most important in controlling acidification dynamics. The objective has been to retain simplicity in the model, while incorporating the physically most relevant processes in its structure.

The model separates the catchment into three distinct water reservoirs (two soil layers and the lake), between which it computes the fluxes of major ions contributing to acidity or alkalinity of surface waters. The mass flow is influenced by the forcing functions as well as by the internal processes taking place in the reservoirs. These processes can be divided into two kinds of reactions: the equilibrium reactions between soil and soil solution that determine the water quality of the runoff; and, second, the flux relationships that may change the chemical state of the soil, thus affecting the equilibrium between soil and soil solution.

Since we are interested in the regional long-term development of lake water quality, we have lumped together some monthly model estimates of lake water acidification to allow the examination of acidification with an annual time step. We consider only those processes that have an effect on the long-term development of lake acidity. These processes are summarized in *Table 9.1*. The overall model structure is presented in *Figure 9.1*, and the model equations are summarized in Appendix 9A.

Table 9.1. Processes considered in the RAINS Lake Module affecting the year-to-year development of lake acidification.

Process	Reference
Forest filtering effect	Mayer and Ulrich (1974)
Lateral flow	Chen *et al.* (1983)
Weathering of base cations	Ulrich (1983a,b)
Cation exchange	Ulrich (1983a,b)
Dissolution of minerals	Christophersen *et al.* (1982)
In-lake sulfate retention	Baker *et al.* (1986)
Dissolution of CO_2	Stumm and Morgan (1981)

In the model the annual sulfur deposition, computed by the air pollutant transport submodel, is transformed into acid load to various sectors of the catchment. To provide a method for fractionating the annual internal flows, a simple two-layer structure is applied (see Christophersen *et al.*, 1984). The terrestrial catchment is vertically segmented into two soil layers (A- and B-reservoir). The annual runoff, computed from the mean monthly precipitation and temperature, is divided into quickflow and baseflow. Physically, the flow from the upper reservoir can be thought of as quickflow, which drains down the hillsides as piped flow or fast throughflow and enters the brooks directly. This water is mainly in contact with humus and the upper mineral layer. The B-reservoir in the model provides the baseflow, which is assumed to come from the saturated zone. The formula for the rate of discharge from the B-layer is simplified from the ILWAS model (Chen *et al.*, 1983). The computation of this flow component is based on rates of precipitation surplus and from catchment characteristics, such as soil thickness, surface slope, and hydraulic conductivity.

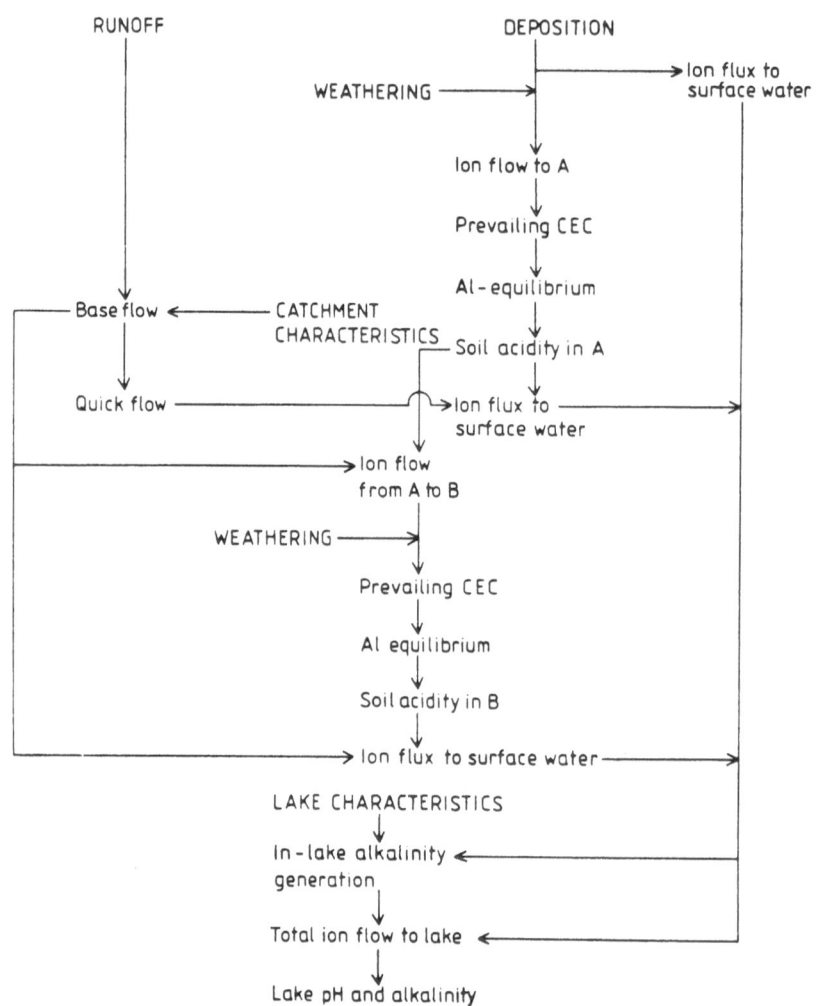

Figure 9.1. Processes and fluxes of ions in RLM.

At the heart of the catchment model are descriptions of long-term chemical soil processes. To compute the ion concentrations of the catchment flows, we use the same analytical approach that is used in the RAINS soil acidification model (see Kauppi *et al.*, 1986). Complete mixing is assumed in the reservoirs, and equilibrium is assumed to be reached according to computed base saturation. The acid load to the soil is assumed to participate in two important equilibrium processes: cation exchange and release of inorganic aluminum species through the dissolution of soil minerals. The net effect of both processes is a buffering of hydrogen ions in the soil solution. The contribution of the soil reservoir to the alkalinity of the surface waters is determined by the amount of cation input, originating from weathering and atmospheric deposition, exceeding the acid load. The leaching of acidity to surface waters is given by the hydrogen ion concentrations in the soil solution discharged from both reservoirs. The base saturation is depleted at a rate corresponding to the acid load minus the total base cation input. The updated soil base saturation then determines the new equilibrium hydrogen ion and aluminum concentrations.

The resulting changes in lake water chemistry are based on well-known equilibrium reactions. The key variable is alkalinity, which originates both from mineral weathering in the soil as well as from in-lake processes, such as the sulfate retention in the lake. The ion fluxes to the lake are assumed to mix within the whole water body. The change in lake acidity is calculated according to equilibrium reactions of inorganic carbon species. The risk for aquatic impacts can be estimated on the basis of simple threshold pH, alkalinity values, or both, but are not part of current calculations. These characteristics are most likely to indicate damage to fish populations and other aquatic organisms.

9.2.2. Regionalization of the model

In an ideal case, if we had correct and complete *a priori* information on the distributions of all parameters (initial conditions as well as on catchment characteristics), and if the model perfectly described all interactions in the catchment, no calibration would be necessary. In simulation models of environmental systems, however, the model structure, the model inputs, and the initial conditions as well as the parameter values all necessarily include uncertainties. The same applies to the water quality observations that are used to constrain the model output.

Spatial variability adds another dimension to the problem. The data available on a large regional scale are characterized by a high degree of heterogeneity and generalization. Even if the actual measurements are reliable and the measurement error is small, the problem of the representativeness of the sample for a regional analysis arises. It has been emphasized in several studies that the analysis of (ecological) models should concentrate on identifying ranges of inputs, rather than on traditional parameter estimation (e.g., Fedra, 1983; Hornberger and Cosby, 1985); this is even more true for a regional analysis.

Our approach to assessing regional surface-water impacts has two distinct levels. At the first level, the catchment model is used to analyze the changes over time in the chemistry of a specific lake. The model can be used for any known system for which the relevant lake, catchment, and soil information is available. The second level consists of regionalizing the model by scaling up from individual catchments to a regional level. In the regional lake acidification assessment, the so-called Monte Carlo Filter Procedure (MFCP) is applied to model regional lake water quality distributions.

The Monte Carlo method is a trial and error procedure for the solution of the inverse problem, i.e., for estimating poorly known input and parameter values from comparing model outputs with available measurements. Monte Carlo methods have been widely used for a variety of purposes: hypothesis testing, prediction, and sensitivity and uncertainty analysis (see Fedra, 1983). In the Monte Carlo simulation, performance criteria (constraints on the output) are formulated to describe the expected satisfactory behavior of the model. Next, probability distributions are defined for all unknown input and parameter values. The Monte Carlo program then randomly samples the parameter vectors from these distributions, runs the simulation model through a selected period of time, and finally tests for violations of the constraints. This process is repeated many times.

In the regional application, the Monte Carlo simulation is used to determine the combinations of inputs and parameters that produce an acceptable distribution of output values observed in the study region. For all inputs and parameters, including those that vary spatially, broad enough ranges are chosen so that any reasonable value for an input can be selected. Monte Carlo simulations are then carried out by randomly selecting a set of input values from these ranges and integrating the equations from a suitable starting time. A subset of all input vectors, producing the actual observed frequency distribution in 1980 in each lake region, is filtered out. The accepted Monte Carlo runs thus produce a set of theoretical lakes that are assumed to be representative of real catchments for that region. These ensembles of input vectors can then be used for scenario analyses of the response of lake regions to different patterns of future deposition.

Mathematically, the Monte Carlo Filter Procedure can be described as follows. The adopted model structure can be represented by a vector function $f = (f_1, \ldots, f_m)$. The arguments of this function are the input and parameter values driving the model – say, $x = (x_1, \ldots, x_n)$ (e.g., x_1 = lake size, x_2 = catchment size, etc.) and time t. With $y = (y_1, \ldots, y_m)$, we denote the output of a model run, e.g., lake water alkalinity, lake water pH, etc.

$$y = f(x, t) \tag{9.1a}$$

or, writing (9.1a) for each component,

$$y_1 = f_1(x_1, \ldots, x_n, t)$$
$$\vdots \tag{9.1b}$$
$$y_m = f_m(x_1, \ldots, x_n, t) \ .$$

Instead of taking fixed input values x and running the model once to obtain the output (prediction) at time t, one allows the input values to vary within an interval, $x_k^{min} \leq x_k \leq x_k^{max}$, $k = 1, \ldots, n$, where the lower and upper bounds are estimated from the catchment characteristics of the region studied. To put it precisely, each input parameter is randomized with a distribution p_k, $k = 1, \ldots, n$, obeying

$$p_k(x_k) = 0 \quad \text{if} \quad x_k < x_k^{min} \quad \text{or} \quad x_k > x_k^{max} \tag{9.2a}$$

and $\int_a^b p_k(x)\,dx$ is the probability that x_k lies in the interval $[a,b]$. Obviously,

$$\int_{x_k^{\min}}^{x_k^{\max}} p_k(x)\,dx \;=\; 1 \quad \text{for} \quad k = 1, \ldots, n \quad . \tag{9.2b}$$

The frequency distributions p_k represent the distribution of the parameters x_k in the region as closely as possible; and in case of a poorly known input parameter, a uniform distribution over $(x_k^{\min}, x_k^{\max})$ is chosen, where the boundaries are wide enough to encompass any feasible value in the region under consideration.

To be able to apply the Monte Carlo procedure, the distributions of the output values y at a certain point in time t_1 (say q_l, $l = 1, \ldots, m$) have to be known from measurements. For the description of the procedure used to solve the inverse problem, i.e., to determine the input parameter distributions for future projections of the model, we assume that the measured distribution at t_1 of the m output variables $y_1, y_2, \ldots, y_m$ is an m-dimensional discrete distribution (I_l ... number of classes of l-th output value, $\eta_{i_l}^{(l)}$ ... class boundaries of l-th output value; $l = 1, \ldots, m$).

$$q(\boldsymbol{y}) \;=\; \begin{cases} q_{i_1, \ldots, i_m} & \text{for } \eta_{i_1-1}^{(1)} < y_{i_1} \leq \eta_{i_1}^{(1)}, \ldots, \eta_{i_m-1}^{(m)} < y_{i_m} \leq \eta_{i_m}^{(m)} \\ & \quad i_l = 1, \ldots, I_l,\, l = 1, \ldots, m \\ 0 & \text{else} \end{cases} \tag{9.3a}$$

with

$$\sum_{i_1=1}^{I_1} \cdots \sum_{i_m=1}^{I_m} q_{i_1, \ldots, i_m} |\eta_{i_1}^{(1)} - \eta_{i_1-1}^{(1)}| \cdots |\eta_{i_m}^{(m)} - \eta_{i_m-1}^{(m)}| \;=\; 1 \tag{9.3b}$$

and

$$\eta_0^{(l)} \;=\; y_l^{\min} \quad \text{and} \quad \eta_{I_l}^{(l)} \;=\; y_l^{\max}, \quad l = 1, \ldots, m \quad . \tag{9.3c}$$

Actually, the assumption of a discrete distribution is not very stringent, since (a) measurements are always given as histograms, and (b) any continuous distribution can be approximated by a discrete one.

To derive "acceptable" input parameter distributions, the model is run many times, each time with a new randomly selected input vector $\boldsymbol{x}$, where the random selection is performed according to the distributions p_k. Let $P = \{\boldsymbol{x}^{(1)}, \ldots, \boldsymbol{x}^{(N)}\}$ be the set of these random vectors $\boldsymbol{x}$ and $Q = \{\boldsymbol{y}^{(1)}, \ldots, \boldsymbol{y}^{(N)}\}$ the set of output values of these runs at time t_1. These N

output values are classified according to the classes defined in (9.3a). Let $N_{i_1, \ldots, i_m}$ be the number of realizations with $\eta^{(1)}_{i_1-1} < y_1 \leq \eta^{(1)}_{i_1}, \ldots,$ $\eta^{(m)}_{i_m-1} < y_m \leq \eta^{(m)}_{i_m}$ with $y = (y_1, \ldots, y_m) \in Q$ (obviously $\sum_{i_1, \ldots, i_m} N_{i_1, \ldots, i_m} = N$). Monte Carlo runs are performed until $N_{i_1, \ldots, i_m} \geq N_0 q_{i_1, \ldots, i_m}$ for all $i_l = 1, \ldots, I_l$ and $l = 1, \ldots, m$, where N_0 is a preselected number of runs to be accepted. In this way a subset $Q_0 = \{y_1, \ldots, y_{N_0}\}$ of Q is selected, so that there are $N_0 q_{i_1, \ldots, i_m}$ output values with $\eta_{i_l-1} < y_l \leq \eta_{i_l}$ ($i_l = 1, \ldots, I_l$ and $l = 1, \ldots, m$) with $y \in Q_0$. To this subset Q_0 corresponds a subset $P_0 = \{x_1, \ldots, x_{N_0}\}$ of P of accepted input vectors x. From this set of accepted input vectors, modified input parameter distributions p_k^0, $k = 1, \ldots, n$ are derived; and these distributions are used for future projections, i.e., for computing y values for $t > t_1$.

9.3. Regional Application to Finland

9.3.1. Regional data

The above initialization of the model for scenario analysis – in other words, the scaling up of the catchment model to a regional level – had several preparatory steps. First, ranges or distributions for unknown parameters were estimated. In the estimation procedure, best available information and best guesses for the input distributions were used. Then – for the model output – a target distribution was specified on the basis of a large number of water quality observations. Finally, the filtering procedure was applied.

For the Monte Carlo Filter Procedure, frequency distributions were estimated independently for 2 driving variables, 17 input variables and parameters, and 2 output variables. They are listed in *Table 9.2*, together with the symbols used for them (see Appendix 9A) and the type of distribution assigned to them in the regional application.

All the relevant lake and catchment information for the five lake regions considered in this application to Finland (see *Figure 9.2*) was collected from the sources discussed in the following subsections.

Driving variables

The spatial variation of the principal driving variable, sulfur deposition, was obtained from the upstream submodels of the RAINS model system. The deposition pattern was calculated with the aid of the EMEP Sulfur Transport Model (Eliassen and Saltbones, 1983; Lehmhaus *et al.*, 1986). For the period from 1960 on, the sulfur emissions data were taken from the RAINS energy-emissions database; for the earlier period, the deposition pattern was derived from historical estimates of sulfur emissions from Europe (see *Figure 9.3*). The time pattern for the base cation deposition was constructed by assigning the present and background values. Between these values the base cation deposition was assumed to

Table 9.2. Model variables and parameters and the type of input assigned to them. "Histogram" and "Uniform" refer to the type of distribution. The hydraulic conductivity is derived from soil texture, and the annual runoff from monthly precipitation and temperature data.

Variables	Symbol	Form of input
Driving variables		
Sulfur deposition	d_{SO_4}	Uniform
Base cation deposition	d_{BC}	Uniform
Input variables and parameters		
Temperature (monthly)	T	Uniform
Precipitation (monthly)	P	Uniform
Lake surface area	A_l	Histogram
Catchment area to lake area ratio	A_c/A_l	Histogram
Lake mean depth	Z_l	Histogram
Mean catchment soil thickness	Z_s	Uniform
Mean surface slope	S	Histogram
Cation exchange capacity	CEC	Histogram
Base saturation in A-layer	Ex_a	Uniform
Base saturation in B-layer	Ex_b	Uniform
Soil moisture content at field capacity	Θ_f	Uniform
Solubility constant for solid Al-phase	K_{so}	Uniform
Partial CO_2 pressure in surface water	p_{CO_2}	Histogram
Forest coverage fraction	f	Histogram
Weathering rate of base cations	wr	Histogram
In-lake SO_4 retention rate coefficient	k_{SO_4}	Uniform
Forest filtering coefficient	φ	Uniform
Hydraulic conductivity at saturation	κ_s	Derived
Annual mean runoff	Q_{tot}	Derived
Output variables		
Lake water pH	$-\log_{10}[H^+]$	Histogram
Lake water alkalinity	$[Alk]$	Histogram

follow the same time development as the sulfur deposition. The present base cation deposition has been derived from measurements of the Finnish deposition monitoring network (e.g., Järvinen, 1986). Background deposition has been assigned the same value (in equivalents) as sulfur. Parallel trajectories for both the sulfur and base cation deposition patterns over the years from 1900 to 2040 were randomly selected from the designated ranges (*Figure 9.4*).

Input variables and parameters

Ranges for the mean monthly temperature and precipitation for each lake region have been derived by interpolating climatic data from about 200 stations in Europe (Müller, 1982). The ranges used in each region are constrained by the minimum and maximum value of that region, thus reflecting the spatial climatic variability within the region.

The *International Geological Map of Europe and the Mediterranean Region* (UNESCO, 1972) was used for assigning distributions for the weathering rate of

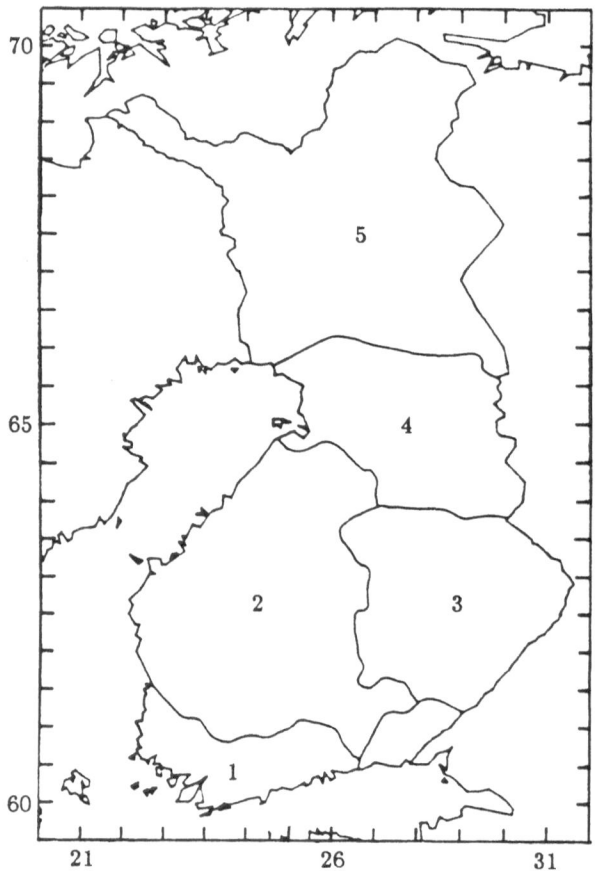

Figure 9.2. The five lake regions considered in the regional application to Finland.

base cations from the silicate parent material. The same classification of different rock types into weathering rate classes was applied as in Kauppi *et al.* (1986). The *Soil Map of the World* (FAO–UNESCO, 1974) provided information on the distributions of typical surface slopes. The *World Forestry Atlas* (1975) was used to derive forest coverage classes for the different lake regions.

Long-term averages of sulfur bulk deposition and of estimated sulfur input to forest floor (throughflow + stemflow) available for several European ecosystems were reviewed (Kämäri, 1986). This information was used for calculating values for the φ parameter, which estimates how much the deposition on forests exceeds the deposition on open land. The bulk deposition was assumed to represent the input to open land. Based on these two sets of measurements, the deposition on forest was calculated to be from 1.1 to 3.9 times larger than the deposition on open land.

The lake depth distribution was obtained from the lake water quality data set by approximating from the observed maximum lake depths. Information on the lake size distribution was provided by the Hydrological Office of the Finnish

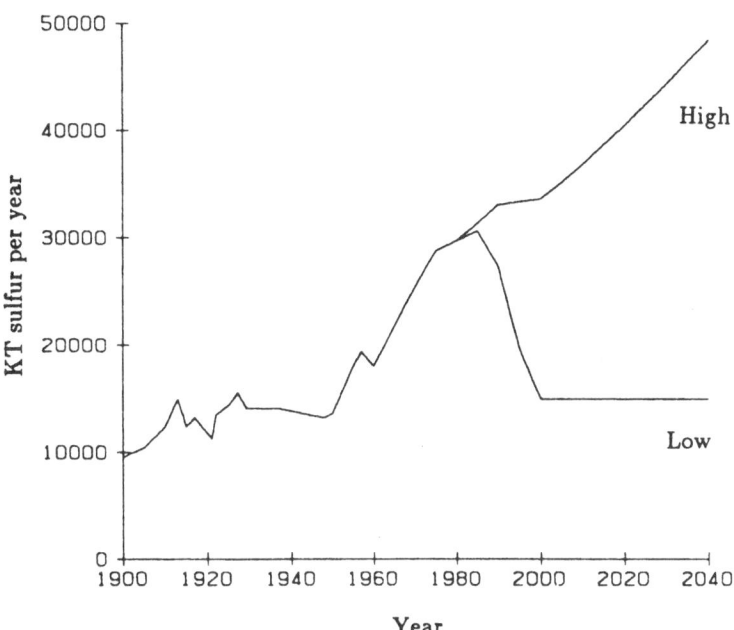

Figure 9.3. Sulfur emissions in Europe (Kt S yr^{-1}) between 1900 and 2040. From 1980 on, the "high" and "low" IIASA–RAINS scenarios are displayed.

National Board of Waters. These data are based on an inventory determining the number and the size distribution of lakes for the whole of Finland. The total number of lakes in Finland (larger than 1.0 hectare) was counted to be 55,000.

All soil data for the Finnish lake regions were obtained from a soil survey of 100 catchments, conducted in 1984 and 1985 by the Geological Survey of Finland. Results of the 1984 survey have been reported by Nuotio *et al.* (1985). Whenever a catchment contained different soil types in its terrestrial catchment areas, samples were taken from all major soil types. Altogether, about 200 samples were analyzed, and the total cation exchange capacity (*CEC*) and the base saturation for both A- and B-layers were calculated for all samples. Frequency distributions for these inputs were formulated by summing up the coverage fractions of each *CEC* and base saturation class within all the catchments.

For some input parameters, there was not enough *a priori* information available to derive frequency distributions. In those cases, the input variables, such as the mean catchment soil thickness and the ratio of lake area to catchment area, were assigned ranges broad enough so that any reasonable value for an input could be selected from these uniform distributions.

Output variables

Lake pH and alkalinity observations were used to constrain the model output. Water quality data on 8,900 lakes from the years 1975 to 1984 were made available by the Finnish National Board of Waters and Environment, Water Quality

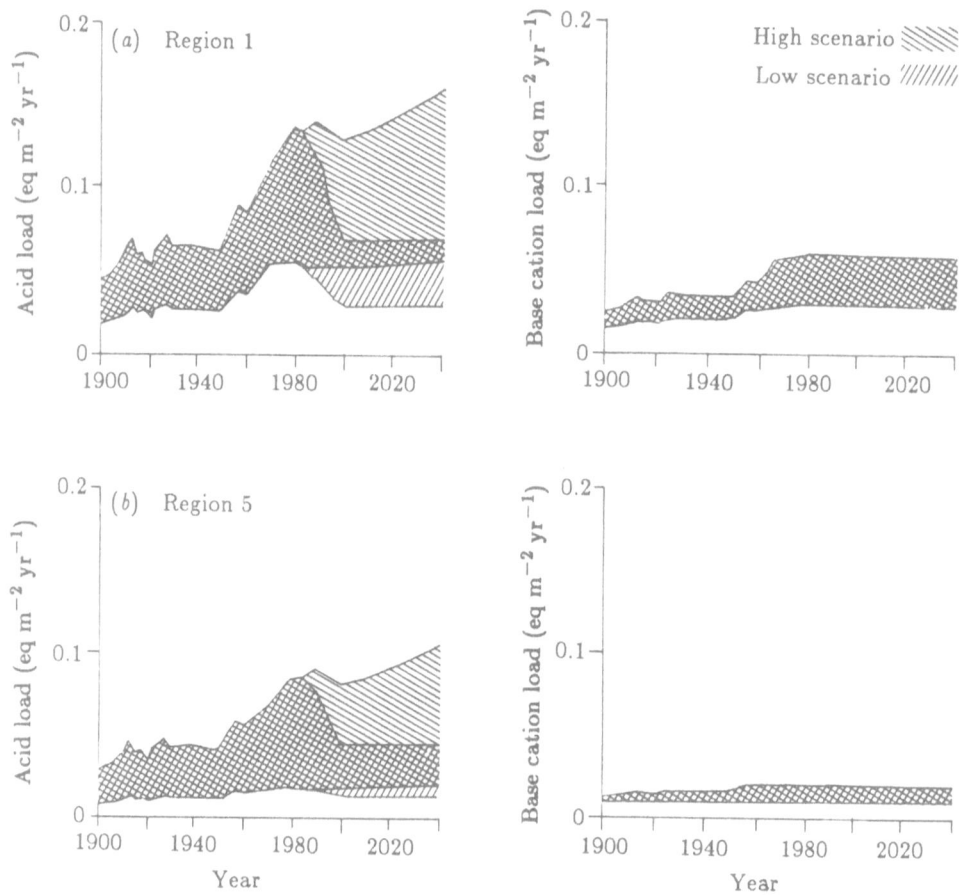

Figure 9.4. (*a*) Maximum and minimum annual acid load and base cation load due to
high and low emission scenarios in (*a*) region 1 and (*b*) region 5.

Data Bank. For a more detailed description of the data, see Forsius (1987). The
lake information was aggregated to form joint lake pH and alkalinity distributions for the five lake regions.

9.3.2. Model results

The MCFP described in the previous section was applied to the five Finnish lake
regions using the 17 distributions for model inputs, initial conditions, and
parameters; the observed (1980) distributions of lake water pH and alkalinity;
and the randomized historical deposition trajectories for sulfur and base cations.
In an earlier version of the model (Kämäri and Posch, 1987), only the observed
pH distribution was used as a constraint in the MCFP. Since, however, in the

routine applications of the lake acidification model within the RAINS framework the user has the option of also displaying the time development of the alkalinity distribution, it seemed necessary to fit the *a priori* input distributions to the joint distribution of pH and alkalinity.

In *Figure 9.5* the pH–alkalinity observations for region 1 (816 clear-water lakes) and region 5 (471 clear-water lakes) are displayed. The observations represent the mean annual conditions in the lakes. If more than one observation was available for a lake, the arithmetic mean was taken. For many lakes, however, the water quality data are based on a single measurement. The curves in *Figure 9.5* show the pH–alkalinity relationship used in the model:

$$[\text{Alk}] = \frac{K_1 K_H p_{CO_2}}{[\text{H}^+]} - [\text{H}^+] - K_{so}[\text{H}^+]^3 \qquad (9.4)$$

where K_1 is the first dissociation constant ($K_1 = 10^{1.5}$ meq l^{-1}); K_H is Henry's law constant ($K_H = 10^{-3.3}$ meq l^{-1} atm); K_{so} is the gibbsite equilibrium constant ($K_{so} = 10^{2.5}$ meq^2 l^{-2}; and p_{CO_2} is the partial CO_2 pressure in water. Curve 1 is drawn with $p_{CO_2} = 1.23 \cdot 10^{-3}$ atm (cf. Wright and Henriksen, 1983), while curves 2 and 3 represent equation (9.4) for $p_{CO_2} = 0.5 \cdot 10^{-3}$ atm and $p_{CO_2} = 10 \cdot 10^{-3}$ atm, respectively. The above pH and alkalinity measurements were then used to derive the joint pH–alkalinity distribution displayed in *Figures 9.6* and *9.7*.

To be able to fit the joint pH–alkalinity distribution, the deterministic relationship of equation (9.4) between pH and alkalinity had to be randomized. To this end the partial CO_2 pressure, p_{CO_2} has been identified as the most uncertain parameter. Therefore, the pH and alkalinity measurements were used to create the p_{CO_2} distribution for each of the five lake regions.

In contrast to an earlier version of the model (see Kämäri et al., 1986), the MCFP was driven with a longer deposition history, starting from 1900 (as opposed to 1960). For different regions, different numbers of runs were required to fill the bins of the joint pH–alkalinity frequency distribution. The percentage of accepted runs varied from 2% to 16%, depending largely on how perfectly the joint distribution was expected to be matched.

Having performed the MCFP and obtained a representative set of parameter combinations, the model can be used to provide estimates of the time patterns of regional lake acidification for any energy-emission scenario between 1980 and 2040. The model was run through the period of 60 years separately for each lake region and for a prespecified number of accepted runs (500 in this exercise).

Estimates of the lake pH and lake alkalinity frequency distributions are produced as the output. In the following paragraphs, hypothetical deposition patterns are evaluated with the regionalized model structure to gain insight into the model behavior under different load patterns. Moreover, the influence on the final outcome of a long versus a short calibration period in the MCFP is evaluated.

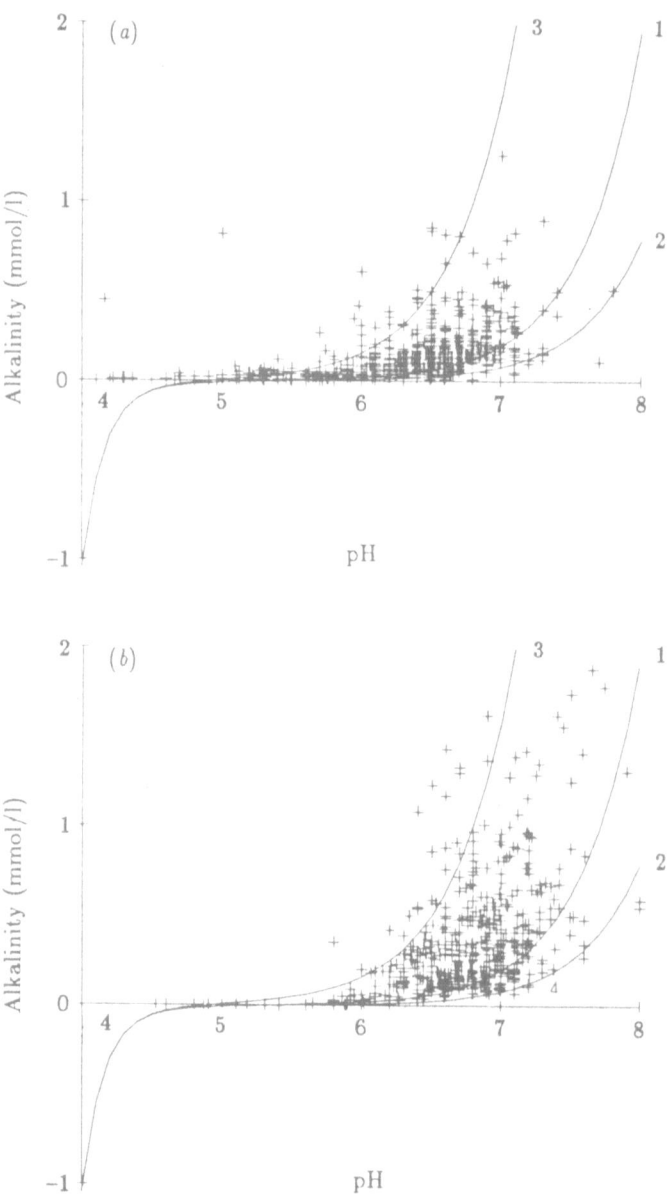

Figure 9.5. Measured pH versus measured alkalinity in (*a*) region 1 (816 clear-water lakes) and (*b*) region 5 (471 clear-water lakes). The curves describe the pH–alkalinity relationship of equation (9.4) for different values of p_{CO_2}.

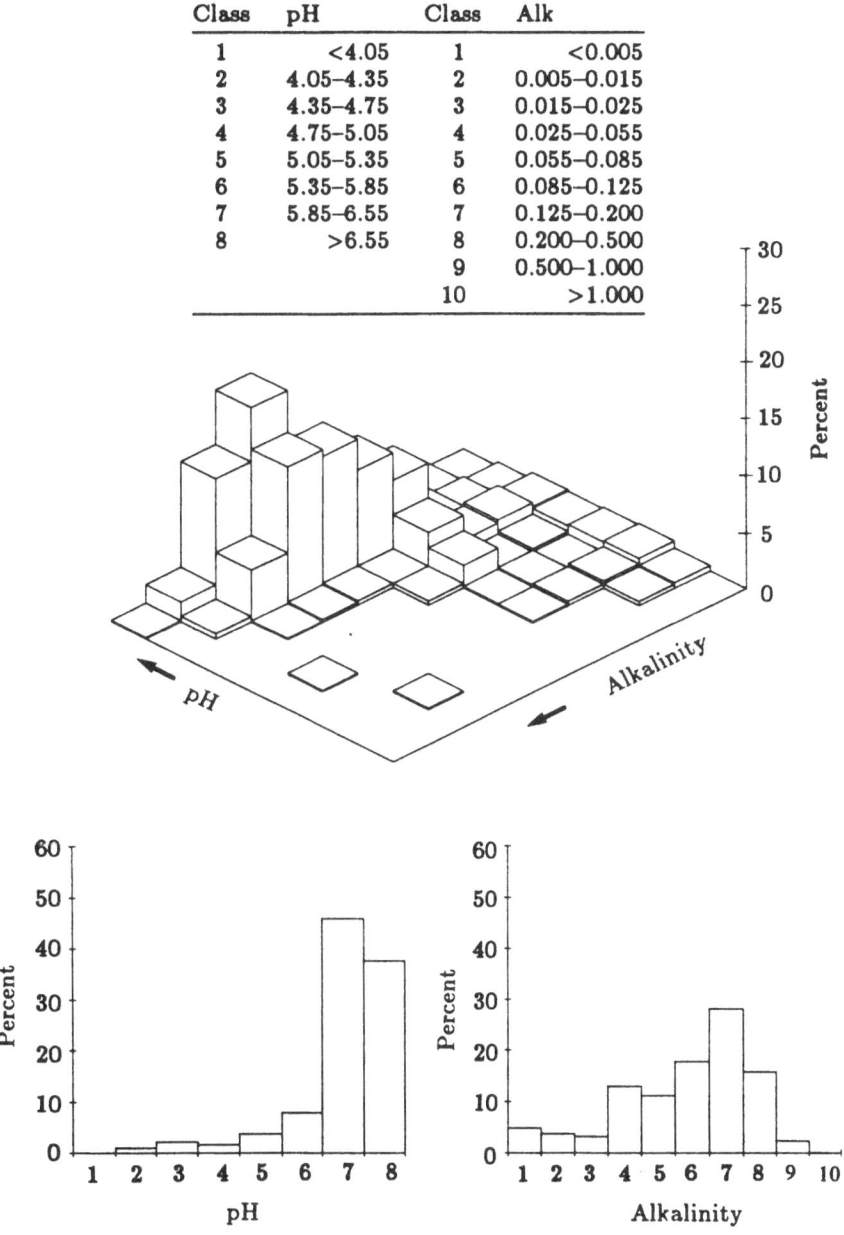

Class	pH		Class	Alk
1	<4.05		1	<0.005
2	4.05–4.35		2	0.005–0.015
3	4.35–4.75		3	0.015–0.025
4	4.75–5.05		4	0.025–0.055
5	5.05–5.35		5	0.055–0.085
6	5.35–5.85		6	0.085–0.125
7	5.85–6.55		7	0.125–0.200
8	>6.55		8	0.200–0.500
			9	0.500–1.000
			10	>1.000

Figure 9.6. Joint pH–alkalinity and marginal pH and alkalinity distributions for region 1.

Class	pH	Class	Alkalinity
1	<4.05	1	<0.005
2	4.05–4.35	2	0.005–0.015
3	4.35–4.75	3	0.015–0.025
4	4.75–5.05	4	0.025–0.055
5	5.05–5.35	5	0.055–0.085
6	5.35–5.85	6	0.085–0.125
7	5.85–6.55	7	0.125–0.200
8	>6.55	8	0.200–0.500
		9	0.500–1.000
		10	>1.000

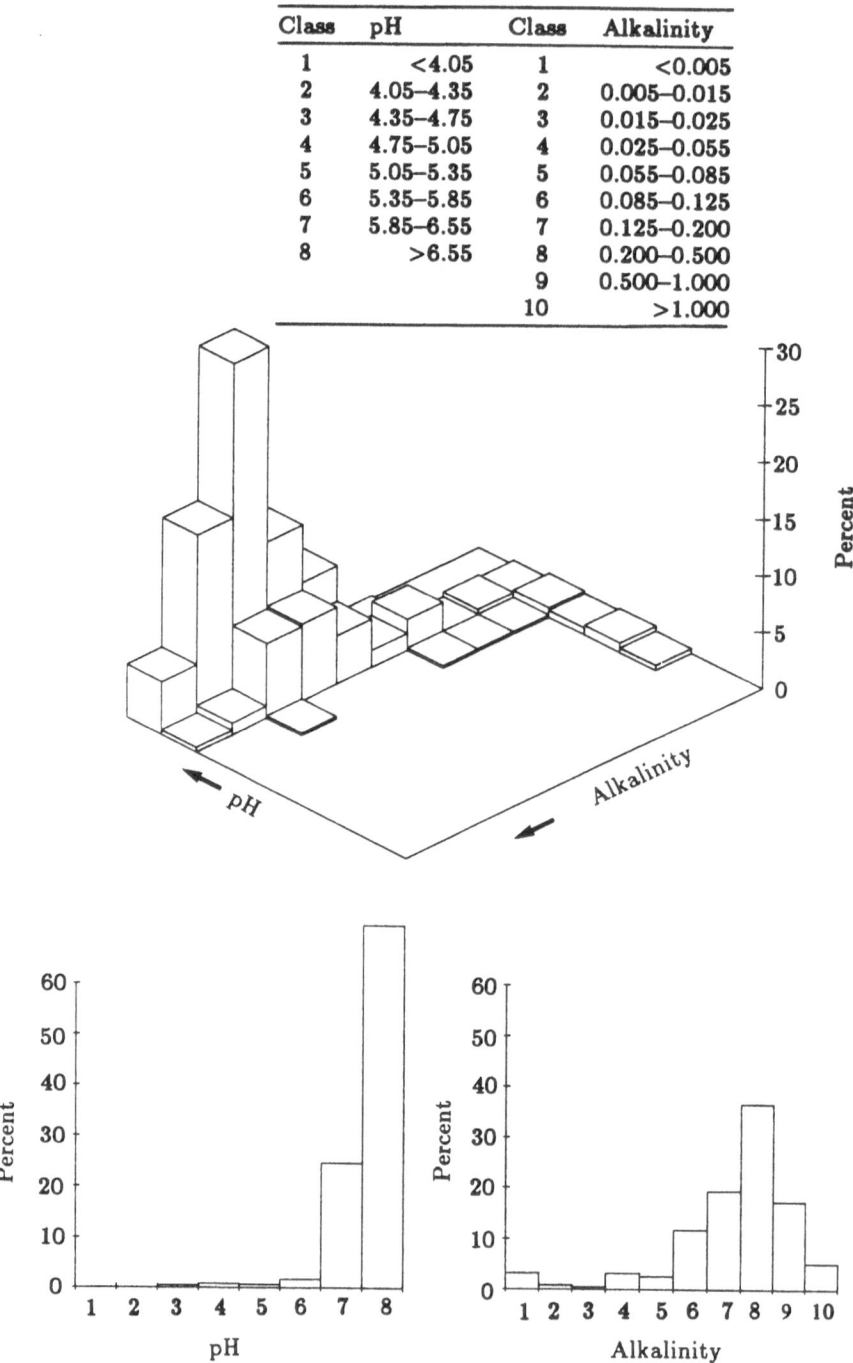

Figure 9.7.   Joint pH–alkalinity and marginal pH and alkalinity distributions for region 5.

Two different scenarios – termed "high" and "low" – from the RAINS energy-emission submodel were used to study the regional lake response. The sulfur emissions of these two scenarios are displayed in *Figure 9.3*, while the spatial variation of the acid and base cation loads from the high and low scenario in regions 1 and 5 are displayed in *Figure 9.4*.

The influence of the length of the calibration period was studied by performing the MCFP twice for both regions – in the first case, using a calibration period of 20 years and in the second case, 80 years. The results, shown in *Figure 9.8*, suggest that the length of the calibration period does not play a major role in controlling the regional behavior of the most acidic lakes. The time pattern for the percentage of acidic lakes (pH < 5.3), produced with the two calibration periods (curves 1 and 2 refer to the 80-year calibration period, while curves 3 and 4 refer to the 20-year period) show very similar results. In fact, the predicted trajectories are so close to each other that they cross several times between 1980 and 2040.

Based on these results, it seems obvious that reductions in European emissions will have a large impact on Finnish lakes. The two energy-emission scenarios give rather different pictures for the distributions of lake pH and alkalinity for any reference year between 1980 and 2040.

Region 1: The response of the lakes to the two scenarios in region 1, displayed with two independent calibration runs, shows that the acidification may actually turn into alkalinization for a large number of lakes if the emissions are reduced drastically. For the high scenario, in turn, the fraction of lakes with pH less than 5.3 is estimated to increase from a level of 6–7% in 1980 to about 13–14% in 2040. For the low scenario, the state of lake acidification in region 1 is predicted to recover, and the fraction of acid lakes (pH < 5.3) to decrease to a level of 4–5%.

Region 5: In region 5 the percentage of lakes with pH < 5.3 rises from the level of about 2% in 1980 to about 4% in 2040 for the high scenario. For the low scenario the percentage of lakes with pH < 5.3 in region 5 reaches in 2040 – after a hardly noticeable increase in the 1990s – the same or a slightly lower level as in 1980. This can be interpreted in such a way that the net acid load of the low scenario is estimated low enough in region 5 to prevent the lakes from further acidification.

9.4. Discussion

Uncertainty in environmental modeling is inevitable. It seems unlikely that any complex environmental system can be well described in the traditional physicochemical sense (Hornberger and Spear, 1981). However, the credibility of any model's results is a key issue in using mathematical models for decision making. An essential aspect of the credibility of the model is how aware the user is of the uncertainties. In regional applications, there remains uncertainty in the accuracy of the data at two levels. First, measurements from the study area,

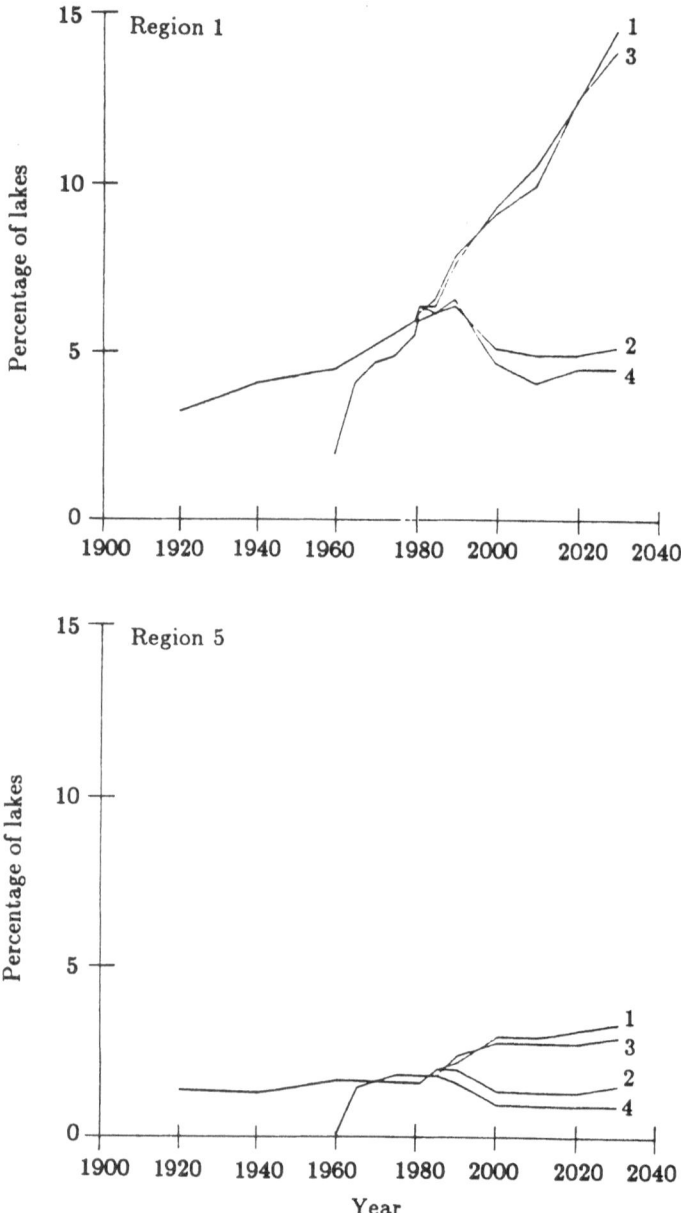

Figure 9.8. Response of the lakes in regions 1 and 5 to the high and low scenarios (pH below 5.3). Curves 1 and 2 refer to the 80-year calibration period, while curves 3 and 4 refer to the 20-year period.

forming the input data used, always include errors. The second level has to do with the interpretation of regional properties. Measurements can only be viewed as samples of the regional system under consideration. It is definitely impossible to sample every one of the catchments in Finland, for instance. The aggregation and interpretation of large-scale information limit the utility of regional data as such. In some cases, measurements are completely absent, and the inputs have to be derived from expert opinion or even guesses. A filtering procedure is chosen, therefore, to avoid unrealistic input ranges by rejecting unrealistic outputs.

The regional application itself forms an additional source of uncertainty, which in fact may result in systematic errors. First test results of MCFP's efficiency in modifying the *a priori* input distributions into the correct direction have been somewhat disappointing (Chapter 12). Nevertheless, in the same tests accurate predictions were produced with MCFP even using inaccurate input distributions. This contradiction deserves further investigation. Moreover, when determining the input ensembles that produced acceptable distributions of output variables, a fixed historical deposition pattern, with a uncertainty band around it, was taken. When this deposition pattern was altered, a new, different set of inputs would be obtained.

Sensitivity and uncertainty analyses of the regional lake acidification model have been performed by Kämäri *et al.* (1986), and the reliability of the regional predictions have been estimated by Gardner *et al.* (Chapter 11). These analyses show that, despite the large uncertainties in some key parameters, the model seems to provide a fair representation of measured pH levels in Finland. Moreover, it has become clear that to improve the results and reduce the uncertainties associated with scenarios, effort should be made to obtain more accurate input distributions for the most critical parameters: the mean catchment soil thickness and the weathering rate of base cations in the region. Relatively little emphasis has been given so far to these two parameters that together largely determine the long-term behavior of the catchments. Current research is, however, continuously expanding our knowledge of these variables, and we expect to be able to incorporate more realistic *a priori* distributions in the future.

The model development is viewed as a stepwise procedure of reducing uncertainties involved in both the model structure and the model application. In the latest version of the RAINS lake module, several additions and modifications were introduced to the model structure. In contrast to the earlier version, the latest version accounts for a long initialization period, filtering of air pollutants by forests, base cation deposition, in-lake sulfate retention, and the use of revised EMEP sulfur transport matrices (Lehmhaus *et al.*, 1986). However, the model has been kept simple, because simplicity is an asset both for model testing and for assisting decision making.

Tests with individual lakes in Fennoscandia showed good agreement between observed (from diatom analyses) and simulated lake responses to the historical deposition pattern (Kämäri *et al.*, 1989). Even better agreement was obtained after including the above-mentioned new process descriptions. Owing to the recent modifications, the regional model response became more stable in its behavior. It suggests much less acidification in the study regions than the earlier applications with comparable scenarios (see Kämäri *et al.*, 1986). A completely new feature is the predicted recovery with the low scenario in southern Finland. As pointed out earlier, the longer calibration period seems not to be responsible for that. However, the complex, dynamic, and nonlinear interactions of the individual model constructs make it difficult to relate the overall performance of the model to any individual process descriptions within the model.

Acidification models assessing long-term responses are extremely difficult to validate. Strict validation of these types of models would require long time series to determine whether the model estimates match the observed catchment responses. Unfortunately, very few, if any, such records exist. Therefore, the question whether the long-term responses estimated by the model are true projections of real systems' responses remains unanswerable (cf. Cosby *et al.*, 1985b). Given the best available data and using parameter values that are within the ranges appropriate for natural soils in Finland, our model seems to produce plausible results. Ultimate validation can never be established. It is our hope, however, that the model structure will be used as a tool for organizing data and for identifying research needs. Even in its present stage, the model may be useful in evaluating policies to combat lake acidification.

Appendix 9A: Model Equations

In this appendix we give a short description of the lake model and present the basic equations for calculating lake water pH and lake water alkalinity.

Acid load: The principal driving variable of the lake acidification model is the total deposition of sulfur, d_{tot}. The throughfall deposition generally exceeds the deposition on open land. The mean deposition on an area is therefore allocated to forests and to open land by assigning the deposition on forested land to exceed that on open land by a factor of φ ($\varphi \geq 1$). The acid load on forest soils of one grid, d_{SO_4}, is then given by

$$d_{SO_4} = d_{tot}\varphi/[1 + (\varphi - 1)f] \tag{9A.1}$$

where f is the fraction of forested land in that grid.

Water fluxes: The basic assumption concerning soil hydraulics is that all of the precipitation surplus, i.e., the annual runoff, Q_{tot} (precipitation, P, minus evapotranspiration, E), infiltrates the A-reservoir. Evapotranspiration is set proportional to the mean temperature, T, above 0 °C, using a evapotranspiration rate coefficient ε (c.f., Christophersen *et al.*, 1984), i.e., $E = \varepsilon \cdot T$.

The water leaves the A-layer as quickflow, Q_a, which is given by

$$Q_a = Q_{tot} - Q_b \tag{9A.2}$$

where Q_b is the baseflow, leaving the B-layer, which is given by

$$Q_b = \frac{\kappa_s S W Z_b}{A_c} \tag{9A.3}$$

where κ_s is the hydraulic conductivity at saturation; S is the surface slope; W is catchment width; Z_b is thickness of the B-layer; and A_c is terrestrial catchment area.

Soil equilibrium: As for the soil chemistry, it is assumed that few equilibrium reactions between soil and soil solution address the water quality of runoff water at a particular time. A nonlinear relationship is assumed between the base saturation, Ex, and the soil pH in the range 4.0 to 5.6. The equation is based on the results of an equilibrium model by Reuss (1983):

$$-\log_{10}[H^+] = 4.0 + 1.6 \cdot Ex^{0.75} \tag{9A.4}$$

where the base saturation is given by

$$Ex = BC/CEC \tag{9A.5}$$

where CEC is the total cation exchange capacity of the soil. If the cation exchange system does not play any role in buffering the inputs to the soil solution, i.e., if BC is zero, it is assumed that the equilibrium with gibbsite controls soil buffering. Aluminum is dissolved or precipitated until the gibbsite equilibrium is reached,

$$[Al^{3+}] = K_{so} [H^+]^3 \tag{9A.6}$$

where K_{so} is the gibbsite solubility constant.

Flux relationships: Over the long term, the flux of acids and bases into and out of the soil change the above chemical state of the soil, thus affecting the soil/soil solution equilibria. In noncalcareous soils the weathering rate of base cations, wr, largely determines the long-term response of the catchments. As long as the input of base cations from weathering, wr, and from base cation deposition, d_{BC}, is larger than the acid load, there will be no change in the H^+ concentration in the soil. If, however, the acid load exceeds the base cation input, the capacity of the cation exchange buffer system is depleted by the rate,

$$\frac{dBC}{dt} = d_{BC} - d_{SO_4} + wr \quad . \tag{9A.7}$$

To calculate lake water pH and alkalinity, the following ion fluxes into the lake are considered. The flux of protons, F_H, is the sum of the convective flow of ions from the two soil reservoirs and the direct input from the atmosphere to the lake

$$F_H = Q_a [H^+]_a + Q_b [H^+]_b + d_{SO_4} A_l \tag{9A.8}$$

where A_l is the lake surface area. The fluxes of bicarbonates contributing to the alkalinity of the lake originate both from the terrestrial catchment, $F_{HCO_3}^{(1)}$ and from the in-lake (internal) alkalinity generation, $F_{HCO_3}^{(2)}$, (cf., Baker *et al.*, 1986):

$$F_{HCO_3}^{(1)} = (d_{BC} - d_{SO_4} + wr) A_c \tag{9A.9}$$

and

$$F_{HCO_3}^{(2)} = \frac{k_{SO_4} d_{tot}}{Q_{tot}/A_l + k_{SO_4}} \tag{9A.10}$$

where k_{SO_4} is the sulfate retention rate coefficient.

Lake equilibrium: The above ion fluxes mix with the lake water and cause a change in the bicarbonate and hydrogen ion concentrations until an equilibrium is reached:

$$[HCO_3^-] = \frac{K_1 K_H p_{CO_2}}{[H^+]} \tag{9A.11}$$

where K_1 is the first dissociation constant, K_H the Henry's law constant and p_{CO_2} the partial CO_2 pressure in water.

CHAPTER 10

Identification of a Direct Distribution Model from a Regionalized Mechanistic Model of Aquatic Acidification

M.J. Small, P.A. Labieniec, and M.C. Sutton

10.1. Introduction

There is a need to predict the effects of future acidic deposition on surface-water quality. The pH and the acid-neutralizing capacity (*ANC*) of surface waters serve as indicators of the conditions faced by aquatic life. Progress has been made in quantifying these effects in single lakes; however, acid rain and its effects occur over larger areas with impacts on many resources. Thus, policy options must take into account acidification not only in a single lake, but also on a regional scale. To accomplish this, we need models capable of predicting the extent of aquatic acidification caused by alternative future deposition scenarios for the full population of lakes in a region.

Existing regional lake acidification models exhibit a wide range of scientific and computational complexity. However, two basic approaches are evident: simple empirical models and mechanistic or process-oriented models. The simple empirical models lack a rigorous physical basis, using lumped weathering factors to represent the effective buffering of acid inputs that occurs throughout a lake and its watershed. As a result of this simplification, empirical models are computationally efficient and easy to apply on a regional scale. In contrast, mechanistic models attempt to represent actual chemical reactions using the kinetic and equilibrium expressions for the geochemical processes involved. They are, however, difficult to apply on a regional scale owing to their extensive data requirements. The Direct Distribution model (Small and Sutton, 1986b; Small *et al.*, 1987) is an example of a model that utilizes the simple empirical approach;

167

J. Kämäri (ed.), Impact Models to Assess Regional Acidification, 167–181.
© 1990 *International Institute for Applied Systems Analysis.*

the regional lake acidification model developed by the modeling group at the International Institute for Applied Systems Analysis (IIASA) is an example of a model that adopts the mechanistic approach (Alcamo *et al.*, 1985).

The Direct Distribution model extends the single lake empirical equilibrium weathering model of Henriksen and Wright (Henriksen, 1979; Wright and Henriksen, 1983) to calculate directly the effects of acidic deposition scenarios on regional distributions of lake *ANC*, pH, and fish populations. Probabilistic methods are used to represent the distribution of weathering factors across a region, and an empirical lag equation is used to represent time delays in the response to acid loadings. The approach is similar to that of Minns (Chapter 5), except that analytical equations are used to represent regional distributions of lake chemistry, whereas Minns uses Monte Carlo methods. Also, an attempt is made in the Direct Distribution model to incorporate empirically temporal lags in the regional lake response. The Direct Distribution model provides a simple method for making regional predictions that is easy to manipulate for sensitivity and uncertainty analysis. It was originally developed for this purpose as part of a linked, integrated assessment model of acid deposition (Marnicio *et al.*, 1985; Rubin *et al.*, Chapter 14). However, it is based on a number of assumptions that have not yet been rigorously tested, including assumptions about the shape of the regional lake *ANC* distribution and the dynamic response profile of lake chemistry.

The lake acidification model developed by IIASA is a component of IIASA's Regional Acidification and INformation Simulation (RAINS) model. The RAINS Lake Model (RLM) was originally developed to simulate the effect of acidic deposition on the pH in a single lake and watershed. The RLM model was subsequently regionalized to simulate lake acidification for a set of statistical lakes representative of the regional distribution of lake pH or *ANC* (Alcamo *et al.*, 1985; Kämäri and Posch, 1987; Posch and Kämäri, Chapter 9).

The ultimate verification of any regional lake acidification model requires the observation and analysis of regional lake chemistry data over time periods during which significant changes in acid deposition occur. Currently, no such data are available. However, as a first step in the evaluation of the Direct Distribution model, we can examine the consistency of the assumptions made in the Direct Distribution model with results derived from a more mechanistic model. The procedure developed here tests this consistency by applying the two models jointly.

The chapter begins with a review of the Direct Distribution and RLM models. This is followed by a description of the method used to determine the weathering characteristics of the RLM model when applied to the southernmost region of Finland. The estimated weathering parameters are incorporated into the Direct Distribution model, and the lake acidification predictions of the resulting Direct Distribution model are compared with the RLM model predictions for a common deposition scenario. The consistencies and inconsistencies between the two models' results are discussed throughout in terms of the key assumptions incorporated into each model.

10.2. The Direct Distribution Model

The Direct Distribution model simulates lake acidification by operating directly on the mean and variance of the regional distribution of ANC. The data required to specify the model are the historical deposition scenario and observed ANC data for a representative sample of the lakes in the region. The regional lake ANC data are fit with an analytical distribution function, usually a three-parameter lognormal. The shape of the ANC distribution is assumed to remain lognormal over time, and because the skewness coefficient of the ANC distribution is assumed to remain constant, the distribution is fully characterized by the calculated first and second moments.

Shifts in the equilibrium distribution of ANC caused by acidic deposition are calculated using regionalized equilibrium weathering models applied to the moments of the fitted distribution. In the current application, a regionalized version of the Henriksen–Wright weathering model is applied.

The Henriksen–Wright model assumes that a constant fraction, F, of the incoming acidity is neutralized in the watershed of each lake (Henriksen, 1979; Wright and Henriksen, 1983). The variation in weathering across a region is characterized by the mean and variance of the weathering factor, μ_F and σ_F, respectively. The resulting mean and variance of the equilibrium ANC distribution in a region is given by (Small and Sutton, 1986b):

$$\mu_{ANC_e} = \mu_{ANC_0} - C(1 - \mu_F) \tag{10.1}$$

$$\sigma^2_{ANC_e} = \sigma^2_{ANC_0} + C^2\sigma_F^2 + 2C\sigma_{ANC_0}\sigma_F\rho \tag{10.2}$$

where

μ_{ANC} = mean of the regional ANC distribution (μeq l^{-1}).
σ_{ANC} = standard deviation of the regional ANC distribution (μeq l^{-1}).
μ_F = mean of the regional weathering factor.
σ_F = standard deviation of the weathering factor.
ρ = correlation coefficient between ANC_0 and the weathering factor.
C = acid concentration of deposition, incorporating the concentrating effects of evapotranspiration (μeq l^{-1}).

Subscript e denotes the equilibrium condition with deposition C; and subscript 0, the initial predeposition condition.

A time lag was incorporated into the Direct Distribution model by Small *et al.* (1987) in recognition of the fact that the equilibrium ANC distribution is not attained instantaneously. There is some delay involved in the approach to equilibrium. The time lag relationship in the Direct Distribution model was developed to consider empirically the following processes:

- The lake's hydraulic retention time controls the time it takes to replace existing water with the new runoff of differing hydrogen ion concentration (Schnoor and Stumm, 1984).
- Soil sulfate adsorption delays the appearance of both hydrogen ions and sulfate anions in the soil solution and surface waters (Reuss and Johnson, 1986; Cosby et al., 1986a).

Very long-term processes involving the loss of buffering capacities are not taken into account.

The Direct Distribution model assumes an exponential shape for the temporal ANC response to a change in deposition, as would be the case of lake flushing controls. That is, for each change in deposition occurring over a time step, the equilibrium condition is calculated, and the approach of ANC to the calculated ANC equilibrium condition is assumed to be exponential with some characteristic lag time, τ. The approach to equilibrium in a given lake can then be expressed as:

$$ANC(t+\Delta t) = ANC(t) + [ANC_e - ANC(t)](1 - e^{-\Delta t/\tau}) \ . \tag{10.3}$$

To approximate the overall response of a region to acidic deposition, a single representative value of τ is selected. The response of the mean and variance of the regional ANC distribution are then given by:

$$\mu_{ANC}(t + \Delta t) = \mu_{ANC}(t)e^{-\Delta t/\tau} + \mu_{ANC_e}(1 - e^{-\Delta t/\tau}) \tag{10.4}$$

$$\sigma^2_{ANC}(t + \Delta t) = \sigma^2_{ANC}(t)e^{-2\Delta t/\tau} + \sigma^2_{ANC_e}(1 - e^{-\Delta t/\tau})^2$$
$$+ 2\sigma_{ANC}(t)\sigma_{ANC_e}e^{-\Delta t/\tau}(1 - e^{-\Delta t/\tau}) \ . \tag{10.5}$$

Equations (10.1), (10.2), (10.4), and (10.5) can be used to calculate the mean and variance of the regional ANC distribution resulting from a deposition scenario. Note that equation (10.5) requires the additional assumption that ANC values at equilibrium are perfectly correlated with ANC values at time t. That is, the rank ordering of ANC values in the lakes of a region remains unchanged over time.

The regional pH distribution is derived using the explicit pH–ANC relationship of Small and Sutton (1986a). The pH distribution can then be used with a fish presence–absence relationship to estimate the fraction of lakes able to support fish populations in the modeled region (Small and Sutton, 1986b; Small et al., 1987).

The steps in estimating and applying the Direct Distribution model can be summarized as follows:

(1)   Data requirements:

- Current ANC and pH data for a set of lakes representative of the region.
- Deposition history.

(2)   Parameter estimation:

- Fit the current ANC data set with a 3-parameter lognormal distribution and calculate μ_{ANC}, σ_{ANC}, and γ_{ANC}.
- Fit the parameters of the pH–ANC relationship to allow the pH distribution to be derived from the ANC distribution.
- Determine a distribution for F and calculate μ_F and σ_F.
- Estimate the correlation coefficient, ρ, between ANC_0 and F.
- Estimate the lag factor, τ.
- Calibrate the model iteratively to determine the moments of the ANC_0 distribution that result in the current ANC distribution moments, given the deposition history [equations (10.4) and (10.5)].

(3)   Model projections:

- Using the moments of the calibrated ANC_0 distribution, calculate the effects of chosen deposition futures on the future regional distribution of ANC [equations (10.1), (10.2), (10.4), and (10.5)].
- Calculate the corresponding future pH distributions using the pH–ANC relationship.
- Calculate the effect on fish viability by integrating the fish survival function with the pH distribution. (In the evaluation presented in this chapter, this step is not included.)

10.3. The RLM Model

The RLM model (Alcamo *et al.*, 1985; Posch and Kämäri, Chapter 9) consists of four submodules:

(1)   A meteorological module to determine the volume and concentration of acid rain falling on the catchment area.
(2)   A hydrologic module to simulate flowpaths through the soil.
(3)   A soil chemistry module to compute the ion concentrations in the soil solution.
(4)   A lake chemistry module.

The first three modules determine the volume and concentration of ions entering the lake; the fourth determines the lake ANC and pH. Relevant soil, lake, and catchment data are required to determine the parameters of the model. Current lake pH data and a deposition history up to the time of the current lake pH data are required to calibrate the model.

A key characteristic of the RLM model is embodied in the soil acidification module, which is based on the model developed by Ulrich (1983a,b). Buffering ranges are identified based on the dominant neutralization reactions that occur in the soil following acidic inputs to the system. These include, in the order of their dominance as the soil is acidified, the carbonate, silicate, cation exchange and aluminum buffering ranges. The capacities of these buffering ranges are assumed to be depleted sequentially as acidification progresses. Buffering or weathering refers to the neutralization reactions that consume hydrogen ions. Each buffering range is assigned a buffering capacity based on the regional soil characteristics. The buffering capacity is the total reservoir of buffering compounds available for the neutralization of the acid inputs in each buffering range. Acid load rates are compared with buffering rates, and accumulated acid load is considered to decrease the buffering capacity of the prevaling buffer range (Kauppi *et al.*, 1986; Posch and Kämäri, Chapter 9). In particular, the base saturation of the soil decreases as long as the acid load is above the silicate buffering rate. This concept of a continuing consumption of buffering capacity is critical for understanding the results of simulations shown later in the chapter.

The RLM model is regionalized by specifying *a priori* parameter distributions from the expected ranges of values for typical lakes in the region. These *a priori* distributions are then sampled using a Monte Carlo procedure, and the model is calibrated by accepting a set of parameter inputs that yields the current distribution of lake pH. Accepted parameter sets are stored and subsequently sampled to simulate the response of the representative set of statistical lakes to specified deposition futures. A detailed description of the RLM model and the Monte Carlo calibration procedure can be found in Kämäri and Posch (1987).

10.4. Estimation of the Direct Distribution Model from RLM

The estimation of the Direct Distribution model from the regionalized RLM model and the use of the Direct Distribution model as a simplified representation of RLM are illustrated by application to a set of lake pH data from the southernmost region of Finland. The RLM model is calibrated for this region and the calibration procedure is verified by comparing predicted and observed pH distributions for the calibration year. The assumptions made in the Direct Distribution model are tested against the RLM model output, and the regional weathering parameters of the Direct Distribution model are estimated based on the behavior of the RLM model. The estimated Direct Distribution model is then executed with the same deposition scenario used in the RLM model, and the results of the Direct Distribution model and the RLM model are compared.

The RLM model was calibrated to the average pH estimates for 816 lakes in the southernmost region of Finland for the year 1980. The raw data were made available by the Finnish National Board of Waters and the Environment, Water Quality Data Bank. The annual pH averages, *a priori* model parameter distributions, and the historical deposition were determined by the modeling group at IIASA (Posch and Kämäri, Chapter 9; Kämäri and Posch, 1987). The calibration procedure determined a parameter set for each of 213 statistical lakes. The pH distribution of the 213 statistical lakes is compared to the pH distribution of the 816 observed lakes in *Figure 10.1*, indicating a successful calibration. The 213 parameter sets associated with the statistical lakes are used for subsequent Monte Carlo sampling to simulate future (post-1980) lake acidification.

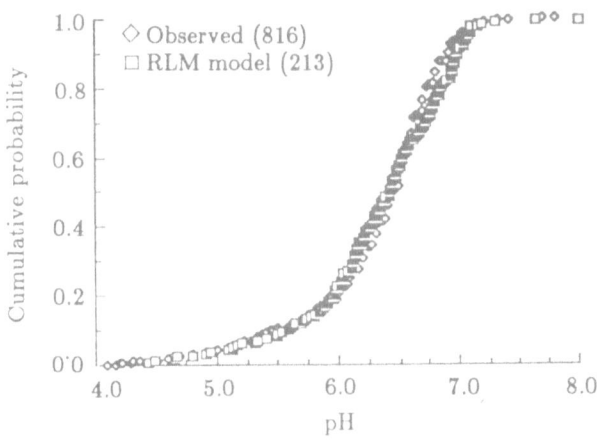

Figure 10.1. The 1980 observed and RLM model lake pH distributions.

Since the RLM model also simulates *ANC* for each statistical lake, a 3-parameter lognormal distribution was fit to the 1980 RLM model *ANC* output using the procedure described in Small and Sutton (1986b). This served to test the assumption made by the Direct Distribution model that the regional lake *ANC* distribution can be fit with an analytical distribution function, and allowed the estimation of the distribution parameters. The lognormal distribution equation and estimated parameters are presented in *Table 10.1*, Section 1. *Figure 10.2* illustrates a good fit for 211 of the 213 RLM model lakes. Two lakes with very high *ANC* biased the estimated distribution parameters and were not included in this analysis. The deletion of 2 lakes out of 213 has a negligible effect (less than a 1% change) on the estimated fraction of acidified lakes (i.e., the estimated fraction of lakes below a given *ANC* or pH).

To compare the RLM and Direct Distribution model predictions of the regional pH distribution, the RLM *ANC* and pH model results for 1980 were used to determine the parameters of the pH–*ANC* relationship presented in *Table 10.1*, Section 5. Nonlinear least squares regression was used to determine the fitted parameter values shown in *Table 10.1*. The RLM model pH–*ANC* values and the fitted pH–*ANC* relationship are compared in *Figure 10.3*. While the fitted equation provides the best functional relationship using the least squares criterion, there is considerable spread around the fitted curve. The implications of this spread for the derived regional pH distribution are discussed below.

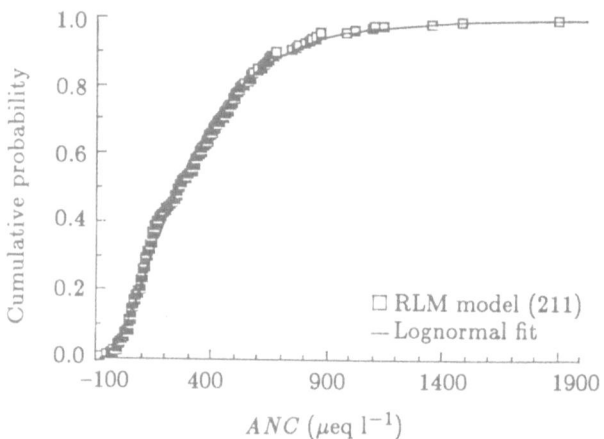

Figure 10.2. The RLM model and the fitted lognormal ANC distributions.

Table 10.1. Direct Distribution model equations and estimated parameters for southern-most Finland.

1.   3-parameter lognormal distribution for current ANC (pdf):

$$f_{ANC}(anc) = \frac{1}{(anc-\theta)\Phi\sqrt{2\pi}} \exp\left\{\frac{-[ln(anc-\theta)-\xi]^2}{2\Phi^2}\right\} ; \quad anc \geq \theta$$

Estimated parameters: $\theta = -140 \ \mu eq/l^{-1}$ $\xi = 5.96$ $\theta = 0.63$

2.   Fitted moments of current ANC distribution:

$$\mu_{ANC_e} = \theta + \exp(\xi + \theta^2/2) = 332 \ \mu eq/l^{-1}$$

$$\sigma_{ANC_e} = [\omega(\omega - 1)\exp(2\xi)]^{1/2}; \quad \text{where } \omega = \exp(\Phi^2)$$

$$= 328 \ \mu eq/l^{-1}$$

3.   Estimated weathering and lag parameters:

$$\mu_F = .79 \quad \sigma_F = .10 \quad \rho = .15 \quad \tau = 30 \text{ years}$$

4.   Calibrated initial ANC moments:

$$\mu_{ANC_0} = 386 \ \mu eq/l^{-1} \quad \sigma_{ANC_0} = 321 \ \mu eq/l^{-1}$$

5.   pH–ANC relationship:

$$pH = a + b \text{ arc } \sin\left|\frac{ANC - d}{c}\right|$$

Estimated parameters: $a = 5.05$ $b = .4343 \ (1/ln\ 10)$
                                       $c = 25.06$ $d = -34.92$

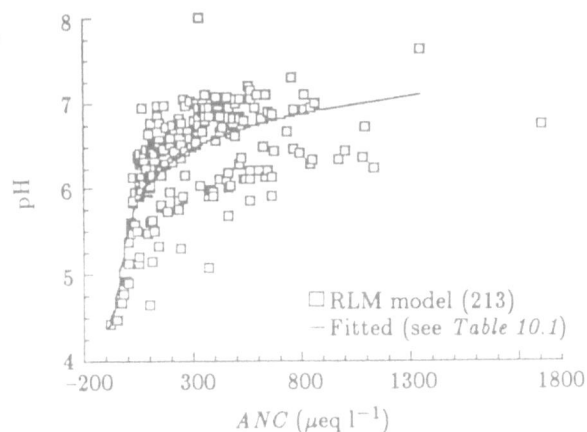

Figure 10.3. The RLM model and the fitted pH–*ANC* relationship.

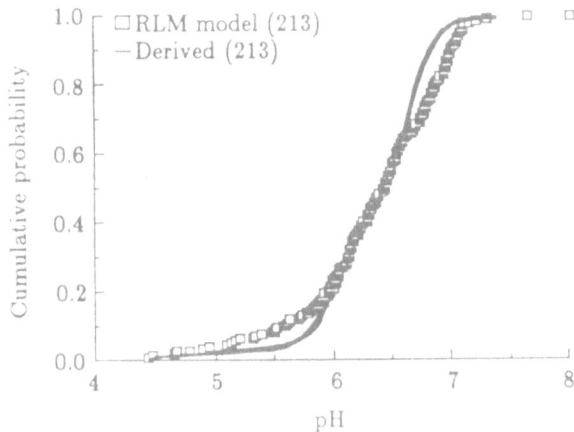

Figure 10.4. The 1980 RLM model and the derived pH distributions.

The pH–*ANC* relationship is used to derive a regional pH distribution from the lognormal *ANC* distribution as described in Small and Sutton (1986b). The resulting distribution is compared with the RLM model pH distribution for 1980 in *Figure 10.4*. The comparison indicates generally good agreement, although the distributions diverge somewhat at the lower and upper ends of the range of pH. Of particular concern for this analysis of lake acidification is the fact that the fraction of lakes with low pH (less than 5.5) is underestimated by the derived distribution. This results largely because of the spread in the RLM model pH–*ANC* predictions (see *Figure 10.3*) for pH less than 5.5 and *ANC* less than 500 μeq/l^{-1}. The derived distribution used in the Direct Distribution model computes all pH values from *ANC* using the fitted line shown in *Figure 10.3*, and is thus unable to account for the RLM model pH predictions that fall below this line. For this reason, results for the RLM and estimated Direct Distribution models were compared for the percentage of lakes with pH less than 6, where the original and derived distributions agree. It is recognized that the divergence indicated in *Figure 10.4* is a potential source of error in the estimated Direct Distribution model. This comparison thus indicates one aspect for which the model simplification procedure is only approximate.

The weathering characteristics of the RLM model were determined by calculating, with the RLM model, the fraction of incoming hydrogen ions that are neutralized in each lake's watershed during each year of the simulation. *Figure 10.5* shows the distribution of F calculated for the 213 statistical lakes at two points in time. The distribution shifts downward from the year 2020 to the year 2040, with a resulting decrease in μ_F. The mean value of F was found to decrease through time for all simulations considered. The consumption of buffering capacity resulting from the neutralization of accumulated acid stress in the soil module is the fundamental cause of this effect.

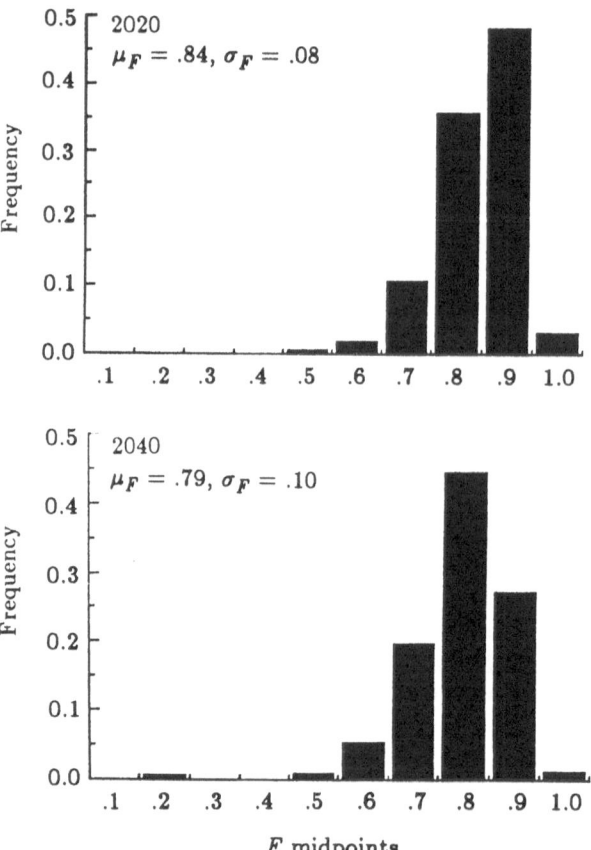

Figure 10.5. Time variation in the frequency distribution of the weathering factor, F, in the RLM model. Results are shown for a particular deposition scenario before (2020) and after (2040) a change in deposition.

The results shown in *Figure 10.5* conflict with the Direct Distribution model assumption that the distribution of F remains constant over time, regardless of the deposition scenario. To match the results of RLM, the Direct Distribution model would have to incorporate a regional value of μ_F that decreases over time and a variable value of σ_F. These capabilities are not currently included in the Direct Distribution model. To demonstrate the implications of this structural difference, constant values of μ_F and σ_F were chosen for use in the Direct Distribution model. Values shown in *Figure 10.5* for the year 2040 – $\mu_F = 0.79$ and $\sigma_F = 0.10$ – were selected. These values are consistent with previous estimates of the moments of F for sensitive regions, as discussed by Small and Sutton (1986b).

A degree of correlation between F and ANC_0 was found to be implicit in the RLM model. The correlation coefficient between F and ANC_0 was determined by correlating the value of F calculated for each of the 213 statistical lakes with the corresponding ANC value predicted for the year 2020. The small positive value of ρ ($\rho = 0.15$) is consistent with the fact that lakes with a higher

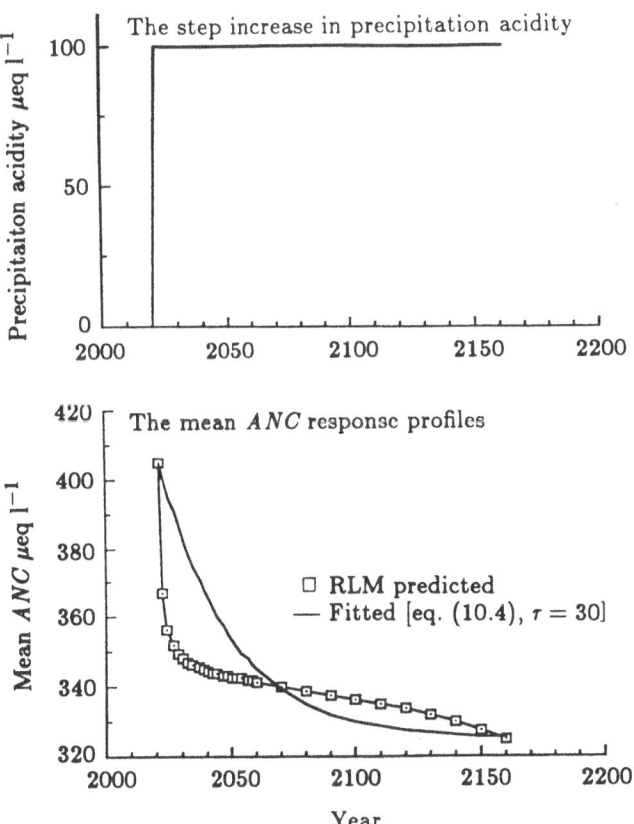

Figure 10.6. RLM predicted and the fitted mean ANC responses to a step increase in precipitation acidity.

initial value of ANC should also tend to exhibit a greater degree of weathering to buffer subsequent acid loads.

The dynamics of the RLM model were assessed by examining the model's response to a step change in deposition. One assumption of the Direct Distribution model confirmed for this simulation is that the rank ordering of ANC remains unchanged, so that equation (10.5) is applicable. The rank order correlation coefficients for ANC values in different years were all greater than 0.985. However, the shape of the mean ANC response of the RLM model is not consistent with the exponential shape assumed in the Direct Distribution model.

Figure 10.6 compares the response of the mean regional ANC predicted by the two models. A distinct difference between the models is indicated. The initial response of the mean ANC to the step increase in deposition in the RLM model is relatively rapid, reflecting processes controlled by the hydraulic residence times of the lakes. This is followed by a slow continuing decay in the mean ANC, even under constant deposition. This again reflects the important structural differences between the RLM model, which predicts long-term continuing losses of buffering capacity, and the Direct Distribution model, which assumes that the weathering capacity remains constant. Nevertheless, an attempt to approximate the overall dynamics of the RLM model was made by selecting value of $\tau = 30$ years for use in the Direct Distribution model, yielding the exponential approximation shown in *Figure 10.6*. The implications of the differences in the response profiles are discussed further in the comparison of the model predictions presented below.

10.5. Comparison of RAINS and Estimated Direct Distribution Model Predictions

The derived parameters presented in *Table 10.1* were used in the Direct Distribution model to simulate acidification in the southernmost region of Finland for comparison with the predictions of the RAINS model. The deposition future for which the models are compared is the "major sulfur reductions scenario" described earlier in the RAINS model study (Kämäri *et al.*, 1985). The common deposition scenario and the comparison of the results for several measures of regional lake acidification are presented in *Figure 10.7*. These measures of acidification include the mean, the standard deviation, the skewness coefficient of the regional ANC distribution, and the percentage of lakes with a pH less than 6.0.

The two key measures of regional acidification – the regional mean ANC and the percentage of lakes with a pH less than 6 – show fairly good agreement until the year 2020. The Direct Distribution model is specifically calibrated to the RLM model predictions for the year 1980, so agreement at this time is assured. The standard deviation and skewness coefficient of the ANC distributions indicate relatively good agreement throughout the simulation period. The skewness coefficient of the RLM model is nearly constant over time, as assumed in the Direct Distribution model. The important differences in the model predictions occur after 2020, when results for the mean ANC and the fraction of lakes

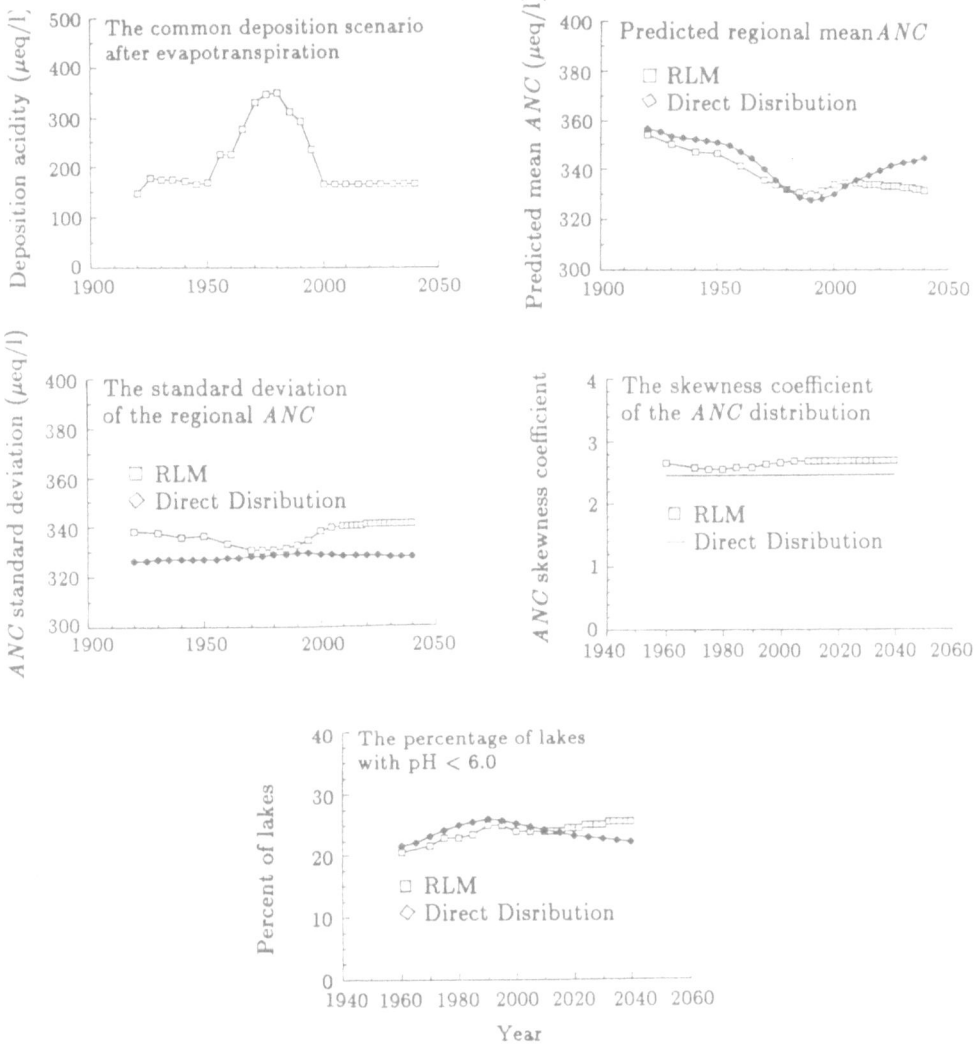

Figure 10.7. Comparison of the RLM model and the estimated Direct Distribution model results for a common deposition scenario.

with a pH less than 6 diverge. The Direct Distribution model results show a recovery of the mean *ANC* and a subsequent reduction in the number of lakes with a pH less than 6 following the decrease in deposition. The RLM model, on the other hand, shows a small recovery followed by continuing acidification, as indicated by the decreasing mean *ANC* and increasing fraction of lakes with pH less than 6.

The results presented in *Figure 10.7* highlight the major fundamental difference between the current Direct Distribution model and the RLM model. The current Direct Distribution model is based on the underlying assumption that the processes affecting lake acidification are completely reversible; there is

no irreversible loss of watershed buffering capacity. The effective weathering factor in the RLM model, however, decays over time, so that the *ANC* distribution
never comes to an equilibrium state even after many years of constant deposition. The loss of watershed buffering capacity continues even after the deposition reduction, as the rate of acid stress still exceeds the rate of primary mineral
weathering. Under the range of deposition reductions considered, the loss in
buffering capacity is thus effectively irrecoverable. This behavior is critical to
the RLM model.

This difference – the effectively irreversible behavior of RLM and the fully
reversible behavior of the current Direct Distribution model – is central to a key
issue in the acid rain policy debate. If acidification processes are to some extent
irreversible, then the timing of acid rain controls is quite important; action to
protect regional lake resources must be taken to prevent further irrecoverable
losses. If the acidification processes are largely reversible, however, then the timing of controls is less critical.

The implications for model development are also clear. If the scientific consensus is such that processes that are, for all practical purposes, irreversible are
indeed important, as represented in the RLM model, then the assumptions and
structure of the Direct Distribution model must be modified to incorporate a
variable (i.e., decaying) regional weathering distribution. This would allow the
Direct Distribution model to summarize the RLM model more effectively,
reflecting the importance of the continued loss of buffering capacity.

10.6. Summary and Conclusions

A procedure is developed for estimating the Direct Distribution model of Small
et al. (1987) – a simple empirical lake acidification model – from the regionalized
mechanistic RLM model. The estimation procedure is illustrated using lakes in
the southernmost region of Finland. The parameters of the Direct Distribution
model are derived from the observed behavior of the RLM model. It is found
that many of the assumptions of the Direct Distribution model are consistent
with the RLM model results. In particular, the regional *ANC* distribution is well
characterized by a 3-parameter lognormal distribution with a constant skewness
coefficient. In addition, the fitted pH–*ANC* relationship allows a representative
pH distribution to be derived, although differences are apparent in the low and
high range.

However, a critical difference between the models is revealed: lake weathering factors decrease over time in the RLM model, whereas the Direct Distribution model assumes a constant distribution of F for the region. The exponential
shape of the response profile assumed by the Direct Distribution model is thus
unable to characterize fully the behavior of the RLM model, which incorporates
both rapid and long-term, effectively irreversible processes.

These differences highlight the conflicting assumptions made by the models.
The current Direct Distribution model assumes that the processes of lake
acidification are completely reversible. The RLM model assumes (1) that
buffering capacity in the soil is lost by the accumulation of acidity and (2) that

this loss is irrecoverable under the range of deposition reductions considered. These differences cause the derived Direct Distribution model to overestimate the regional *ANC* recovery resulting from a future decrease in deposition and to underestimate the corresponding fraction of lakes with pH less than 6.0.

The shape of the response profile and the assumed constancy of the weathering factor in the Direct Distribution model require further examination. To accomplish this, we must look at the behavior of other regional mechanistic models, such as the MAGIC model developed by Cosby *et al.* (Chapter 7). This examination may help to illuminate the relative roles of reversible and irreversible processes in other mechanistic lake acidification models. Clarifying these relative roles is essential to developing a consensus in the prediction of the long-term effects of acidic deposition on distributions of regional lake chemistry. The estimation procedure presented in this chapter can be beneficial for examining and comparing more complex models for this purpose in a common framework, and for applying the results of these models in an integrated assessment framework.

Acknowledgments

This research was supported by the Claude Worthington Benedum Foundation and the US National Science Foundation (PYI, ECI – 8552772). We would like to thank the researchers at IIASA for making this work possible by providing access to the RAINS model and the input data and *a priori* distributions for southernmost Finland.

PART III

Model Reliability
and
Model Use

CHAPTER 11

Estimating the Reliability of Regional Predictions of Aquatic Effects of Acid Deposition

R.H. Gardner, J.-P. Hettelingh, J. Kämäri, and S.M. Bartell

11.1. Introduction

The reliability of environmental predictions is directly related to the accuracy of the model used and the uncertainties associated with model parameters (Bartell *et al.*, 1986; O'Neill and Gardner, 1979). Models that forecast broad-scale effects as a function of site-specific estimates (e.g., regional models) are subject to an additional source of error, owing to extrapolation of site-specific results to the region.

Because extrapolation errors can be critical for predicting the broad-scale effects of acid deposition on aquatic systems, it is important that research in this area address a number of concerns, including:

(1) Current regional boundaries are based on historical, pragmatic, or political precedents. How do these arbitrary boundaries affect prediction errors? Can predictions be improved by developing regional definitions based on the homogeneity of processes within the region?

(2) Variability in the quality and quantity of regional data implies that simple models may be more useful than detailed, complex models. Is it possible to determine, *a priori*, the optimal level of model complexity that maximizes the use of information and still minimizes prediction errors?

(3) Regional predictions may require the extrapolation of information from one site to another. Can methods be developed to test models against existing

J. Kämäri (ed.), Impact Models to Assess Regional Acidification, 185–207.
© 1990 *International Institute for Applied Systems Analysis.*

data, and can these results be used to extrapolate from one region to another?

The RAINS Lake Model (RLM) (Kämäri *et al.*, 1985), and the extensive data that describe the lakes in Fennoscandia (*Table 11.1*) and document the process of lake acidification, combine to provide an opportunity to examine these problems. The purpose of this chapter is to use the RLM and data from Fennoscandia to: (1) identify and quantify errors associated with extrapolation of site-specific information; (2) test and quantify our abilities to perform useful extrapolations; and (3) establish guidelines and limits for the role of data and models in regional assessments.

Table 11.1. Lakes sampled by region in Finland, Norway, and Sweden[a].

Country/ Region	Total number of lakes	Number of[b] lakes sampled	Percentage of total lakes sampled
Finland			
1	2,833	1,590	56.1
2	13,579	647	4.8
3	12,146	2,061	16.9
4	9,644	1,537	15.5
5	17,841	512	2.9
Norway			
1	unknown	382	unknown
2	unknown	65	unknown
3	unknown	67	unknown
Sweden			
1	1,194	8,000	15% lakes 1–9 ha. 50% lakes 10–99 ha.
2	5,887	8,000	
3	7,580	8,000	2% of all Swedish lakes
4	15,308	8,000	have been sampled
5	9,943	8,000	
6	47,500	8,000	

[a]See *Figure 11.2* for geographical location of regions in Fennoscandia.
[b]The Finnish lakes were sampled under supervision of the Finnish National Board of Waters and the Environment between 1975 and 1983. The Swedish morphological data were obtained from the Swedish Morphological Institute (SMHI), and the information on the lake acidification was obtained from the National Swedish Environmental Protection Board from samples taken between 1977 and 1980. The data we have used here were obtained from reports that aggregated the information into frequency distributions in which details about the sample sizes were made explicit. The Norwegian data were sampled by the Norwegian Institute for Water Research (NIVA) between 1974 and 1977.

11.2. Sources of Errors in Regional Models

The sources of errors that affect model predictions have been partitioned in many ways (e.g., O'Neill and Gardner, 1979; Gardner *et al.*, 1980); however, the effects of different sources of error on regional models have not been clearly specified. Three general sources of errors seem relevant:

(1) Errors due to uncertainties associated with estimates of model parameters (ε_1).
(2) Errors due to the extrapolation of model predictions from one region to another (ε_2).
(3) Residual errors due to assumptions and simplifications in the model structure and the computer implementation of the model (ε_3).

In general it is not easy to distinguish the effects of these errors. Residual errors (ε_3) are clearly dependent on the adequacy of parameter estimates. Likewise, extrapolation errors (ε_2) will be affected by changes in ε_1. Nevertheless, it is important to characterize the effects of these errors and, where possible, quantify their effects if we ever expect to make reliable regional predictions.

Various methods have been developed for analyzing the effect of parameter uncertainties (ε_1) on model predictions (e.g., O'Neill and Gardner, 1979; Gardner et al., 1980). Monte Carlo methods are generally preferred for complex models because simple analytical solutions are difficult or impossible to obtain. The Monte Carlo method is implemented by selecting a random set of model parameters from prespecified frequency distributions; performing model simulations to obtain the set of model predictions that corresponds to the set of random parameter values; and analyzing the combined set of parameters and predictions to characterize model uncertainties and identify the parameters that most affect model results. Monte Carlo methods have been applied to the RAINS Lake Model (Kämäri et al., 1986), and this information was used to rank, for each output, the most influential parameters.

This chapter uses Monte Carlo methods to quantify the effect of ε_1 on predictions of the RAINS Lake Model (RLM). Errors due to extrapolation, ε_2, of RLM to another region cannot be so easily studied. Minimum effects are expected to occur when parameter uncertainties are relatively low, data characterizing important differences between regions are available, and the behavior of sites in the target region is similar to that for sites to which the model has been calibrated. Our approach is to characterize the joint effect of ε_1 and ε_2 by varying the entire set of model parameters and comparing different extrapolation procedures. Thus, errors other than those caused by ε_1 can be attributed to ε_2. The effect of ε_3, the residual model error, is assumed to be important when reductions of ε_1 and ε_2 – by improvements in model calibration, reduction of parameter uncertainties, or additional measurements that characterize critical regional attributes – do not result in a reduction of prediction errors.

The adequacy of model predictions can be quantified by comparison of predictions against independent data sets (Mankin et al., 1977). Identification of model inadequacies due to structure (ε_3), inappropriate extrapolation of results to other regions (ε_2), or improper estimation of model parameters (ε_1) should be directly dependent on the quality and quantity of data. Section 11.3 describes the RLM and the data that are available to test this model, and Section 11.4 describes the methods used for comparison of model results against these data.

11.3. The RAINS Lake Model (RLM) and the Data of Fennoscandia

11.3.1. Model description

The RAINS Lake Model (RLM) simulates the site-specific changes in soil and lake chemistry that occur as a result of sulfur deposition to the watershed (Posch and Kämäri, Chapter 9; Kämäri *et al.*, 1985). The RLM is subdivided into several modules, each simulating separate components of lake acidification (*Figure 11.1*):

- The meteorologic submodule calculates the amount of water and total sulfur directly entering the soil or lake each month.
- The soil acidity submodule accounts for the soil solution chemistry and estimates ion concentrations in the soil by computing the carbonate, silicate, and the aluminum buffer intensity (Kauppi *et al.*, 1986).
- A hydrologic submodule simulates the movement of water through soils so that convective flow of ions can be estimated.
- The lake submodule calculates lake acidity according to the equilibrium reactions of inorganic carbon species.

(The version of the RLM used here does not consider recent modifications that include effects on alkalinity due to in-lake retention of sulfur.) The RLM parameters subject to uncertainty are listed in *Table 11.2*.

11.3.2. Sulfur deposition scenarios

The input to a particular watershed is computed from the total European emissions multiplied by the source–receptor matrix of Eliassen and Saltbones (1983). The European emissions figures from 1960 to 2040 are based on the Energy/Emission model of RAINS, in which different emission scenarios can be defined (Alcamo *et al.*, 1987; Hettelingh and Hordijk, 1986). Levels of total (wet and dry) deposition in the RLM are computed from the combination of the scenario and the source–receptor matrix. Site-specific deposition values are randomly drawn from the range of deposition values in 1960 computed for the particular region. This random factor is then used to compute the site-specific relative magnitude of deposition for all years after 1960.

11.3.3. The Fennoscandian data

An extensive set of measured pH and alkalinity values exists for lakes in Finland, Sweden, and Norway (*Table 11.1*). These data have been arranged into five districts for Finland, six for Sweden, and three for Norway (*Figure 11.2*). For each lake region, a set of corresponding parameter distributions was developed, based on measurements of the physical characteristics of lakes and catchments.

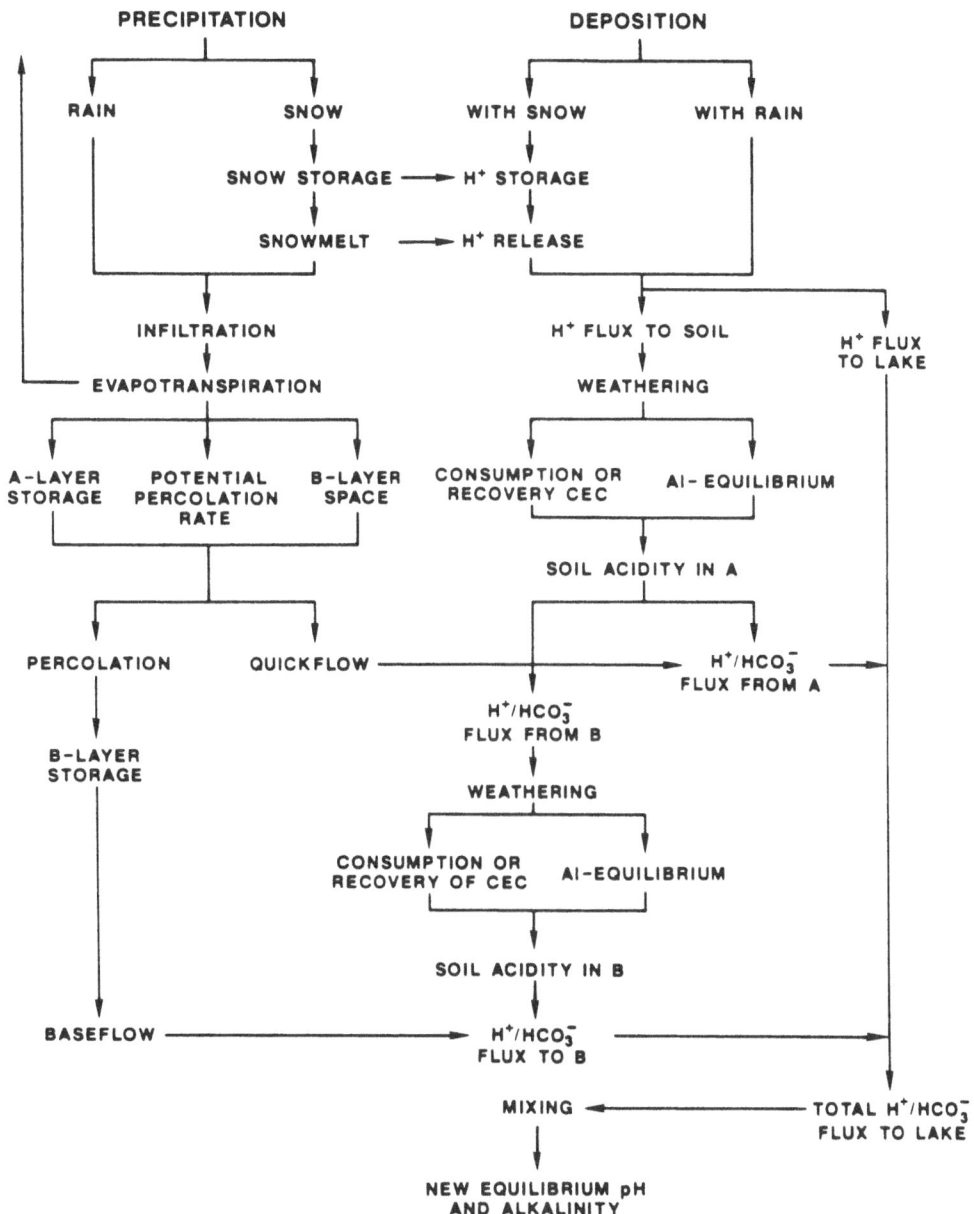

Figure 11.1. Flow diagram of the chemical and physical processes modeled by the RAINS Lake Module (RLM).

The quality and quantity of data on which these input distributions were based are not consistent for all districts of the three countries. Sample sizes were larger in regions where deposition levels are high (region 1 of each country and region 2 in Sweden), and smaller in regions receiving relatively low levels of

Table 11.2. RLM input parameters[a].

No.	Abbreviation	Explanation
1	*AREAL*	Lake surface area
2	*RATCL*	Ratio of lake to catchment area
3	*LDEPTH*	Lake depth
4	*SOILT*	Total soil thickness (sum of A and B layer)
5	*SLOPE*	Mean surface slope
6	*SIBR*	Silicate weathering rate
7	*CEC*	Total cation exchange capacity
8	*BASEA*	Base saturation of A-layer
9	*FCAP*	Soil moisture content at field capacity
10	*BSATB*	Base saturation of B-layer
11	*INPH*	Measured pH in 1980
12–23	*TEM01–TEM12*	Monthly mean air temperature
24–35	*PRE01–PRE12*	Monthly mean precipitation
36	*STFAC*	Factor of deposition on forest vs. open land
37	*FSATU*	Soil moisture content at field saturation
38	*RZO*	Depth of lake mixing layer
39	*EPSEV*	Evapotranspiration coefficient
40	*TLOW*	Threshold temperature of snow
42	*THIGH*	Threshold temperature of precipitation
43	*BRATE*	Melting rate coefficient
44	*CALK*	Alkalinity/H^+ constant: Henry's law
45	*COND*	Hydraulic conductivity at saturation

[a]See Kämäri *et al.* (1985) for detailed explanation of the RLM model and parameters.

sulfur. *Table 11.1* gives an overview of the differences in sampling intensities among countries and regions.

The frequency distributions of the input parameters were derived from the sampled data. When the data did not provide adequate information for estimating parameter means, variances, and the shape of the distribution, a uniform distribution was assumed, with minimum and maximum values derived from available information. This approach means that differences in data accuracy are implicitly taken into account when simulating the aquatic properties of the Fennoscandia lakes.

11.4. Comparing Predictions with Data

Monte Carlo simulations of RLM have been used to generate regional estimates of the shifts in the frequency distribution of lake pH over time (Kämäri *et al.*, 1986; Posch and Kämäri, Chapter 9; Sutton, Chapter 12). This approach is appropriate when the statistical distributions of key parameters in the model reflects the observed regional variability.

Because extensive data sets exist for Fennoscandia lakes, it is possible to test the adequacy of model predictions by direct comparison with the data. The problem is that traditional methods of goodness-of-fit, such as the χ^2 or Kolmogorov–Smirnoff test, are not well suited to this purpose because they may

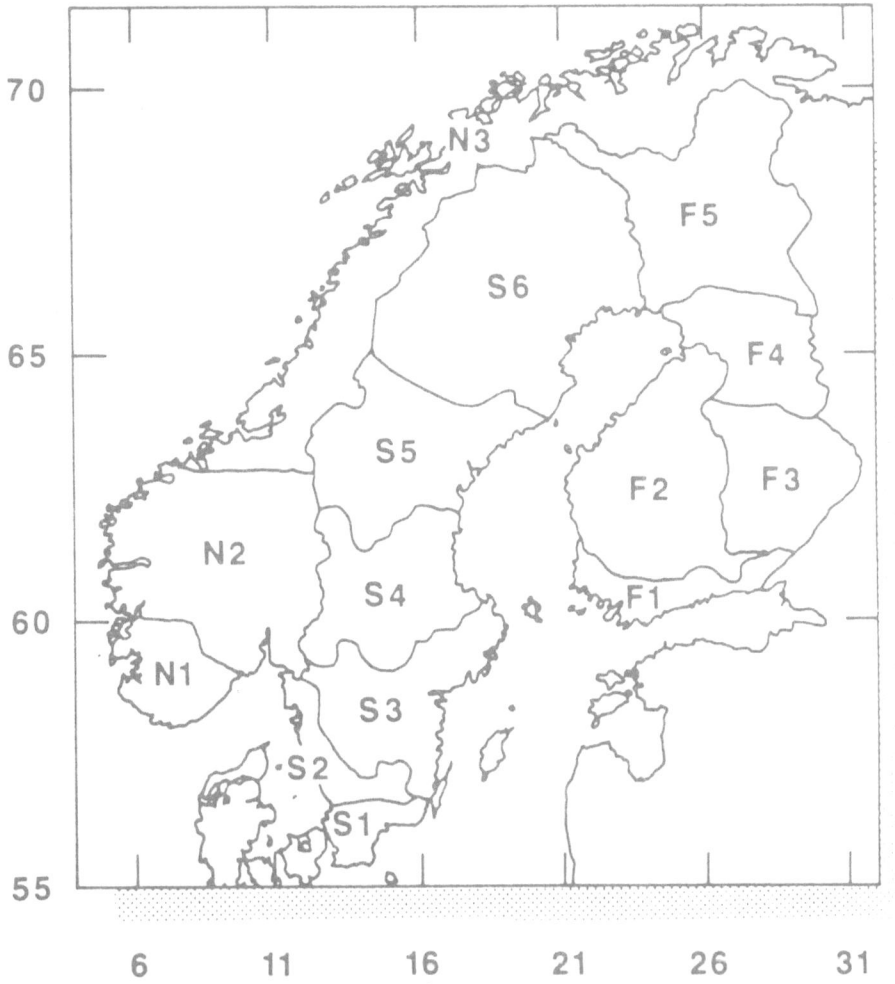

Figure 11.2. Map illustrating the regional boundaries within Finland (5), Sweden (6), and Norway (3).

identify significant differences that have little effect on the capability of the model to predict regional lake acidification (e.g., inadequate estimates of the frequency of lakes with pH higher than 8.0). Therefore, in the next two sections we present a method to compare models and data that is appropriate for estimating regional lake acidification, a method of model calibration that allows extrapolation of results to other regions, and a final section that outlines the application of this model for estimation of errors and uncertainties.

11.4.1. Method of comparison

Not all differences between models and data are of equal interest, nor are they equally critical in assessing a model's adequacy for predicting regional shifts in lake acidity due to sulfur deposition. The comparisons of interest for this study are the differences between the RLM and the Fennoscandian data for the mean, variance, and range of predicted pH values.

If P is the mean of the predicted value, O is the mean of the measured values, and S_o is the standard deviation of the measured values, then relative bias, rB, can be identified as $(P - O)/S_o$. Relative bias measures model accuracy by quantifying the mean difference between the model and data in units of standard deviations of the data. If S_p^2 is the variance of the predicted effects and S_o^2 is the variance of the measured effects, then F, the ratio of the variances $(F = S_p^2/S_o^2)$, indicates the relative spread of the model predictions compared to the data. For F values < 1, the model predictions have a narrower distribution than the data; while for $F > 1$, the converse is true.

For any combination of rB and F, a probability, p, for which the distributions are the same, can be estimated from the appropriate cumulative distribution function. By assuming that the model predictions and data are normally distributed, it is possible to express the relationships between rB, F, and p as a contour plot (*Figure 11.3*). If the model and data were drawn from identical distributions, then rB will equal 0.0, F will equal 1.0, and p will equal 1.0. The result of this procedure quantifies the degree of confidence that can be placed in model predictions as an adequate representation of the data.

The use of the normal distribution is not critical to the method, but is used here as a matter of convenience because it allows calculation of p by knowing only the means and variances. It is possible to generalize this procedure and make the results free of distributional assumptions. For instance, any percentile of the cumulative distribution can be substituted for the mean and the results expressed in terms of the coincidence of two empirical frequency distributions.

Two recent studies have used rB and F to compare models with available data. Bartell *et al.* (1986) found that the relative bias of an aquatic model was low but varied with time, while the variance of model predictions underestimated the variability of the data. Comparisons of a forest succession model with forest inventory data (Dale *et al.*, 1989) were similar: the relative bias of model predictions was low, but the high variability of the data was underestimated by the model. Dale *et al.* recommended that the uncertainties associated with the measured values should be reduced before significant efforts are made to improve model predictions.

11.4.2. The calibration method

We designed a model calibration procedure for this study that was simple to apply, resulted in adequate model prediction of the frequency distribution of 1980 lake pH, and produced parameter sets that could be used to make predictions in other regions.

Regions 1 of Finland, Sweden, and Norway (*Figure 11.2*) were chosen to calibrate the model because these regions receive some of the highest levels of

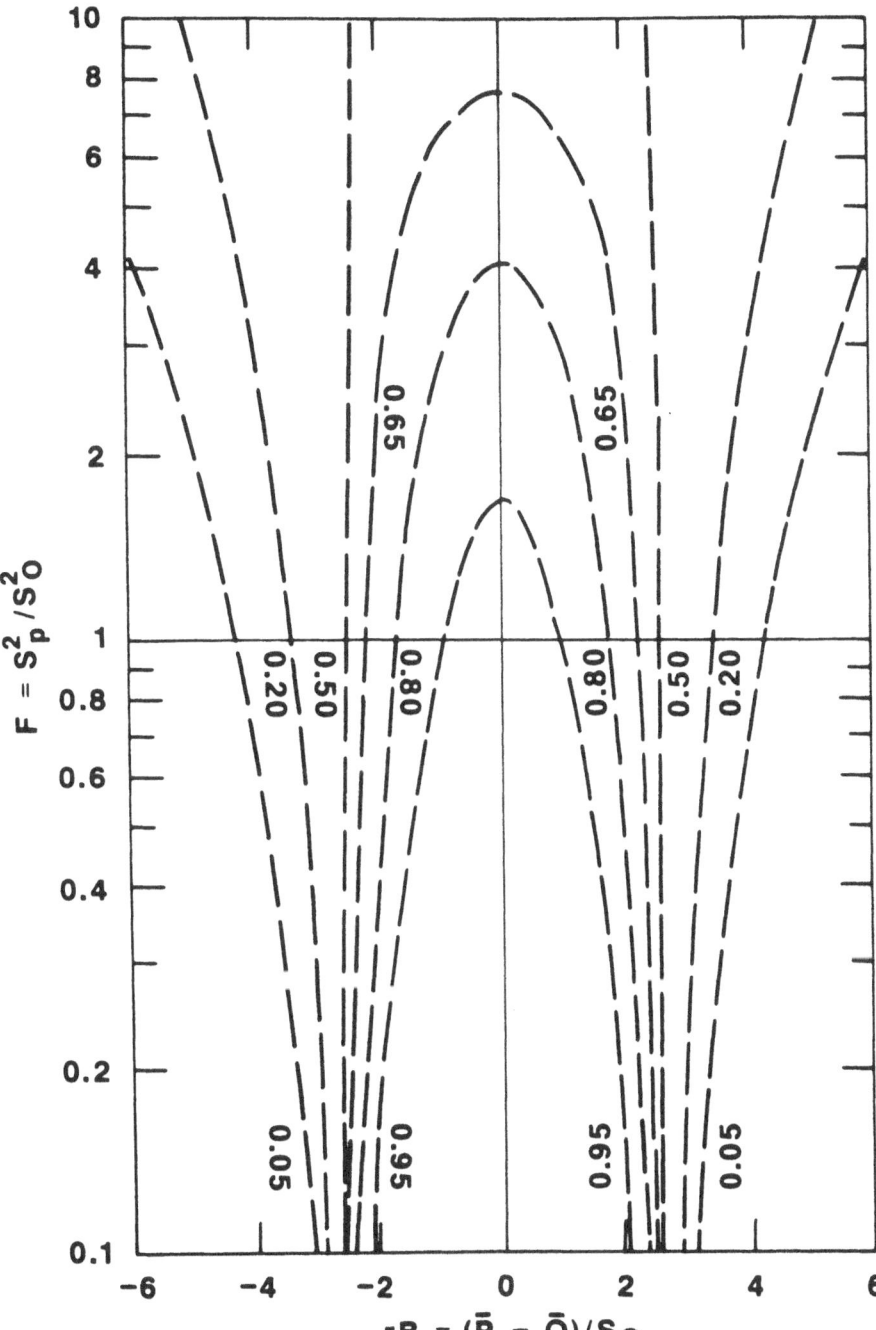

Figure 11.3. Contour plot of the relationship between the relative bias (rB) and the ratio (F) of variance in model predictions (S_p) to the variance in observations (S_o) for different degrees of overlap (p) between two normal distributions.

Table 11.3. Parameter uncertainty analysis for Monte Carlo simulations of region 1 in Finland, Sweden, and Norway[a].

Parameter	Finland Rank	%	Norway Rank	%	Sweden Rank	%
SOILT	1	32.4	1	49.1	1	25.9
SIBR	2	7.3	2	16.8	2	9.1
RATCL	3	4.8	3	8.3	6	2.4
COND					3	4.8
FCAP	4	3.4				
BASEA			4	3.2		
CEC			5	3.1		
AREAL	5	2.9	6	2.4	8	1.8
TEM12					5	3.2
PRE07	6	2.9				
LDEPTH	7	2.6				
TEM11			7	1.8		
STFAC					7	2.0
SLOPE	8	1.9				
EPSEV			8	1.5		
BSATB	9	1.2			4	4.5
TEM06			9	1.5	10	1.1
PRE04					9	1.5
TEM08	10	1.1				
PRE12			10	1.1		

[a]Parameter ranks are based on the percentage effect that each parameter has on predicted pH values in 1980. The percentage effect is estimated from multiple regression methods as (partial sum of squares / total sum of squares * 100). Values for the uncertainty analysis are determined by simultaneous variation of parameters from prespecified frequency distributions. Further details concerning the Monte Carlo methods may be found in Kämäri *et al.* (1986).

sulfur deposition in Fennoscandia. Previous analysis of model uncertainties (Kämäri *et al.*, 1986) has shown that the most important model parameters for Finland are *SOILT* (the average soil thickness of the catchment) and *SIBR* (the weathering rate of silicates). When this analysis was repeated for region 1 of Finland, Norway, and Sweden, *SOILT* and *SIBR* accounted for a total of 39.7%, 65.9%, and 36.0% (*Table 11.3*), respectively, of the variability of simulated pH values. Based on these results the frequency distributions of *SOILT* and *SIBR* for region 1 of each country were iteratively adjusted until:

(1)　The standardized difference (rB) between the predicted and measured mean pH was minimized.
(2)　The ratio of the variances (F) of the measured and predicted pH was approximately equal to 1.0.
(3)　The difference in the ranges of predicted and measured pH values was as small as possible.

The initial pH values in 1960 were assumed to be neutral (pH of 7), while more recent measurements (*Table 11.1*) were assumed to represent 1980 pH values.

The values of *SOILT*, the average soil thickness of the catchment, have not been directly estimated for all sites; therefore, *SOILT* was initially set to a uniform distribution. Iterative adjustments of *SOILT* resulted in a final distribution, which was triangular. This use of the triangular distribution does not imply that the actual distribution of *SOILT* in a region is triangular, but rather that the effective soil thickness and associated buffer capacities must show a strong central tendency in order to make the mean and variance of model predictions match those of the data. The converse is also true: although the uncertainty analysis has shown that other parameters are relatively unimportant, and therefore were not adjusted during the calibration process, this does not imply that they have been correctly estimated.

The calibration procedure for region 1 of Finland resulted in a slight improvement in model predictions. The uncalibrated model produced values of rB and F of 0.21 and 1.27, respectively, while the calibrated model produced an rB of 0.07 and an F of 1.17. The frequency of model predictions beyond the limits established by the data (ν_r) was 8% before calibration and 7% afterward. The calibration process did not alter the value of p from the original value of 0.98.

Region 1 of Sweden was more difficult to calibrate. The uncalibrated model produced an rB of 0.57, an F of 1.04, with 11% of the model predictions beyond the limits of the data. The model calibration process reduced ν_r to 6%, but increased rB to 0.82 and changed F to 0.86. The uncalibrated value of p for region 1 of Sweden was 0.98, and the calibrated value was 0.97.

The most difficult calibration was for region 1 of Norway. The uncalibrated values of rB and F were 3.19 and 4.67, respectively, producing a value of p of 0.38. The calibration process improved rB to 0.04 and reduced F to 2.97, resulting in a value of p of 0.86. Neither the uncalibrated nor calibrated model produced any predictions beyond the limits established by the data.

The difficulty in calibrating Sweden and Norway may be due to the interaction of high sulfur depositions in Sweden and Norway and differences in biogeochemical processes that are not accounted for by the parameters of the RLM. In both cases adjustments to *SOILT* and *SIBR* were not entirely adequate, indicating the existence of residual model error (see Section 11.2).

11.4.3. Procedure for regional comparison

We developed the procedure for a regional comparison of RLM's adequacy to produce results that could be compared across different sulfur deposition scenarios and through time. The method was intended to compensate for errors intrinsic to the model (i.e., the residual errors defined in Section 11.2), and make explicit the errors associated with the extrapolation process. Our final objective was to test the adequacy of model predictions and identify where improvements in the model, the data, or both will be most useful.

The effect of parameter uncertainties (ε_1) and residual model errors (ε_3) was characterized by superimposing the calibration results (see Section 11.4.2) of the most influential parameters (*SOILT* and *SIBR*) of region 1 on all other

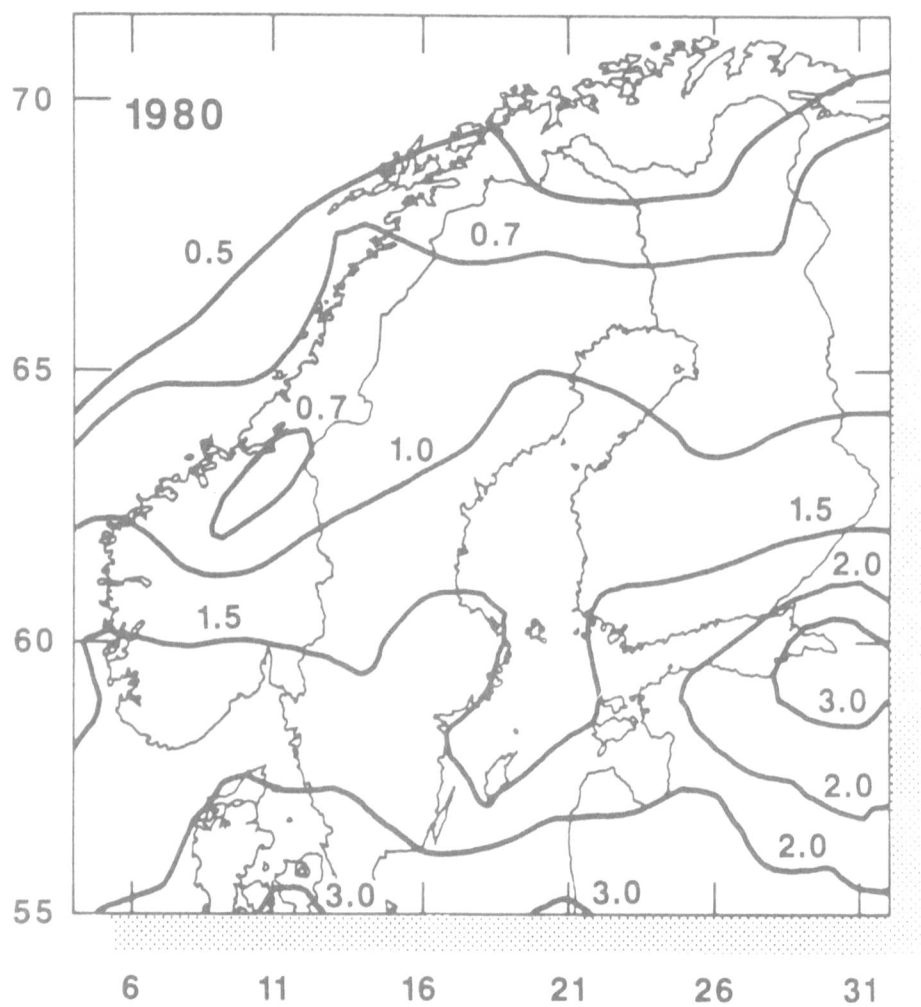

Figure 11.4. Isopleths of the simulated 1980 total sulfur deposition ($gm^{-2}yr^{-1}$).

regions within the same country. This extrapolation with its associated errors was made for all but the first region within each country and is referred to as the intra-country comparison and intra-country error. Additional error owing to the extrapolation process (ε_2) was investigated by superimposing the results of the calibration process and all other input parameters (see *Table 11.2*) from region 1 in Finland to all other regions in Fennoscandia. Region 1 of Finland was chosen for this extrapolation because of the good agreement between the model and data. Such inter-country extrapolation assumes that all other regions "behave" as region 1 of Finland, except for the pattern and quantity of sulfur deposition (*Figure 11.4*). This second type of extrapolation with its associated errors is referred to as the inter-country comparison and inter-country error.

In summary, the procedure for regional comparison is:

(1) A sensitivity and uncertainty analysis of RLM to identify the minimum set of important model parameters.
(2) Calibration of the model by adjustment of these parameters to region 1 of Finland, Sweden, and Norway.
(3) Extrapolation of the calibrated model within each country (intra-country comparison) and characterization of model and parameter estimate errors (ε_1) by comparison with available data.
(4) Extrapolation using the model calibrated to region 1 of Finland for all regions in Fennoscandia (the inter-country comparison) and characterization of model errors and bias (ε_2) by comparison with available data.
(5) Comparison of model results through time by simulation of different sulfur deposition scenarios.

11.5. Results

11.5.1. Intra-country comparisons

Table 11.4 summarizes the results of the intra-country comparisons of the RLM simulations with available data. The model simulations for the intra-country comparisons are based on the calibration of the two parameters (*SOILT* and *SIBR*) to region 1 of Finland, Norway, and Sweden (see *Figure 11.2*) and the use of these values for simulations of lake pH for other regions within each country (see Sections 11.4.2 and 11.4.3 for further details). The first four columns of *Table 11.4* give the mean and coefficient of variation (*CV*) of lake pH estimated from data and model simulations. The columns labeled *d* and % are the absolute and percentage deviations of the mean of the model simulations from the mean of the data. The statistics *rB*, *F*, and *P* are, respectively, the relative bias, the ratio of the variances, and the probability of coincidence (see Section 11.2 for a complete description).

The results in *Table 11.4* show that, in general, the method of model extrapolation used in the intra-country comparison provides a satisfactory estimate of the mean and variance of measured lake pH values. Except for region 1 of Norway, the values of *p* are uniformly high (i.e., greater than 0.95), indicating that the cumulative frequency distribution (*cfd*) of model results is similar to the *cfd* of the data. The values of *rB* are all less than 1.0, indicating that the means of the model predictions are always less than one standard deviation from the data. The ratio of the variances, *F*, is less than 1.0 in Finland regions 2 through 5, near 1.0 in Sweden regions 2 through 6, and above 1.0 in Norway regions 2 and 3. The pattern of *rB*, *F*, and *P* is illustrated in *Figure 11.5*.

An inspection of *Figure 11.5* shows that an underestimation of the variance by the model (i.e., values of *F* less than 1.0) will produce a higher value of *P* than the corresponding overestimation of the variance. For instance, if the *F* ratio is 0.5 (model variance is one-half that of the data), then *P* equals 0.999; but

Table 11.4. Intra-country comparisons of 1980 measured pH values against model predictions[a].

Country/ Region	Data		Model						
	Mean	CV	Mean	CV	d	%	rB	F	P
Finland									
1	6.24	13.8	6.30	14.8	0.06	1.0	0.07	1.15	0.98
2	6.31	11.3	6.81	8.4	0.50	7.9	0.70	0.63	0.99
3	6.47	12.1	6.83	9.1	0.37	5.7	0.47	0.64	0.99
4	6.58	11.6	6.94	8.1	0.36	5.5	0.48	0.53	0.99
5	6.79	9.4	6.93	5.9	0.14	2.1	0.21	0.40	0.99
$\Sigma|d|/5$					0.29				
Sweden									
1	5.72	17.5	6.55	14.2	0.83	14.5	0.82	0.86	0.97
2	5.30	22.1	5.84	17.5	0.54	10.2	0.46	0.76	0.99
3	6.33	18.3	6.26	14.7	0.07	1.1	−0.06	0.62	0.99
4	5.84	12.3	6.41	12.9	0.58	9.9	0.80	1.33	0.94
5	6.36	11.0	6.51	11.8	0.14	2.3	0.21	1.19	0.98
6	6.61	12.1	6.60	9.5	0.01	0.1	−0.01	0.62	0.99
$\Sigma|d|/6$					0.36				
Norway									
1	4.96	8.9	4.98	15.3	0.02	0.4	0.04	2.97	0.86
2	5.54	15.2	5.48	15.7	0.07	1.3	−0.08	1.03	0.99
3	5.96	8.9	6.29	11.6	0.33	5.5	0.61	1.88	0.91
$\Sigma|d|/3$					0.14				
$\Sigma|d|/14$					0.29				

[a] Intra-country comparisons are based on model simulations from parameter sets that were adjusted to obtain a satisfactory agreement between the means and variances of the observed
and predicted values for the region 1 of each country. The mean values for the data were tak-
en from information described in *Table 11.2*; 500 Monte Carlo simulations were used to calculate the means for the model. The CV is the relative variability calculated as the (standard
deviation / mean) X 100. d is the absolute value of the difference between the mean of the
data and the mean of the model. The mean absolute values of d for each country and for the
entire table are also listed. The % column gives d as a percentage of the mean of the data. rB
is the relative bias and is equal to the difference between the means divided by the standard
deviation of the data. F is the ratio of the variance of the data to the variance of the model.
P is the probability of coincidence of the two distributions.

if the F ratio is 2.0 (model variance is twice that of the data), then P equals
0.932. This potential problem does not affect the behavior of rB or d because
these statistics are not related to the model variance. However, persistent
underestimation by a model of the true variability of a system will affect estimates of extreme percentiles of the cumulative frequency distribution (cfd).
Therefore, an analysis that focuses on extreme percentiles of the cfd (i.e., the
lower range of simulated pH values) should first calibrate the model to the
desired percentile and then test predictions against the appropriate data.

The relative bias, rB, is a standardized measure allowing uniform comparison with data, but the absolute deviation from measured values, d, is also of
interest. The values of d for region 1 in Finland, Sweden, and Norway are 0.6,

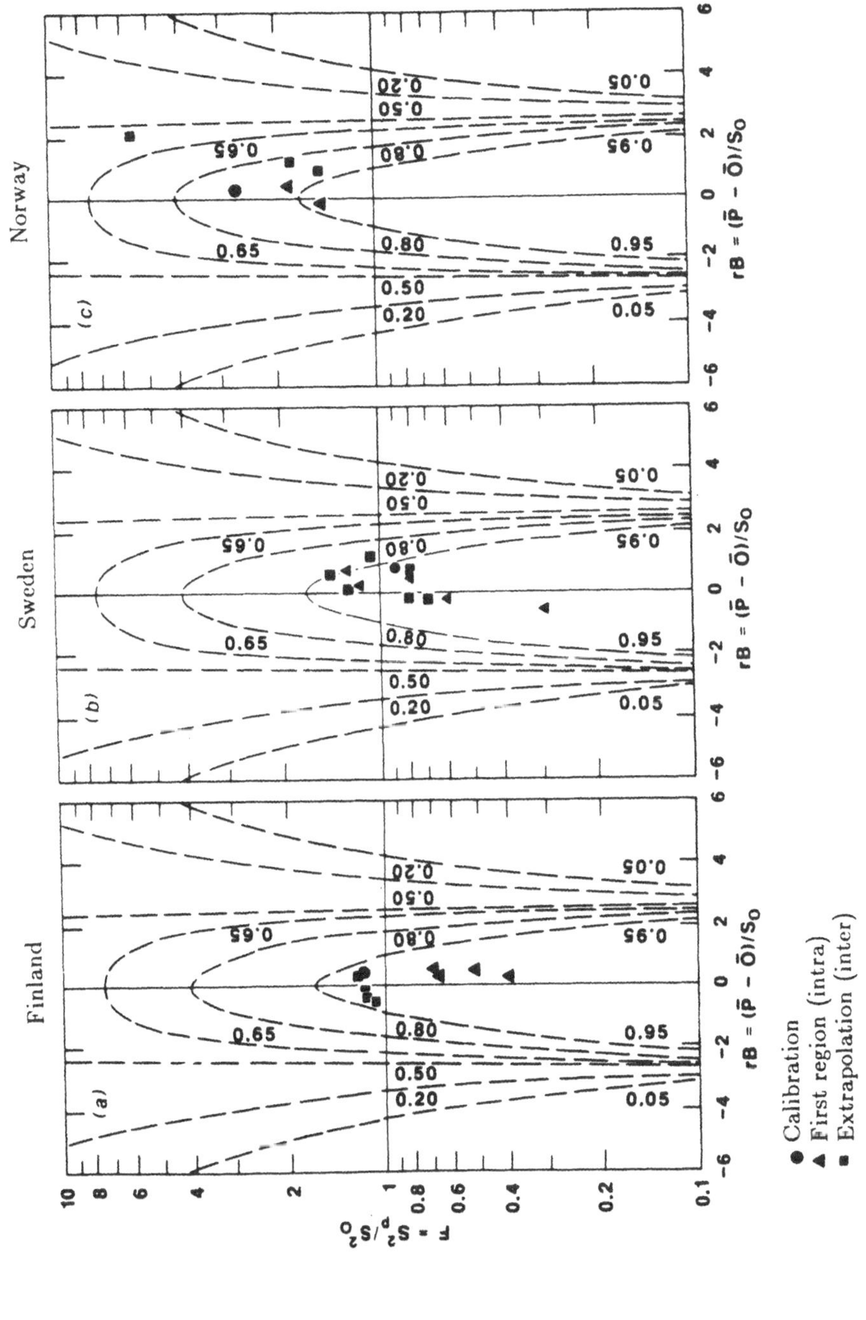

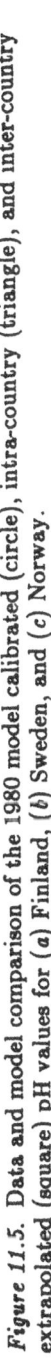

Figure 11.5. Data and model comparison of the 1980 model calibrated (circle), intra-country (triangle), and inter-country extrapolated (square) pH values for (*a*) Finland, (*b*) Sweden, and (*c*) Norway.

0.83, and 0.2 (*Table 11.4*), respectively, with an average for these regions of 0.30. The average value of d for all comparisons is 0.29, indicating that the extrapolation process produces no increase in absolute bias.

Inspection of the maximum percentage deviation (column labeled % in *Table 11.4*) shows that these errors are less 8%, 11%, and 6%, for extrapolation regions in Finland, Sweden, and Norway, respectively. The maximum percentage deviation is thus smallest in Norway, while the standardized error $(1 - P)$ is less than 2, 4, and 14%, respectively. Thus, the maximum standardized prediction error is largest in Norway. These two maxima indicate that parameter uncertainty (ε_1) is smallest in Norway, whereas bias (ε_2) is largest there. The opposite conclusion may be stated for Finland and Sweden, which have the same level of both error types.

11.5.2. Inter-country comparisons

Table 11.5 summarizes the results of the inter-country comparisons of the RLM simulations with available data. The inter-country comparisons are based on the extrapolation of all parameters from region 1 of Finland (*Figure 11.2*) to other regions within Fennoscandia, with the only differences accounted for by different patterns and amounts of sulfur deposition (*Figure 11.4*). The data used for comparison are the same as those for the intra-country comparison (see Section 11.3.3).

Table 11.5 shows that the inter-country extrapolation for Finland is generally better than the intra-country (*Table 11.4*) comparison (recall that only *SOILT* and *SIBR* were extrapolated in the intra-country comparison). The relative bias of results, rB, is less; and the ratio of the variances, F, more closely estimates the variability of the data. Although these differences do not affect the value of P, *Figure 11.5* shows that the pattern of simulations is closer to the ideal values of zero bias and an F ratio of 1.0.

The results of the inter-country extrapolation for Sweden were expected to produce larger errors and lower values of P. However, the average of d for the inter-country comparisons (*Table 11.5*) are the same as the intra-country comparisons (*Table 11.4*). In fact, the simulation of region 1 in Sweden with all parameters taken from region 1 in Finland produced better results than the intra-country calibration of *SOILT* and *SIBR*. Although these results appear to be spurious (the improvement in the prediction of region 1 corresponds to a poorer prediction in region 2), the difficulties in predicting the regional behavior of lakes in these regions may be due to interaction of high rates of sulfur deposition (*Figure 11.4*) and differences in geochemical processes that were not accounted for by the parameters of the RLM.

The inter-country extrapolations are rather poor for the three regions in Norway. Both the absolute (d) and relative (rB) bias are high, and the variance is poorly estimated for region 1 ($F = 5.08$, *Table 11.5*). In general, the quality of the inter-country extrapolations is best in the northernmost districts of Fennoscandia.

Table 11.5. Inter-country comparisons of 1980 measured pH values against model pre-dictions[a].

Country/ Region	Data		Model		d	%	rB	F	P
	Mean	CV	Mean	CV					
Finland									
1	6.24	13.8	6.30	14.8	0.06	1.0	0.07	1.15	0.98
2	6.31	11.3	6.41	13.1	0.10	1.6	0.14	1.37	0.97
3	6.47	12.0	6.39	13.3	0.08	1.2	−0.10	1.20	0.98
4	6.58	11.6	6.43	12.7	0.15	2.2	−0.19	1.14	0.98
5	6.79	9.4	6.55	10.7	0.24	3.5	−0.37	1.21	0.97
$\Sigma|d|/5$					0.13				
Sweden									
1	5.72	17.5	6.15	17.1	0.43	7.5	0.43	1.10	0.98
2	5.30	22.1	6.18	16.7	0.89	16.8	0.76	0.77	0.98
3	6.33	18.3	6.25	16.6	0.08	1.2	−0.07	0.71	0.99
4	5.84	12.3	6.37	13.6	0.53	9.2	0.74	1.45	0.93
5	6.36	11.0	6.44	12.6	0.08	1.2	0.11	1.33	0.97
6	6.61	12.1	6.54	11.0	0.07	1.1	−0.09	0.81	0.99
$\Sigma|d|/6$					0.35				
Norway									
1	4.96	8.9	6.22	16.1	1.26	25.4	2.85	5.08	0.44
2	5.54	15.2	6.44	12.5	0.90	16.2	1.07	0.91	0.94
3	5.96	8.9	6.56	10.7	0.60	10.1	1.12	1.72	0.86
$\Sigma|d|/3$					0.92				
$\Sigma|d|/14$					0.39				

[a]Intra-country comparisons are based on model simulations from parameter sets that were adjusted to obtain a satisfactory agreement between the means and variances of the observed and predicted values for the region 1 of each country. The mean values for the data were taken from information described in *Table 11.2*; 500 Monte Carlo simulations were used to calculate the means for the model. The CV is the relative variability calculated as the (standard deviation / mean) X 100. d is the absolute value of the difference between the mean of the data and the mean of the model. The mean absolute values of d for each country and for the entire table are also listed. The % column gives d as a percentage of the mean of the data. rB is the relative bias and is equal to the difference between the means divided by the standard deviation of the data. F is the ratio of the variance of the data to the variance of the model. P is the probability of coincidence of the two distributions.

11.5.3. Time-dependent comparisons

Model simulations of the mean pH values in the year 2040 for two sulfur deposition scenarios were performed with the intra- and inter-country parameter sets. Because these predictions cannot be verified by data, the results will concentrate on the differences due to the interaction of the extrapolation procedure with the sulfur deposition scenario. Section 11.3.2 describes the two deposition scenarios, and *Figures 11.6* and *11.7* illustrate the resulting geographic pattern of sulfur deposition.

The difference, Δ, between the intra- and inter-country simulations for each scenario is given in *Table 11.6*. There is less difference between intra- and

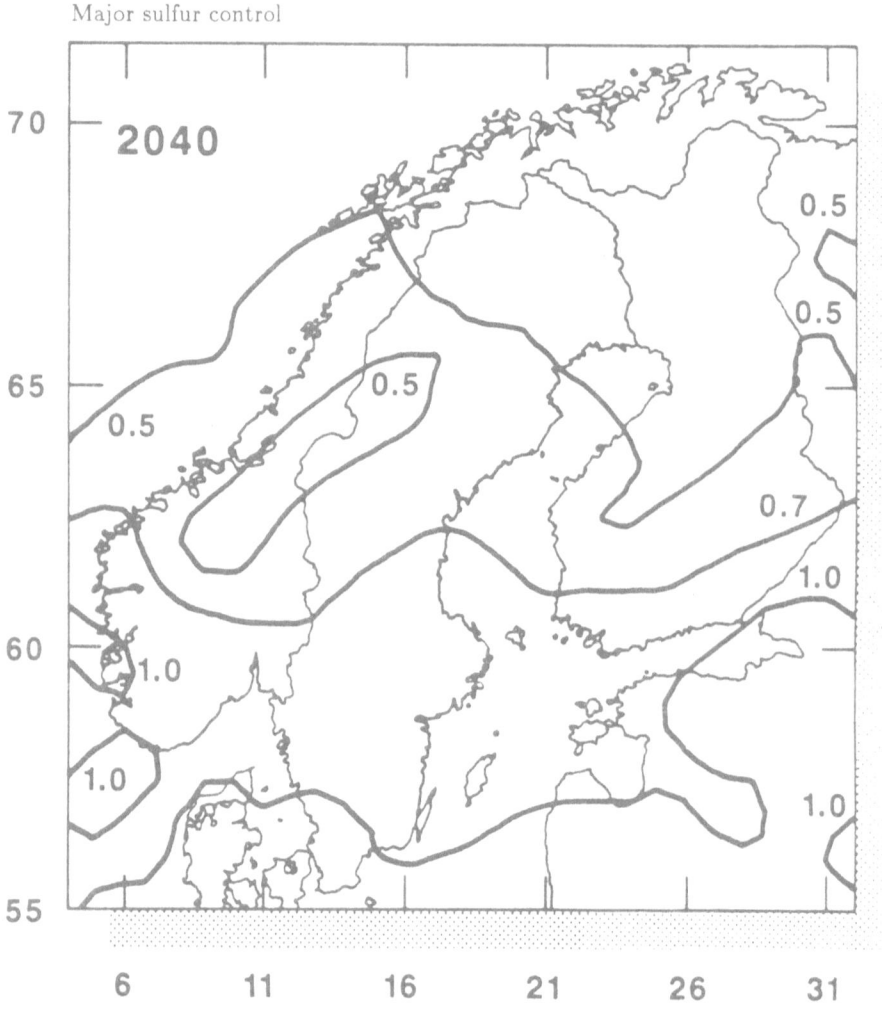

Figure 11.6. Isopleths of the simulated total sulfur deposition in year 2040 for the low-emission scenario $(gm^{-2}yr^{-1})$.

inter-country comparisons from 1980 to 2040 for the high deposition case for Finland (from 0.34 to 0.41 vs. 0.34), marginal differences for Sweden (from 0.15 to 0.18 vs. 0.17), and somewhat larger differences for Norway (from 0.82 to 0.97 vs. 1.03). The difficulties of calibrating and predicting lake pH for regions 1 and 2 in Sweden are still evident by inspection of Δ for year 2040, as well as the extrapolation error for Norway (*Table 11.6*).

The mean predicted pH in 2040 for the calibration regions (solid circles), intra-country extrapolations (solid triangles), and inter-country extrapolations (solid squares) are illustrated in *Figure 11.8*. The simulations of the sulfur high-deposition scenario and low-deposition scenario are connected by a solid line, with the high-deposition scenario always predicting a lower mean pH.

Linear extrapolation of ECE trends

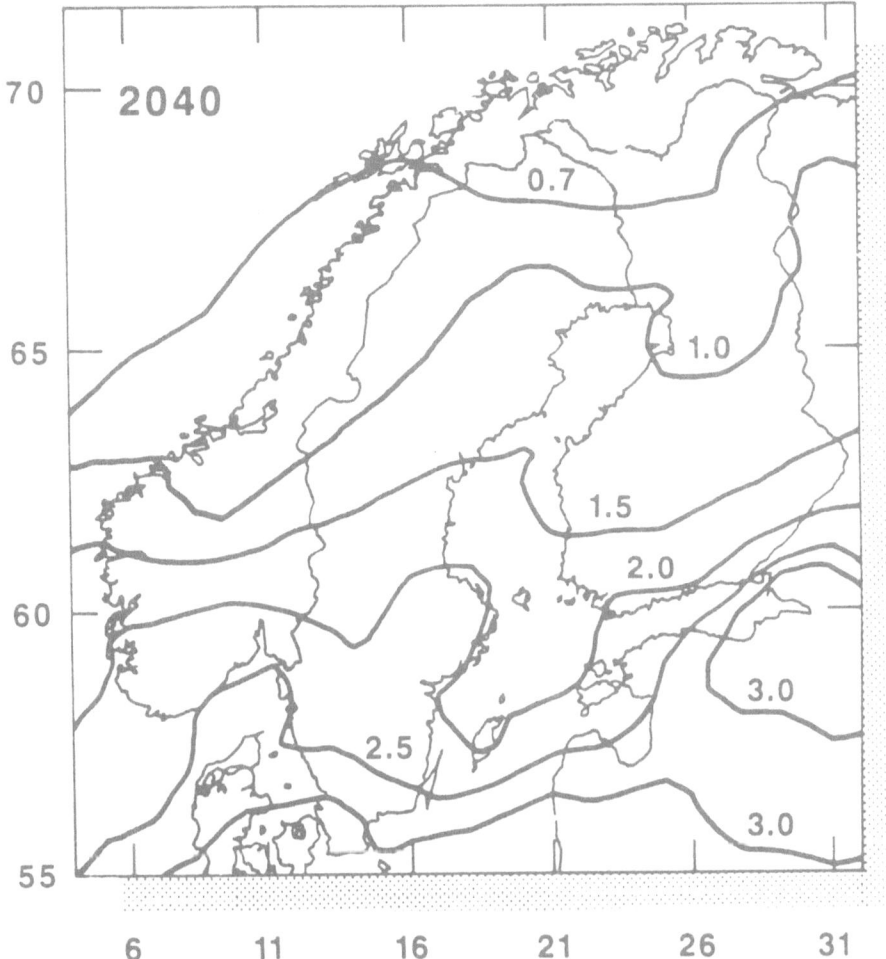

Figure 11.7. Isopleths of the simulated total sulfur deposition in year 2040 for the high-emission scenario $(\mathrm{gm}^{-2}\mathrm{yr}^{-1})$.

Each point illustrated in *Figure 11.8* represents 500 independent Monte Carlo simulations. The large number of Monte Carlo samples ensures that differences between sets of model simulations illustrated in *Figure 11.8* are reliable (i.e., a repeated set of Monte Carlo simulations will produce the same pattern of results). On average, the mean difference between scenarios ranges from 0.18 to 0.28 pH units.

The pattern of results shown in *Figure 11.8* indicates:

(1) The differences between scenarios were consistent, no matter which method of model calibration was used (*Table 11.6*).

Table 11.6. Differences between inter- and intra-country comparisons for the high and low sulfur deposition scenarios[a].

Country/ Region	1980		Scenario[b] for year 2040			
			Low		High	
	Δ	%	Δ	%	Δ	%
Finland						
1	0.00	0.0	0.00	0.0	0.00	0.0
2	0.40	6.3	0.40	6.2	0.41	6.1
3	0.44	6.5	0.43	6.2	0.48	7.1
4	0.51	7.5	0.84	6.8	0.49	7.1
5	0.38	5.5	0.37	5.3	0.36	5.2
$\Sigma\|d\|/5$	0.34		0.41		0.34	
Sweden						
1	0.40	6.1	0.42	6.3	0.44	6.9
2	0.34	5.8	0.35	5.9	0.42	7.6
3	0.01	0.2	0.05	0.8	0.08	1.3
4	0.04	0.6	0.02	0.3	0.02	0.3
5	0.07	1.1	0.12	1.8	0.05	0.8
6	0.06	0.9	0.14	2.1	0.03	0.5
$\Sigma\|d\|/6$	0.15		0.18		0.17	
Norway						
1	1.24	25.0	1.58	33.3	1.43	31.1
2	0.96	17.5	1.13	20.9	1.34	26.7
3	0.27	4.3	0.19	2.9	0.31	5.0
$\Sigma\|d\|/3$	0.82		0.97		1.03	
$\Sigma\|d\|/14$	0.37		0.43		0.42	

[a]Δ is the absolute value of the difference between sets of simulations calibrated to the first region of each country (*Tables 11.4* and *11.5*) and sets of simulations calibrated to region 1 of Finland and extrapolated to other countries (*Table 11.6*). The % column gives the values of Δ as a percentage of the mean predicted value of *Tables 11.4* and *11.5*. The mean absolute values of Δ for each country and for the entire table are also listed.
[b]The low- and high-deposition scenarios in 2040 are illustrated in *Figures 11.6* and *11.7*, respectively.

(2)   A comparison of the 1980 versus 2040 results shows that differences are the same for Finland (0.34 and 0.34), nearly the same for Sweden (0.15 vs. 0.17), and inconsistent for Norway owing to the large bias (ε_2) in this country.

(3)   The pattern of differences between the intra-country and inter-country comparison is similar to those in *Tables 11.4* and *11.5*, suggesting that the errors of parameter estimation (ε_1) and bias (ε_2) are dependent on the differences between the emission scenarios.

These results indicate that if the RLM is used as a measure of Δ (the net difference in pH due to the different sulfur deposition scenarios), then it is not important what method of model calibration is used. If the accuracy of the simulations is more important, then because errors do not increase with time,

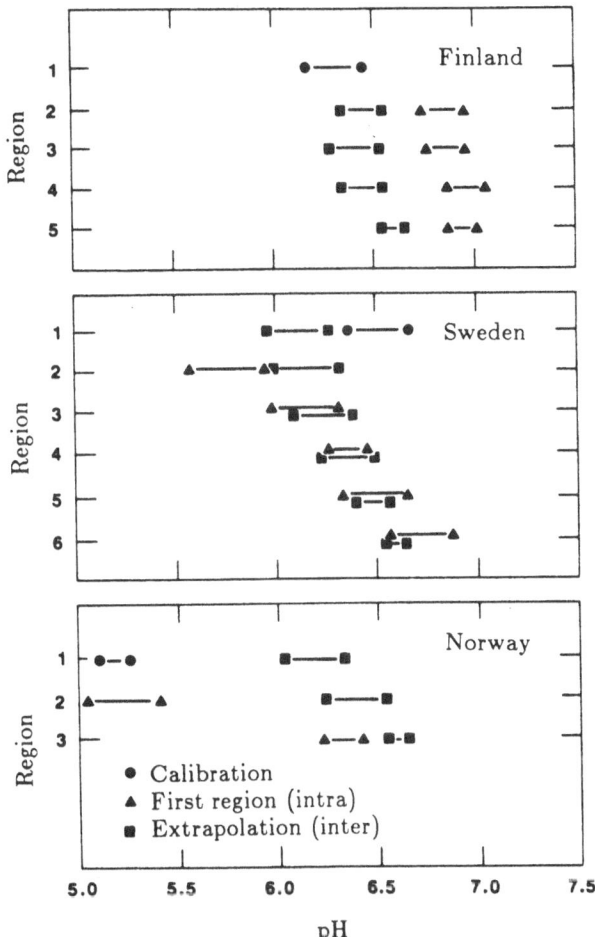

Figure 11.8. Comparison of model predictions for the low (*Figure 11.6*) and high (*Figure 11.7*) sulfur deposition scenarios. Circles indicate predictions for the calibration regions, triangles indicate the intra-country extrapolations, and squares indicate the inter-country extrapolations.

the results in *Tables 11.4* and *11.5* should be used to select the most suitable set of model parameters.

11.6. Conclusions

Studies of the broad-scale consequences of sulfur emission, transport, and deposition on aquatic resources have resulted in the development of extensive data bases (e.g., Eilers *et al.*, 1983; Linthurst *et al.*, 1986; Cook *et al.*, 1988; Rosen *et al.*, 1988), empirical models and multivariate statistical analyses (e.g., Hunsaker *et al.*, 1987; Rapp *et al.*, 1985; Schnoor *et al.*, 1985), and several complex process models (Cosby *et al.*, 1985a; Kämari *et al.*, 1986; Chen *et al.*, 1983; Schnoor *et al.*, 1984). In spite of the success of these studies, the temporal and spatial extent of sulfur deposition and effects exceeds our abilities to gather critical information in time and space. Therefore, this chapter has investigated the problem of applying information from regions that have been well studied and where local processes have been adequately modeled to regions where less data may be available and where local processes may not be as well understood. The central thesis is that the errors associated with this process may be used to determine the limits of measurement and prediction and thus define regions within which the patterns of sulfur effects may be reliably assessed.

The Fennoscandian data set of measured lake pH values (see *Table 11.1*) allowed the systematic calibration and testing of the RAINS Lake Module (RLM) to: (1) identify the major sources of errors associated with the regional application of RLM; (2) establish possible geographic limits for model and data extrapolation by looking at resulting parameter uncertainty (ε_1) and bias (ε_2); and (3) identify situations where further improvements in data, models, or both will most improve predictions. Although many statistics may be of interest (e.g., extreme pH, seasonal dynamics in pH, alkalinity), the first two moments of the cumulative frequency distribution of lake pH were evaluated here because they can be reliably estimated from existing data and easily compared with model results. The methods are intended to be general and can be applied to other studies whose prediction objectives differ from this application.

The results show that simple procedures can be used to calibrate the RLM and quantify the prediction errors (*Tables 11.4* and *11.5*). Sensitivity and uncertainty analyses identified *SOILT* (the soil-thickness of the catchment) and *SIBR* (the silicate buffer rate) as the key model parameters (*Table 11.3*). When *SOILT* and *SIBR* were calibrated to region 1 of Finland, the cumulative frequency distribution (*cfd*) of predicted pH values adequately matched the mean and standard deviation of the measurements. This result is important because the regional application of a complex site-specific model, such as RLM, involves many unknowns. The identification of a few critical values greatly simplifies the process and indicates that detailed information for all variables is not required for regional application of this model.

The use of the parameters from regions 1 of Finland, Sweden, and Norway for other regions within each country (intra-country comparison) and the use of the parameters from region 1 of Finland for all regions in Fennoscandia (inter-country comparison) indicate that broad geographic areas exist where the errors of extrapolation have relatively little effect on the accuracy of model predictions (e.g., all regions in Finland and regions 3–6 of Sweden). However, regions 1 and 2 of Sweden and all of Norway were difficult to calibrate (*Figure 11.5*), indicating practical limits to the regional applicability of RLM.

The predicted change in lake pH from 1980 to 2040 shows that the variance of model predictions does not increase with time (*Table 11.6*), and the pattern of predicted pH change is consistent with the simulated level of sulfur deposition (*Figure 11.8*). The results of Cosby *et al.* in Chapter 7 show rapid initial changes in Norwegian lakes due to acid deposition, with slower rates of change with continued acid input. Assuming that the RLM adequately simulates system response for sulfur deposition, the relative error associated with the extrapolation process can be quantified by comparison of the 1980 results with data. If the accuracy of simulations is the most important objective, then the results in *Tables 11.4* and *11.5* can be used to select a satisfactory set of parameters. Because the net difference between model predictions for different scenarios appears to be insensitive to the method of parameter calibration, relative differences might be employed to reduce the problem of residual model error.

Three reasonable options seem to exist for applying RLM to areas in Europe and North America where the information is not as extensive as in Fennoscandia. Option 1 is to model all systems as if they were similar to region 1 in Finland (the inter-country extrapolation) and use the net difference between scenarios as an indication of system effects. The advantages of this approach are that Finnish lakes are sensitive to sulfur deposition, the RLM model has been well calibrated to this region and adequately predicts system response, and little additional effort needs to be made for this application. The disadvantage of option 1 is that unique responses to other regions will never be detected by this type of extrapolation process. The second option is similar to option 1, except additional information critical to each region is gathered for key system variables. The difficulty with option 2 is that of identifying and obtaining, *a priori*, information for these key variables. However, some indication of the change or variance of important parameters across a region will be useful for identifying the possibility of unique regional responses to sulfur deposition. The third option is to develop methods that will identify the geographical distribution of key variables and thus define – on the "first principles" of geology, topography, land-use history, etc. – what the regional limits of model extrapolation may be.

The issues discussed here are not limited to the problems of acid deposition, but are equally applicable to other broad-scale environmental problems, such as atmospheric CO_2 (Gardner and Trabalka, 1985), pest outbreaks (Eager, 1984), and toxic chemical fate and effects (Bartell *et al.*, 1983; Bartell *et al.*, 1986). The challenge of using complex site-specific models for broad-scale predictions will be to identify processes that are critical to a region and thus define, on first principles, the natural boundaries of system behavior. Methods that identify these boundaries may allow model predictions to have errors that are small relative to the natural variability of the system. Progress in the solution to this problem must recognize the interplay between the complexity of site-specific processes, the regional extent of model application, and the eventual limitations of regional data.

Acknowledgments

The financial support from IIASA that facilitated the collaboration with the Environmental Sciences Division, Oak Ridge National Laboratory, is gratefully acknowledged. Research at Oak Ridge National Laboratory was supported in part by the Office of Health and Environmental Research, US Department of Energy, under contract No. DE–AC05–840R21400 with Martin Marietta Energy Systems, Inc.; and in part by the Ecosystem Studies Program, National Science Foundation, under Interagency Agreement BSR 8516951 with the US Department of Energy.

CHAPTER 12

Evaluating the Performance of Monte Carlo Calibration Procedures

M.C. Sutton

12.1. Introduction

The development of computer models to simulate environmental systems is a costly and time-consuming business. Owing to constraints on time and money, model development is generally an exercise in the distribution of scarce resources. Modelers must decide how their resources will be distributed among the model, data collection, and calibration – a decision complicated by their interrelationships. For example, the model's design guides data collection; the calibration procedure and data are used to tune the model; and the model's usefulness is influenced by the conceptual model's validity, the calibration procedure's performance, and the amount and quality of data available. Consequently, it is often very difficult to determine which of the above is responsible for a particular imperfection in the final model's performance.

With respect to modeling the impacts of acid deposition on lakes, studies of single watersheds have provided us with relatively good conceptual models of the many chemical and physical processes involved (de Grosbois *et al.*, 1986; Baker *et al.*, 1986; Cosby *et al.*, 1985a), and our computers are powerful enough to perform the necessary computations. However, when these processes are linked together to simulate the behavior of even one watershed, let alone all those in a region, the amounts of data needed to provide inputs, estimate parameter values, and validate the model are such that we can rarely do more than scratch the surface. Even with the extensive surface-water surveys already completed (Linthurst *et al.*, 1986; Overton *et al.*, 1986; Kaniciruk *et al.*, 1986), such data are relatively scarce owing to the large number of lakes in acid-sensitive regions, the costs involved, and the physical difficulty of obtaining data. This scarcity is

J. Kämäri (ed.), Impact Models to Assess Regional Acidification, 209–232.
© 1990 *International Institute for Applied Systems Analysis.*

amplified by the spatial variability of conditions within and among watersheds, which make precise characterization difficult even with unlimited resources. An additional factor is that while the response times for some of the ion exchange and weathering processes can be as long as decades or even centuries (Galloway *et al.*, 1983; Schnoor and Stumm, 1984; Cosby *et al.*, 1985b, 1985c, 1986), acidification has only been studied closely for the last 15–20 years. The net result is that we have what amounts to a low-resolution still picture of a complex and dynamic situation.

The above characteristics and the need for models to predict the regional impacts of alternative acid deposition policies have led some modelers to employ stochastic techniques for the purpose of regionalizing models designed to predict the acidification of individual lakes. The resultant models are expected to generate predictions that represent the distribution of chemistry for the population of lakes within a given region (Alcamo *et al.*, 1987; Cosby *et al.*, Chapter 7; Small and Sutton, 1986; Small *et al.*, 1988). While the approaches taken in these models differ in detail, some characteristics are common to all: They add a level of abstraction and complexity to an already complex set of models, taking us from the more traditional world of environmental modeling, in which models are calibrated and validated with respect to identifiable real world systems, to a world where models are predicting the behavior of a set of theoretical lakes that are assumed to be representative of the population. In this latter world, we need ways to evaluate the performance and validity of the calibration–regionalization procedures that do not depend on the validity of the models being calibrated.

Performance of models and calibration procedures can be categorized as in *Figure 12.1*: both perform sufficiently well; neither performs sufficiently well; one, but not both, performs sufficiently well. While most modelers prefer to find out that their model falls into the first category, knowing which of the other three categories it falls into is also valuable information as it allows the modeler to focus future efforts where they are needed the most. Making such a classification requires information on both the model and the calibration procedure, independent of one another. For regional-scale lake acidification models, information can often be provided by comparisons of the single-watershed versions with available data. However, obtaining information on the calibration procedure requires an evaluation whose results are independent of the model's realism, since there will always be some uncertainty about its validity.

A method for using synthetically generated systems to evaluate calibration procedure performance, independent of model validity, is proposed and presented in this chapter. This approach does not require a large amount of real data from the system to be modeled (although the more, the better), because the model itself is used to generate any data not actually available. Indeed, the term true is used throughout this chapter to refer to the system defined by the modeler, as opposed to a system that actually exists. To implement the method the modeler specifies the true parameter values and model inputs; defines true system behavior by running the model with these parameters and inputs; makes one or more of the parameter distributions different than the true distributions to see how well the calibration procedure calibrates them; calibrates the model to its own self-defined true behavior; evaluates performance of the calibration

Model performance

GOOD   POOR

		GOOD	neither needs improvement	model needs improvement
	POOR		calibration procedure needs improvement	both need improvement

Calibration procedure performance (left vertical label)

Figure 12.1. Categorization of the performance of models and calibration procedures: both perform sufficiently well; neither performs sufficiently well; one, but not both, performs sufficiently well.

procedure by comparing true and calibrated parameter values along with true and predicted system behavior.

Use of this synthetic system to evaluate calibration procedure performance is demonstrated on the Lake Module of the Regional Acidification INformation and Simulation (RAINS) model, developed at the International Institute for Applied Systems Analysis (IIASA), Laxenburg, Austria (Alcamo *et al.*, 1987; Kauppi *et al.*, 1986; Kämäri *et al.*, 1985, 1986; Kämäri and Posch, 1987; Posch and Kämäri, Chapter 9). The results of this demonstration are used to evaluate the performance of the calibration method incorporated in RAINS Lake Module (RLM), suggest ways of improving the RLM calibration method, and illustrate some potential applications of synthetically generated systems.

12.2. Calibration of RLM by Monte Carlo Filtering

The RLM model makes use of Monte Carlo techniques and statistical filtering to calibrate and regionalize simultaneously a single-watershed acidification model (*Figure 12.2*). The resulting model is expected to predict the spatial frequency distribution of lake pH. This is done by assigning – a priori – spatial frequency distributions, based on a synoptic survey of surface-water chemistry (see Hornberger *et al.*, Chapter 13; Kämäri *et al.*, 1986), to model parameters that exhibit spatial variability. These distributions are used to generate randomly a large number of parameter sets, the model is run once with each set, and the predictions are saved. The predicted and observed pH distributions are compared; and model predictions – and their associated parameter sets – are rejected so as to make the predicted pH distribution match the observed (see *Figure 12.3* and the Appendix 12A).

The *a posteriori* (accepted) parameter sets constitute a set of theoretical lakes that are assumed to be representative of the lake population. Predictions about the regional distribution of lake pH are made by specifying future acid deposition rates and running the model once with each of the *a posteriori* parameter sets. Because the comparison–rejection step consists of filtering the

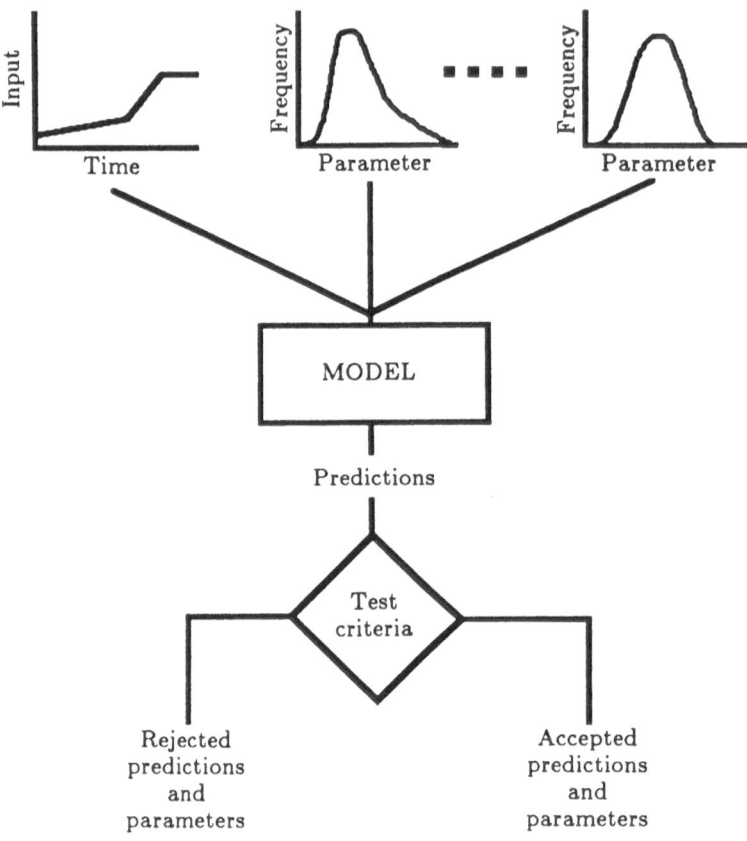

Figure 12.2. Parameters with spatial variability are assigned frequency distributions, and these distributions are used to generate random sets of parameters. The model is run once with each parameter set, and the predictions are saved. Predictions and their associated parameter sets are then judged as either unacceptable or acceptable, in which case they are kept for further analysis.

randomly generated sets of parameters, the procedure is referred to as the Monte Carlo Filter Procedure (MCFP). The mathematics of MCFP are described in detail in the Appendix to this chapter.

The mechanics of the MCFP consist of first rejecting any predictions that lie outside the range of observed values. The range of pH values is divided into intervals (*Figure 12.3*), making the intervals shortest where the simulated or observed cumulative distribution functions (*cdfs*) of pH are the most nonlinear, as this allows the greatest flexibility in the areas most likely to need it. The system of equations, which must be solved to determine the number of predictions that must be rejected from each pH interval to make the *cdfs* match at the interval boundaries, is given by:

$$(1 - cdf_i) \sum_{j=1}^{i} r_j - cdf_i \sum_{k=i+1}^{n} r_k = N_i - cdf_i N_{\text{tot}} \qquad i = 1,2, \ldots, n \quad (12.1)$$

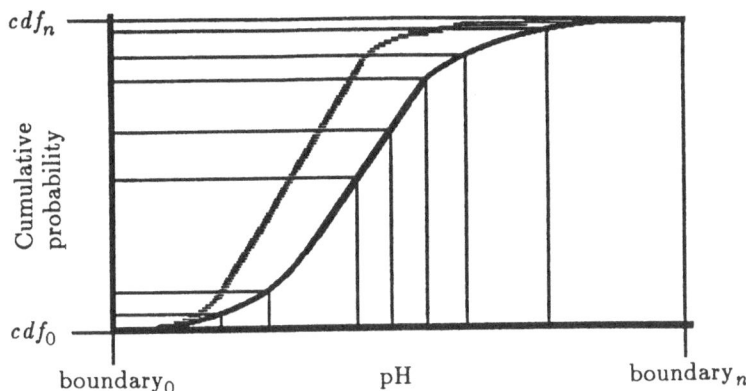

Figure 12.3. The observed (lower) and predicted (upper) cumulative distribution functions (*cdfs*) of pH are divided into intervals. The observed *cdf* is characterized by values $cdf_0 = 0, \ldots, cdf_n = 1$.

subject to

$$r_i \geq 0 \quad \text{and} \quad \sum_{j=1}^{n} r_j < N_{\text{tot}} \quad ,$$

where N_{tot} is the total number of predictions; N_i is the number of predictions between the $(i-1)$th and ith pH boundary, where the boundaries are numbered from 0 to n; r_i is the number of predictions to be rejected between the $(i-1)$th and ith pH boundaries; and cdf_i is the observed cumulative probability at the ith pH boundary. The nth equation is an identity $(1 = 1)$, so that there are only $(n-1)$ independent equations and n unknown r_i; therefore, one of the r_i must be specified by the modeler. Within each interval, the predictions to be rejected are chosen randomly on the basis of a uniform distribution.

12.3. Use of a Synthetic Region to Evaluate Calibration Procedure Performance

12.3.1. Design of a synthetic region

The synthetic region for evaluating RLM was created by specifying the annual precipitation and temperature patterns as well as the historical and future deposition trajectories; assigning true spatial frequency distributions to each randomized parameter; using the true distributions to generate randomly a large number of parameter sets and introduce correlations, if so desired (Iman and Conover, 1982); define true regional behavior – historic and future – by running the model once with each parameter set and deposition scenario; simulating a lack of *a priori* parameter data by altering the true parameter distributions; calibrating the model to the true regional behavior and then running the calibrated model with the future deposition scenarios; and, finally, assessing the performance of the calibration procedure by comparing the true and calibrated model behavior and parameter distributions.

Some advantages of using a synthetic region to evaluate the calibration procedure's performance are that:

(1) True regional behavior, parameter distributions, and parameter correlations are known with certainty, thus simplifying the task of evaluating performance.

(2) The ability of the calibration procedure to compensate for different types of imperfections in the *a priori* data can be tested by systematically varying the type, number, and magnitude of imperfections.

(3) It is possible to study the degree to which additional observations – in either space or time – may improve model calibration.

Design of the synthetic region is a major factor in determining how relevant the results will be. Credibility of the results obtained will be questionable if important characteristics of the synthetic and real regions are not at least qualitatively similar. Characteristics likely to be important are the upper and lower bounds of parameters and output variables, the parameter distribution shapes, and the correlations between parameters. As it is unlikely that there will be enough data to determine these things with much certainty in the real system, it will probably be necessary to design a set of synthetic regions that span the range of possibilities.

The variables and parameters that need to be included when creating a region are determined by the model being used. The RLM model has three exogenous inputs (acid deposition, precipitation, and temperature) and 16 randomized parameters: partial pressure of CO_2 in surface waters, soil depth, silicate weathering rate, base saturation in the upper soil layer (Base-A), base saturation in the lower soil layer (Base-B), cation exchange capacity, soil moisture content at field capacity (Kauppi *et al.*, 1986; Kämäri *et al.*, 1985 and 1986), sulfate in-lake retention coefficient (Baker *et al.*, 1985), lake surface area, the ratio of catchment area to lake area, mean lake depth, watershed surface slope, fraction of watershed area covered by forest, forest filtering coefficient, temperature coefficient, and precipitation coefficient. Both the temperature and precipitation coefficients are uniform random variables (0, 1) used to select temperature and precipitation values between the minimum and maximum values shown in *Figures 12.4* and *12.5*.

The synthetic region created for the demonstration has temperature (*Figure 12.4*), precipitation (*Figure 12.5*), and acid deposition profiles (*Figure 12.6*) similar to those of southern Finland. The true parameter distributions of the four most influential parameters are displayed by the dashed lines in *Figures 12.11–12.15*. The judgment of which parameters are influential was based on a combination of sensitivity–uncertainty analysis (Kämäri *et al.*, 1986), Regional Sensitivity Analysis, and correlation coefficients.

Once the inputs and parameters have been specified, the model can be used to produce the true regional pH distributions, which are used to calibrate the

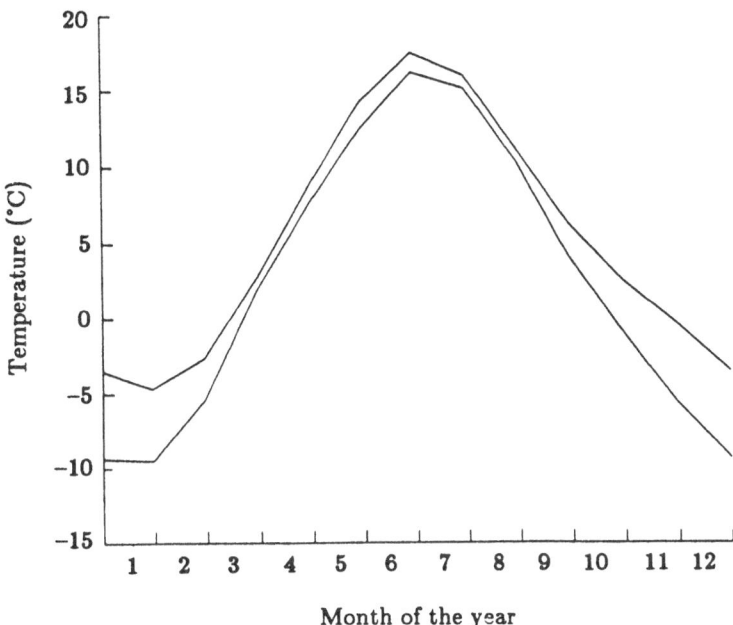

Figure 12.4. Range of monthly temperatures for the synthetic region.

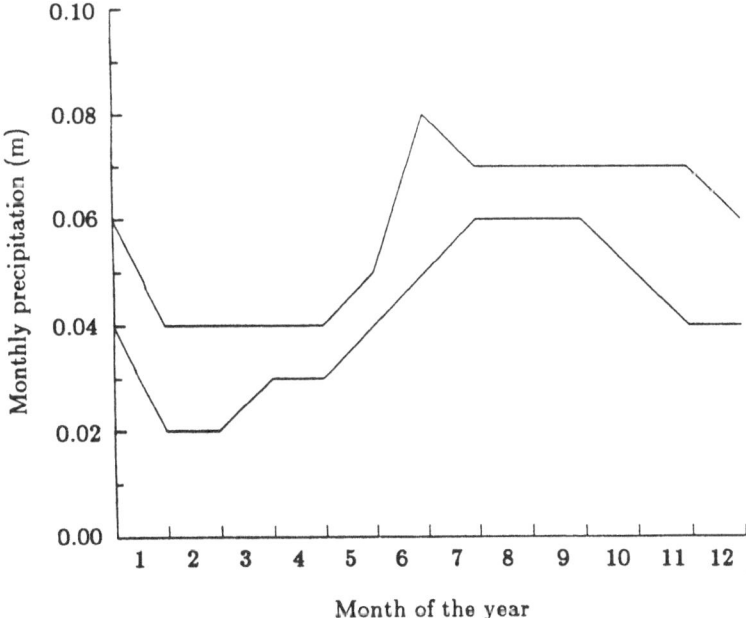

Figure 12.5. Range of monthly precipitation rates for the synthetic region.

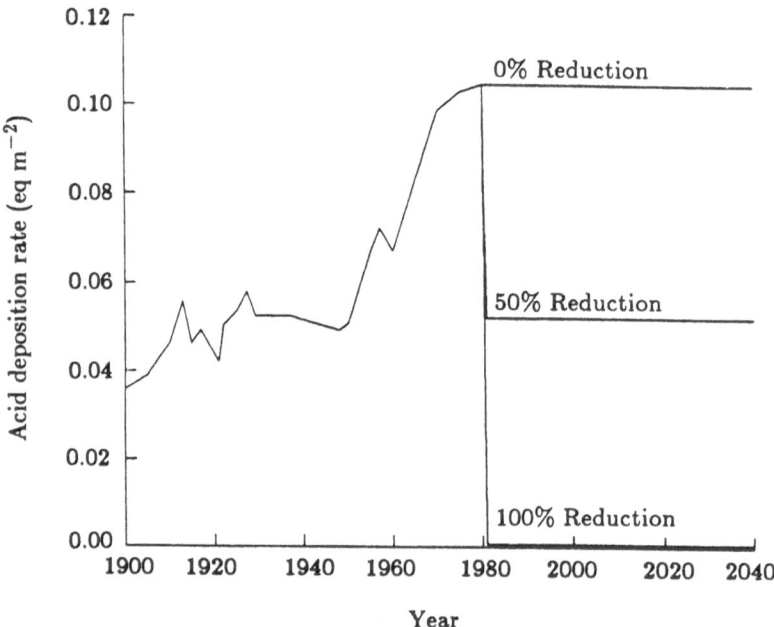

Figure 12.6. Historic acid deposition and the three future (post-1980) deposition scenarios: no reduction, 50% reduction, and 100% reduction.

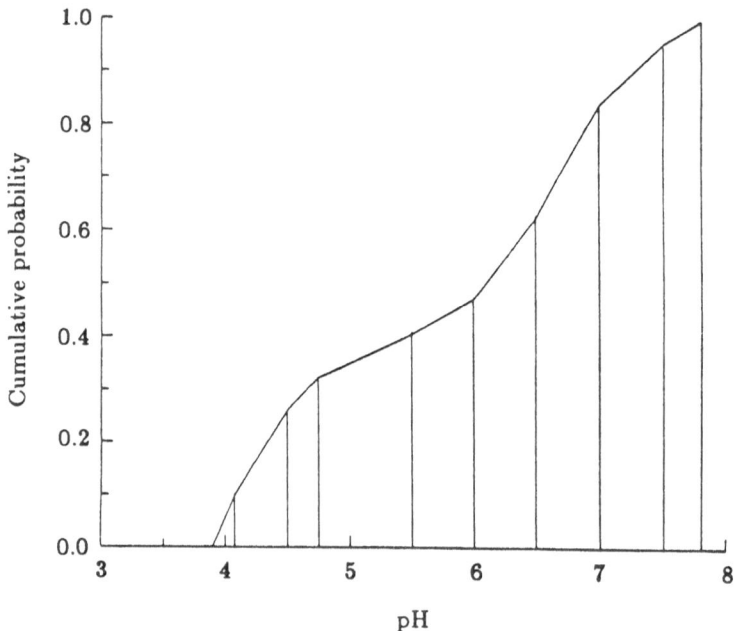

Figure 12.7. The true (Case 1) cumulative distribution function of pH for the synthetic region and the pH interval boundaries used by the MCFP.

model versions having poorly known parameters and to perform the tests by which the performance of the calibration procedures will be evaluated. The RLM model was calibrated using the cumulative probability distribution (cdf) of pH in the year 1980, which was produced by running the RLM model without filtering. The 1980 pH cdf is shown in *Figure 12.7* along with the interval boundaries used by the MCFP. True regional pH behavior was generated by running the RLM model from the year 1900 to 2040 once with each randomly generated set of parameters, without filtering, using the three deposition scenarios depicted in *Figure 12.6*.

12.3.2. Simulation of poorly known parameters

The next step in using a synthetic region is to simulate a lack of data for one or more of the parameters. There are a variety of ways in which this may be done for the parameter distributions used in the RLM model, although the actual method(s) chosen should be based on the types and quality of data available for the real region to be modeled.

In the following demonstration, poorly known parameters were simulated by substituting uniform distributions for the true distributions, but keeping the same upper and lower bounds. This method of simulating poorly known parameter distributions implies that the range of possible parameter values is known – perhaps from our understanding of the processes involved – but that there are few, if any, data from the real region to be modeled. The only deviation from this method occurs when the partial CO_2 pressure (pCO_2) is simulated as being poorly known in Case 2. The very long upper tail of the true pCO_2 distribution means that using a uniform distribution with the same bounds greatly increases the expected value of pCO_2. This resulted in the rejection of so many predictions that it was found necessary to lower the upper pCO_2 bound in Case 2 from 0.3 to 0.03 to keep calibration times reasonable.

The MCFP was evaluated with respect to its ability to calibrate the RLM model for criteria in which only one of the four most influential parameter distributions is poorly known (Cases 2–5) and when all four are poorly known (Case 6):

Case 1: All parameter distributions known; used to define true regional behavior.

Case 2: All parameter distributions known except for pCO_2.

Case 3: All parameter distributions known except for soil depth.

Case 4: All parameter distributions known except for silicate weathering rate.

Case 5: All parameter distributions known except for base saturation in the upper soil layer.

Case 6: Distributions for pCO_2, soil depth, silicate weathering rate, and base saturation in the upper soil layer are poorly known.

RLM was calibrated six times to the 1980 pH distribution of Case 1 (*Figure 12.7*), using a different random seed each time (6 Cases × 6 random seeds = 36 calibrations). The *a posteriori* parameter sets and predictions from each calibration were used to calculate correlation coefficients and parameter *cdf*s. Predictions were made for each of 36 calibrations by running the RLM model out to the year 2040 once with each of the accepted parameter sets, using the three future (post-1980) deposition scenarios shown in *Figure 12.6*.

12.3.3. Testing calibration procedure performance

The choice of test criteria to be used in judging performance of the calibration procedure depends on what is expected of the model as well as the characteristics of the model and calibration procedure being tested. The criteria to be used in this demonstration consist of:

(1)   The Smith–Satterthwaite (Devore, 1982) two-sample test applied to the predicted distribution of pH (for the year 2020) to test the alternative hypothesis that the mean, standard deviation, and percentage of lakes with pH < 5 are not equivalent to the corresponding true values.
(2)   Comparison of the true, *a priori*, and *a posteriori* parameter *cdf*s.
(3)   Examination of the true and *a posteriori* parameter correlation coefficients.

The mean and standard deviation of the pH distribution were chosen to serve as general indications of the overall fit of the regional pH distribution, which is what the MCFP actually works with. The percentage of lakes with pH < 5 was chosen because it is more likely to be the sort of metric upon which decisions will be based. The Smith–Satterthwaite test (level of significance 0.05) was chosen because it cannot be assumed that both populations are normal or their variances equal.

Comparison of the parameter *cdf*s will provide some insights about how and what the MCFP actually calibrates and whether we might expect the calibrated model to behave like the true region for only some or most deposition scenarios. When frequency distributions are being calibrated, it is possible for one part to be improved while another part deteriorates. For the sake of simplicity, a variant of the Kolomogorov–Smirnov test is used here, which focuses on deterioration in parameter fits, rather than on improvements. When the *a priori* and true parameter distributions are the same, the statistic (D) is equal to the greatest vertical distance between the true and *a posteriori* cdfs $(D = |cdf_{true} - cdf_{a\ posteriori}|)$. When the *a priori* differs from the true distribution, the statistic is equal to the greatest vertical distance between the *a priori* and a *posteriori* cdfs, at a point where the *a posteriori* cdf is further away than the *a priori* cdf from the true cdf.

When simulated and true predictions are similar, the examination of the *a posteriori* parameter characteristics can indicate whether the good predictions are the result of a realistic calibration or a combination of parameters having little in common with the true region. When simulated predictions are a poor representation of true regional behavior, the *a posteriori* parameters can be useful for identifying where and how the MCFP has failed.

Examination of the correlation coefficients among the parameters and pH provides some insight into the interaction between different parts of the model and the MCFP.

12.4. Results

Results from the model calibrations and predictions are presented in both graphs and tables. The graphs give a broad picture of what the MCFP did, while the tables contain the performance of the MCFP with respect to the specific test criteria, described in Section 12.3.2.

Predictions of the mean regional pH are found in *Figure 12.8* and *Table 12.1*; predictions of the regional pH standard deviation are found in *Figure 12.9* and *Table 12.2*; predictions of the percentage of lakes with pH < 5 are found in *Figure 12.10* and *Table 12.3*; correlation coefficients between parameters and $[H^+]$ are found in *Table 12.4*; changes in the parameter distributions are found in *Figures 12.11–12.15* and *Table 12.5*.

Quickly summarizing the results from Test Cases 2–6 before discussing them in greater detail: When pCO_2 is poorly known (Case 2), predictions are inaccurate, parameter fits are poor, and the MCFP creates significant parameter correlations between two different parameter pairs. When soil depth is poorly known (Case 3), the predictions are quite close to the true values even though they do not pass the test criteria; parameter fits are quite good, and no significant parameter correlations are found. When silicate weathering rate is poorly known (Case 4), predictions are inaccurate and parameter calibration is poor, although no significant parameter correlations were created by MCFP. When base saturation in the upper soil layer is poorly known (Case 5), predictions are quite close to the true values, but the MCFP did very little to improve – or alter in any way – the parameter fits. This suggests that the accuracy of the predictions is primarily due to the unimportance of the base saturation parameter in the model. Correlations were created between two parameter pairs. When all four parameters were poorly known (Case 6), predictions were inaccurate.

Table 12.1. Is the predicted mean regional pH equivalent to (EQUAL) or different from (-----) the true regional pH, according to the Smith–Satterthwaite test at a 0.05 level of significance?

Test criteria	Deposition after 1980		
	Constant	50% reduction	100% reduction
Case 2	-----	-----	EQUAL
Case 3	-----	-----	-----
Case 4	-----	-----	-----
Case 5	EQUAL	-----	-----
Case 6	-----	-----	-----

Table 12.2. Is the predicted regional pH standard deviation equivalent to (EQUAL) or different from (-----) the true pH standard deviation, according to the Smith–Satterthwaite test at a 0.05 level of significance?

	Deposition after 1980		
Test criteria	Constant	50% reduction	100% reduction
Case 2	-----	-----	EQUAL
Case 3	-----	EQUAL	-----
Case 4	-----	EQUAL	-----
Case 5	EQUAL	-----	-----
Case 6	-----	-----	-----

Table 12.3. Is the predicted percentage of lakes with pH < 5 equivalent to (EQUAL) or different from (-----) the true percentage of lakes with pH < 5, according to the Smith–Satterthwaite test at a 0.05 level of significance?

	Deposition after 1980		
Test criteria	Constant	50% reduction	100% reduction
Case 2	-----	-----	EQUAL
Case 3	-----	-----	-----
Case 4	-----	-----	-----
Case 5	EQUAL	-----	-----
Case 6	-----	-----	-----

Table 12.4. Means of the major correlation coefficients between parameters and 1980 hydrogen ion concentration.

	A priori test cases					
Parameter	True	2	3	4	5	6
pCO_2	+0.10	+0.25	+0.06	+0.05	+0.10	+0.70
Soil depth	-0.70	-0.70	-0.68	-0.70	-0.70	-0.70
Sil weath rate	+0.10	-0.15	-0.10	+0.05	-0.05	+0.05
Base-A	-0.14	-0.10	-0.20	-0.10	-0.10	-0.10

Table 12.5. Fit of the *a posteriori* parameter distributions with respect to *a priori* distribution: only improvement seen (GOOD); no change (-----); the maximum amount of vertical deterioration seen in the parameter's *cdf* (*D*).

	Fit of the calibrated parameters			
Poorly known parameter(s)	pCO_2	Soil depth	Sil weath rate	Base-A
pCO_2	GOOD	-0.10	-0.08	-0.05
Soil depth	-----	GOOD	-----	-0.08
Sil weath rate	-0.05	+0.12	GOOD	-----
Base-A	-----	+0.03	-----	GOOD
All above	GOOD	-0.04	-0.10	GOOD

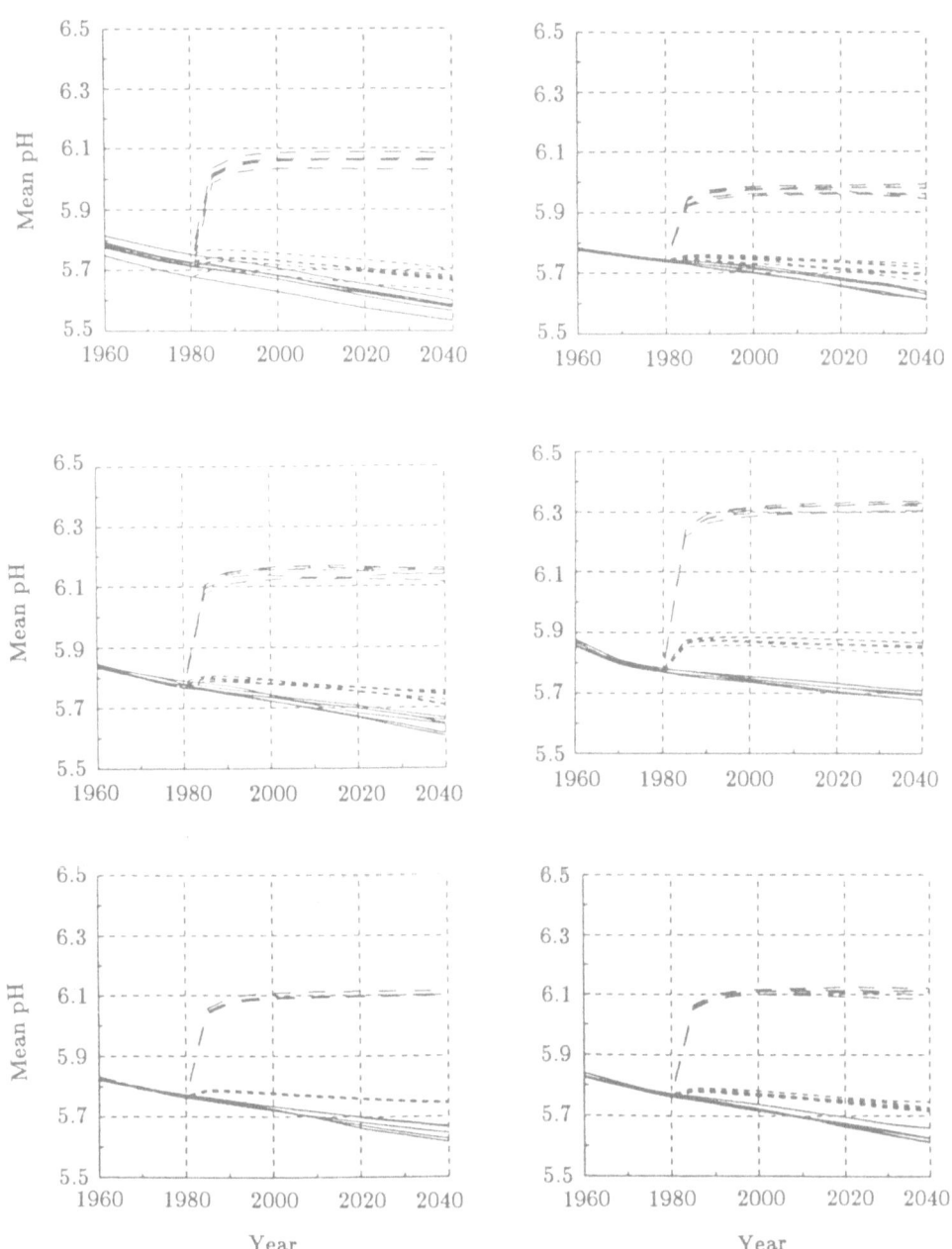

Figure 12.8. Predicted mean regional pH for no reduction, 50% reduction, and 100% reduction scenarios. Results are shown for six model replications using a different random number seed each time.

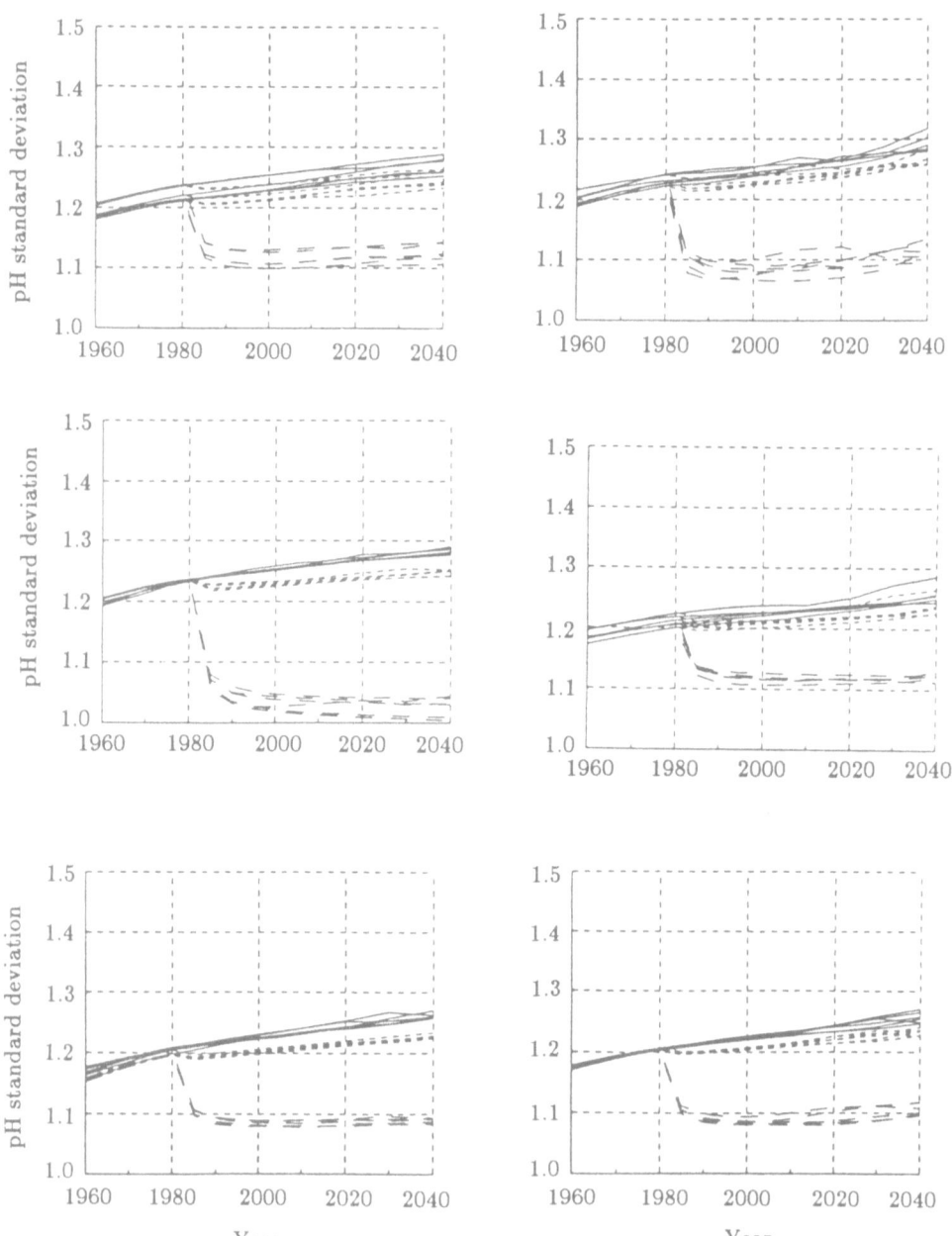

Figure 12.9. Predicted standard deviation of regional pH for no reduction, 50% reduction, and 100% reduction scenarios. Results are shown for six model replications using a different random number seed each time.

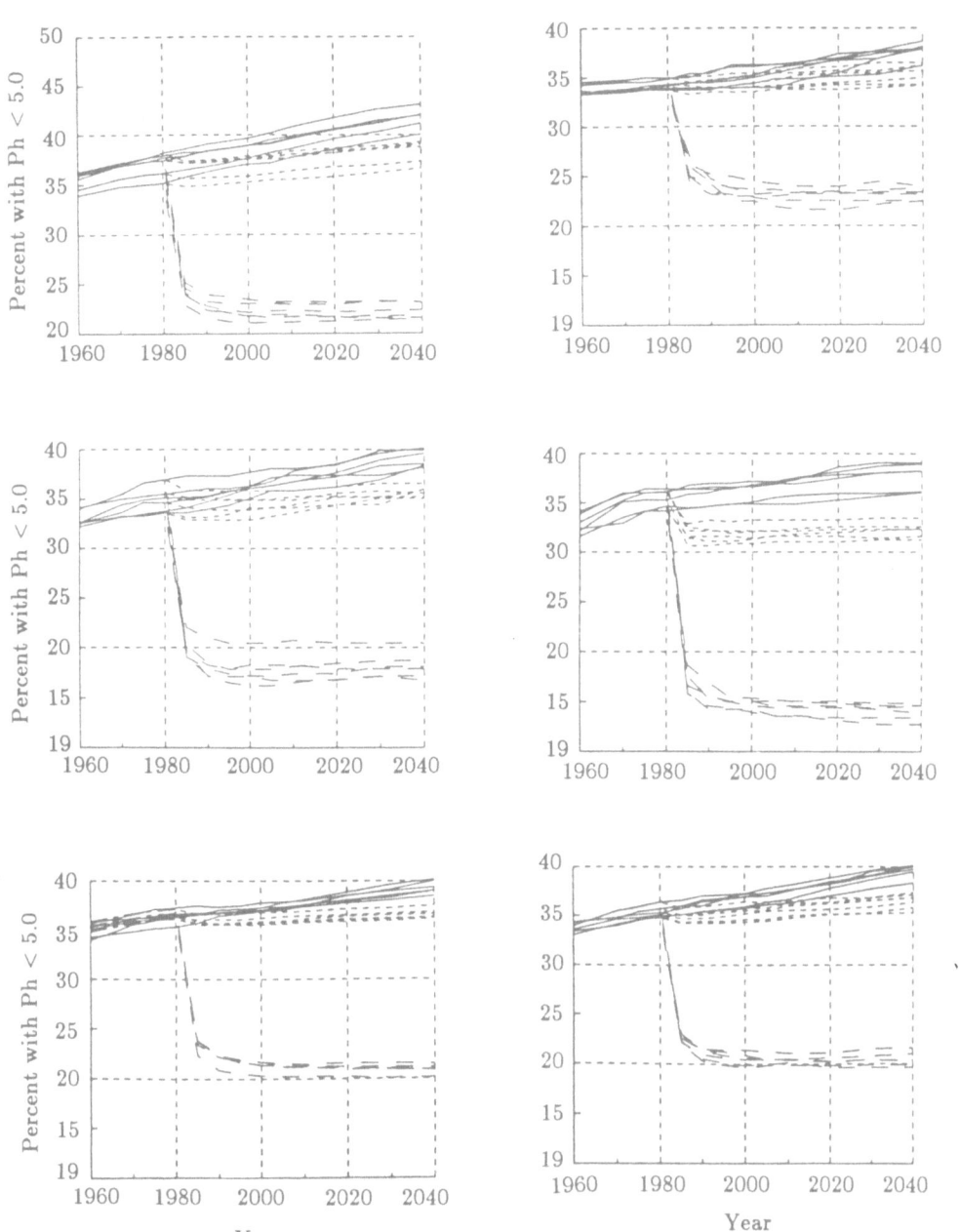

Figure 12.10. Predicted percentage of lakes with pH < 5.0 for no reduction, 50% reduction, and 100% reduction scenarios. Results are shown for six model replications using a different random number seed each time.

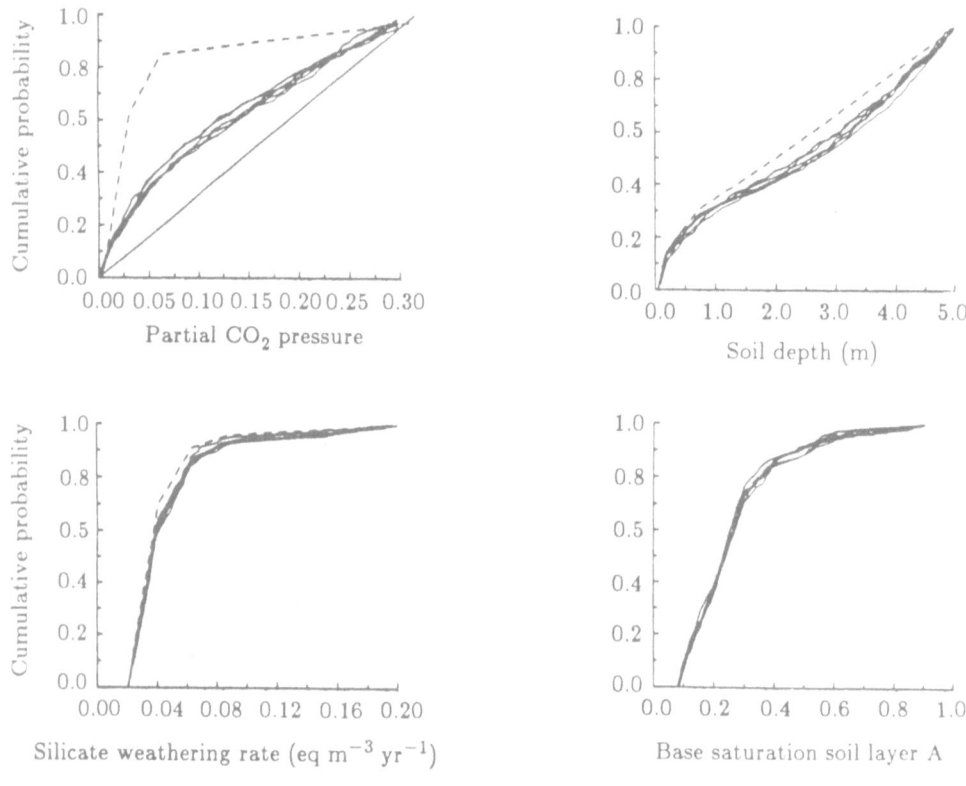

Figure 12.11. Case 2: pCO_2 poorly known. Comparison of the true (- - -) and *a posteriori* (——) parameter *cdfs*. The uniform *a priori* pCO_2 distribution is shown by the straight line. Results are shown for six model replications.

Examination of the parameter distributions shows that the MCFP improved some while worsening others, but did not create any significant parameter correlations. This suggests that the accuracy of the predictions may be attributed to errors canceling out each other. A tentative conclusion from these results is that the MCFP performs best when there is only one poorly known, and very influential, parameter.

Case 1: True behavior

Case 1 was used to define true system behavior. Unlike Cases 2–6, the parameter sets and predictions were not subjected to any filtering. The 1980 pH distribution of this Case (*Figure 12.7*) was used by the MCFP to calibrate the other *a priori* data Cases, and its predictions were used to evaluate the performance of the MCFP in Cases 2–6. Each randomized parameter was generated independently of the others. However, when the t-test (0.05 level of significance; Benjamin and Cornell, 1970) was used to test the null hypothesis that the mean correlation coefficients between parameters are equal to zero, it was rejected for 7 parameter pairs. Therefore, when considering the correlation coefficients of

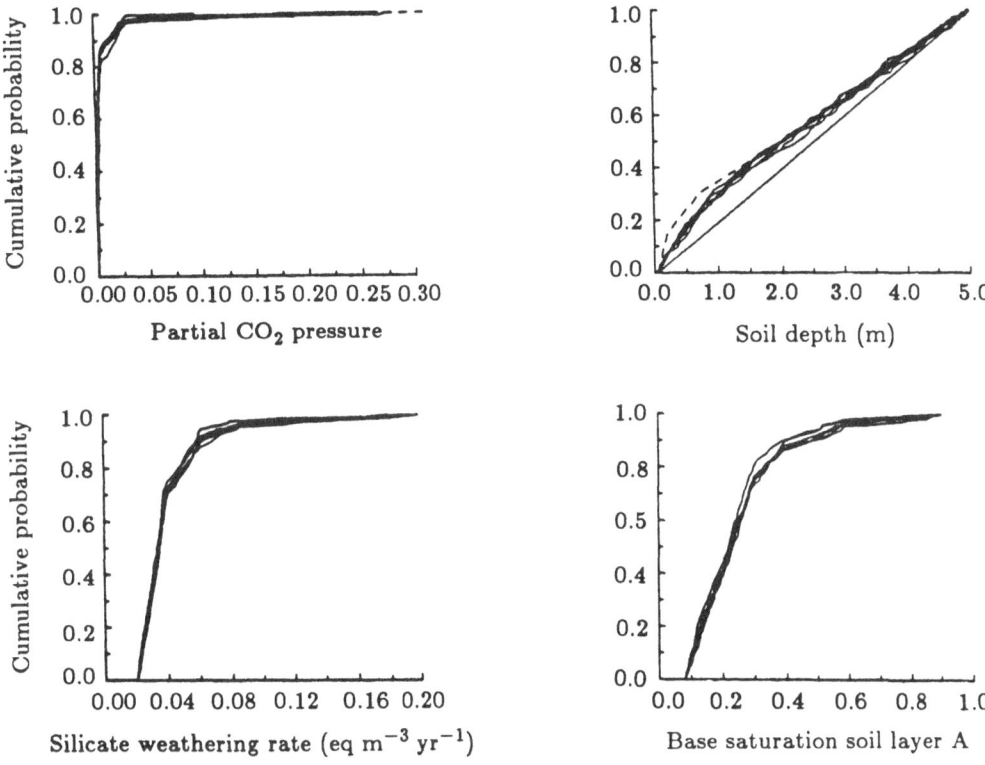

Figure 12.12. Case 3: soil depth poorly known. Comparison of the true (- - -) and *a posteriori* (——) parameter *cdfs*. The uniform *a priori* soil depth distribution is shown by the straight line. Results are shown for six model replications.

Cases 2–6, values of the same magnitude as those in Case 1 were considered to be insignificant.

The ranking of parameters with respect to their influence on the predicted $[H^+]$ is an important factor to consider when interpreting the indications of the parameter fits and correlations. Uncertainty analysis was used by Kämäri *et al.* (1986) to estimate parameter influence in an earlier version of the RLM model. RLM has been modified since then; and because the model is being applied to a different region, the results of Kämäri *et al.* were confirmed by the use of Regionalized Sensitivity Analysis (RSA; Hornberger and Cosby, 1985; see also *Figure 12.15*) and examination of the correlation coefficients between the parameters and $[H^+]$ (*Table 12.4*). The results of these analyses indicated that soil depth, silicate weathering rate, pCO_2, and base saturation in the upper soil layer are the most influential parameters in the RLM model.

Case 2: pCO_2 poorly known

As mentioned earlier, this Case is different from the others in that it was necessary to decrease the upper bound of the pCO_2 distribution to achieve calibration

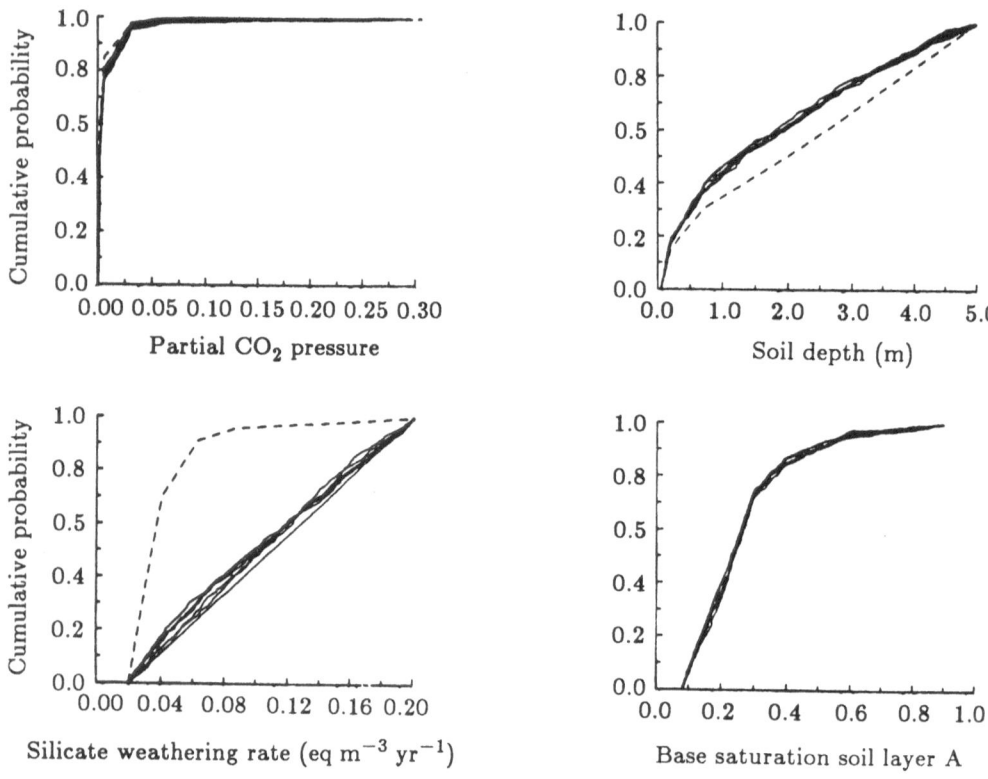

Figure 12.13. Case 4: silicate weathering rate poorly known. Comparison of the true (------) and *a posteriori* (——) parameter *cdfs*. The uniform *a priori* silicate weathering rate distribution is shown by the straight line. Results are shown for six model replications.

without an excessive number of models runs being required. Even with this change, the expected value of the *a priori* pCO_2 distribution was much higher than the true value.

When the predictions of the calibrated model were tested to determine if they were equivalent to the true regional values in the year 2020, only the predictions for a 100% deposition reduction passed the test criteria (*Tables 12.1, 12.2, and 12.3*). Also, while the MCFP did improve the pCO_2 distribution, it also created errors in the distributions of soil depth and silicate weathering rate, and correlations – not present in the true system – of 0.12 between soil depth and pCO_2 and –0.14 between soil depth and silicate weathering rate. The net effect of these correlations is to reduce the acidity of predictions, as the result of the MCFP compensating for the too-high values of pCO_2. Considering that only one deposition scenario passed the test criteria and the problems with the parameter distributions, the one successful prediction is more likely the result of carbonate equilibria being the dominant acid-base process in the absence of acid deposition, rather than the MCFP.

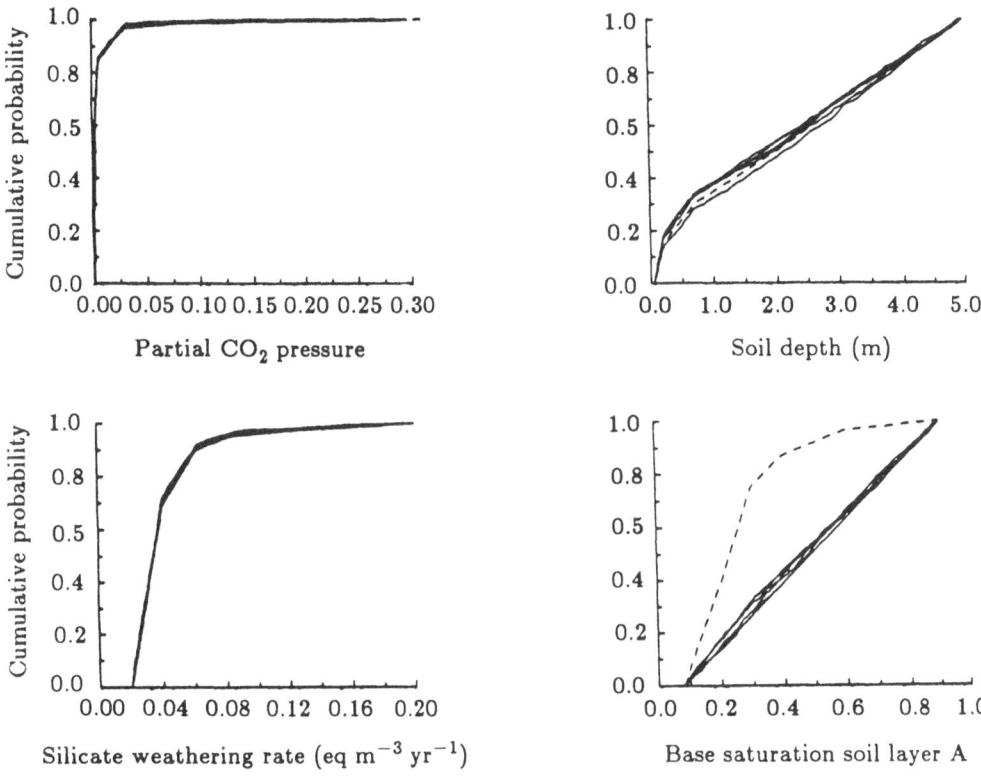

Figure 12.14. Case 5: base saturation in the upper soil layer poorly known. Comparison of the true (- - -) and *a posteriori* (——) parameter *cdfs*. The uniform *a priori* base saturation distribution is shown by the straight line. Results are shown for six model replications.

Case 3: Soil depth poorly known

Even though the prediction for pH standard deviation with a 50% deposition reduction was the only one to pass the test criteria (*Tables 12.1, 12.2,* and *12.3*), the MCFP turned in its best overall performance in this Case. The other predictions were quite close to the values seen in the true region and would have been judged equivalent under slightly less stringent criteria.

The parameter fits (*Figure 12.12*) were also good, with substantial improvements in the soil depth distribution and only a small amount of deterioration in the base saturation distribution. Also, no significant correlations were created between any of the parameters.

Case 4: Silicate weathering rate poorly known

If the results in *Tables 12.1, 12.2,* and *12.3* were the only basis for judgment, the performance of the MCFP in this Case would be judged equivalent to that in Case 3. However, examination of the model predictions in *Figures 12.8, 12.9,*

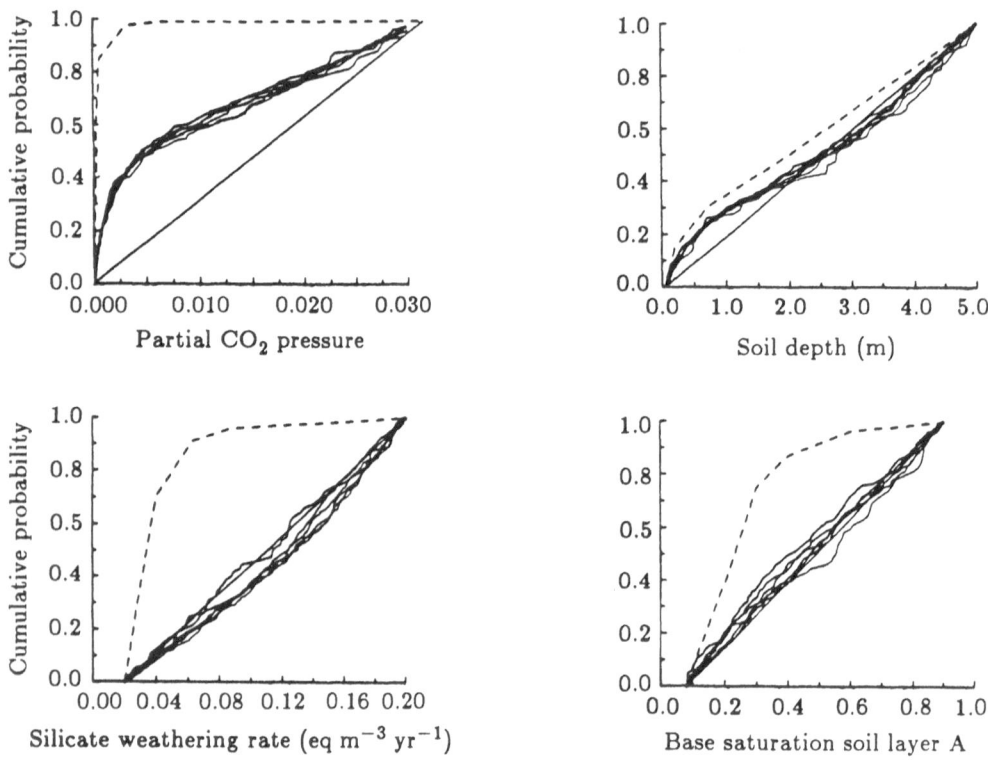

Figure 12.15. Case 6: all four parameters poorly known. Comparison of the true (- - -) and *a posteriori* (——) parameter *cdfs*. The uniform *a priori* distributions are shown by the straight lines. Results are shown for six model replications.

and *12.10* reveals that the predictions were much less accurate than those when soil depth was the poorly known parameter. Parameter calibration was also poor (*Figure 12.13*), with the silicate weathering rate distribution showing only marginal improvements, while the distributions for soil depth and silicate weathering rate both became substantially worse (*Figure 12.13* and *Table 12.5*). There were no significant correlations introduced between any of the parameters.

Case 5: Base saturation in the upper soil layer poorly known

Based solely on the accuracy of model predictions, the MCFP would appear to have put in its best performance in this Case, even though predictions only passed the test criteria when there was no change in deposition (*Tables 12.1, 12.2,* and *12.3*). However, the absence of significant change in any of the parameter distributions suggests that the accuracy of the predictions is due to the lack of influence base saturation has on predicted pH, or because base saturation is defined as the fraction of the cation exchange capacity being occupied by base cations and as long as the base saturation remains above 0.05–0.10 (as it did in the true system), soil pH will remain between 4.2 and 5.0.

In this Case the expected value of the *a priori* base saturation distribution was substantially greater than the true expected value so that, with no change in deposition, the calibrated model should accurately simulate the true system. However, because the MCFP did make some small changes in the soil depth distribution, which affects the total amount of silicate buffering, model predictions differed from true regional behavior when deposition was greatly reduced. The MCFP did create correlations of 0.13 between soil depth and silicate buffer rate and −0.28 between soil depth and pCO_2, both of which serve to increase acidity and are the result of the MCFP compensating for the higher than true expected value of base saturation.

Case 6: All four parameters poorly known

The MCFP performed badly when all four parameters were poorly known, and none of the predictions passed the test criteria (*Tables 12.1, 12.2*, and *12.3*). A comparison of the *a posteriori* and true parameter distributions (*Figure 12.15*) reveals that the MCFP improved the pCO_2 and soil depth distributions, but worsened that of silicate weathering rate. While the MCFP did not create any significant parameter correlations, the correlation between $[H^+]$ and pCO_2 (*Table 12.4*) is much larger than in the true Case (0.7 as opposed to 0.1). This is largely the result of the MCFP compensating for the higher – than true – values of soil depth, silicate weathering rate, and base saturation.

12.5. Conclusions

At first glance, the MCFP does not appear to have done very well in the above evaluation. When calibrated by the MCFP, the RLM model predictions consistently failed the test criteria. The MCFP was as likely to "calibrate" the model by introducing errors into initially correct parameter distributions as it was to improve inaccurate distributions. In defense of the MCFP, the tests applied to the model predictions were rather stringent – more stringent than might be needed by some users of such a model. An alternative and somewhat more forgiving null hypothesis to test would be that the mean pH, pH standard deviation, and the percentage of lakes with pH < 5 are assumed to be within a given percentage of the true values.

The examination of parameter–parameter correlation coefficients along with the comparisons of the true, *a priori*, and *a posteriori* parameter distributions has revealed a serious weakness in the MCFP as applied to the RLM model: the MCFP has many degrees of freedom, but only the distribution of pH at one point in time to serve as a constraint. As a result, the MCFP is able to choose sets of parameters that produce a fit for the 1980 pH distribution, but may exhibit response characteristics and neutralization capacities much different from those of the true system. When soil depth   the most influential of the parameters – was the only poorly known parameter, the MCFP was able to produce good parameter fits and the model predictions came close to true regional behavior. Whereas when pCO_2 – a less influential parameter – was poorly known, the MCFP matched the 1980 pH distributions by adjusting the pCO_2 distribution as well as the already correct soil depth distribution. This suggests a number of changes that might improve the MCFP's performance:

- Having at least one output variable for, and strongly correlated with, each poorly known parameter.
- Assigning single values to poorly known parameters that do not have such a corresponding output variable.
- Using pH distributions, or any output variable, from more than one point in time (e.g., 1980 and 1986) to filter the predicted pH distributions.

The basic objective of these changes is to equalize the number of degrees of freedom with the number of independent constraints, by eliminating parameters (degrees of freedom) whose values cannot be clearly linked to a particular aspect of the model's predictions. Having the MCFP filter of the model's output distributions at more than one point in time should increase its ability to discriminate between parameters associated with equilibrium and rate-based processes.

Use of a synthetically generated system to evaluate calibration procedure performance has a number of applications. One such use would be to estimate the amount of predictive error inherent in the calibration procedure, which could also be thought of as maximum achievable model accuracy given perfect information. Knowing the maximum accuracy a particular calibration procedure is likely to achieve can be helpful in estimating the value of collecting additional data. Identifying the situations in which the calibration procedure can be expected to succeed or fail can guide future development of the calibration procedure, the acquisition of data to be used for calibration, or both.

Acknowledgments

This research was made possible by resources provided by the International Institute for Applied Systems Analysis (IIASA) and the National Science Foundation (ECE-8552712) through Carnegie-Mellon University. Discussions with and comments from M.J. Small, J. Kämäri, M. Posch, and J.-P. Hettelingh were of great assistance. The views or opinions expressed have received only limited review at IIASA and do not necessarily represent those of IIASA or its National Member Organizations.

Appendix 12A: Optimal Filtering of Monte Carlo Simulations

M. Posch and M.C. Sutton

The primary objective of the filtering procedure is to reject a number of model simulations from each pH interval so that the resultant *cdf* of the model pH simulations will match the observed *cdf* at the interval boundaries. The secondary objective is to reject as few Monte Carlo simulations as possible in achieving this match.

To achieve these objectives, the *cdf* of the observations is first divided by $(n+1)$ boundaries (*Figure 12.3*). Let N_i be the number of simulations falling between the $(i-1)$th and the ith boundary, and the interval between these two boundaries will be referred to as the ith bin. The approach is to derive a formula for the maximum number of *accepted simulations* (a_i) in each bin, such that the simulated and observed *cdf*s are the same. This requires that:

$$cdf_1 = \frac{a_1}{A}$$

$$cdf_2 = \frac{a_1 + a_2}{A}$$

$$cdf_3 = \frac{a_1 + a_2 + a_3}{A} \tag{12A.1}$$

$$\cdot$$
$$\cdot$$
$$\cdot$$

$$cdf_n = \frac{a_1 + \ldots + a_n}{A}$$

where

$$A = \sum_{i=1}^{n} a_i$$

is the total number of accepted simulations. There are n equations for n unknowns $a_1, \ldots, a_n$, but the nth equation in equation (12A.1) is an identity $(1 = 1)$, so that there are in effect only $(n-1)$ equations. This means that there is one degree of freedom to fix. It is clear from the definitions that the constraint on the values a_i exist:

$$0 \le a_i \le N_i, \quad i = 1, \ldots, n \quad . \tag{12A.2}$$

Next the $(i-1)$th equation is subtracted from the ith equation in (12A.1) and, after rearranging, the following result is obtained:

$$x_i = \frac{a_i}{cdf_i - cdf_{i-1}} = A, \quad i = 1, \ldots, n \tag{12A.3}$$

which introduces a new unknown, x_i. Equation (12A.3) can also be written as

$$x_1 = x_2 = \ldots = x_n = A \tag{12A.4}$$

showing that all of the x_is are identical to the total number of accepted runs. The constraints for the x_is read [compare with equation (12A.2)]

$$0 \leq x_i \leq \frac{N_i}{cdf_i - cdf_{i-1}}, \quad i = 1, \ldots, n \quad . \tag{12A.5}$$

Now let M be the minimum of the upper bounds for x_i:

$$M = \min \left\{ \frac{N_i}{cdf_i - cdf_{i-1}} \right\}, \quad i = 1, \ldots, n \quad . \tag{12A.6}$$

Then it follows from equation (12A.4) that

$$0 \leq x_i \leq M \quad \text{for all } i \quad . \tag{12A.7}$$

Now the surplus degree of freedom of the a_is is fixed in such a way that the a_is (number of accepted simulations) is maximized. From the definition of the x_is [see equation (12A.3)], it follows that the x_is have to be equal; and from equation (12A.7), it follows that

$$x_i = M \quad \text{for all } i \tag{12A.8}$$

which means that the optimal a_is are given by [see equation (12A.3)]

$$a_i = (cdf_i - cdf_{i-1})M, \quad i = 1, \ldots, n \tag{12A.9}$$

with M given by (12A.6).

The generalization to more dimensions is straightforward and is mainly a matter of notation and proper interpretation of the one-dimensional result. Note that $cdf_i - cdf_{i-1} = \Delta(cdf_i)$ is the number of observations in bin i divided by the total number of observations $\sum_{i=1}^{n} cdf_i = 1$. Therefore, the following result is obtained for the d-dimensional case. The maximum number of accepted simulations for bin $(i_1, \ldots, i_d)$ is given by

$$A_{i_1, \ldots, i_d} = \Delta(cdf_{i_1, \ldots, i_d}) M \tag{12A.10}$$

where

$$M = \min \left\{ \frac{N_{i_1, \ldots, i_d}}{\Delta(cdf_{i_1, \ldots, i_d})} \right\}, \quad i_k = 1, \ldots, n_k, \quad k = 1, \ldots, d \quad . \tag{12A.11}$$

Here $N_{i_1, \ldots, i_d}$ denotes the number of simulations for bin $(i_1, \ldots, i_d)$ and n_k is the number of classes of variable k.

CHAPTER 13

A Regional Model of Surface-Water Acidification in Southern Norway: Uncertainty in Long-Term Hindcasts and Forecasts

G.M. Hornberger, B.J. Cosby, and R.F. Wright

13.1. Introduction

Mathematical models of surface-water acidification, used in conjunction with data from regional surveys of water chemistry at one point in time to obtain long-term hindcasts and forecasts for an entire region, can be an important tool in assessing the effects of atmospheric acid deposition on aquatic systems. We have developed a Monte Carlo-based procedure that permits such a regional application of conceptual water quality models. The procedure selects those "acceptable" simulations from a sequence of Monte Carlo trials (model input parameters selected from prespecified distributions) whose output variables fall within "windows" defined by the survey data. The accumulated ensemble of acceptable simulations (those simulations that are consistent with the observed data) can be used to "calibrate" the model for the region of interest (Cosby *et al.*, Chapter 7; Hornberger and Cosby, 1985b; Hornberger *et al.*, 1986a, 1986b, 1987). The procedure can also be used to estimate input parameter sensitivities in regional applications of the model (Hornberger *et al.*, 1986a; Hornberger and Cosby, 1985a).

The regionalization procedure we use is similar in many respects to modeling protocols that have been applied to other studies in earth sciences and ecology. In essence, the procedure substitutes space for time as the dimension along which data are collected to provide constraints in calibrating the model (cf. Chapter 7). The procedure uses spatially distributed survey data taken at a

J. Kämäri (ed.), Impact Models to Assess Regional Acidification, 233–252.
© 1990 *International Institute for Applied Systems Analysis.*

single point in time instead of a temporal sequence of data taken at a single site. No attempt is made to preserve a one-to-one correspondence between any particular simulation and any particular system in the survey. Rather, the regionalization procedure attempts to match the bulk statistical properties of the simulation ensemble to those of the survey data set.

Given that the intent of the procedure is not to simulate any *particular* system in the survey set, individual simulations can be judged "acceptable" only if all variable values for that simulation lie completely within the range of values observed in the survey. The decision to retain or discard any individual simulation is binary (yes or no) and can be made only by reference to the maxima and minima of the observed data. Comparing a simulation result with the means and variances of the observed data, for instance, does not provide an unequivocal decision of acceptability. Only when a large number of simulations have been retained can the *distributions* of variables for the acceptable ensemble be compared with the observed distributions.

In some instances, the "fit" of the distributions produced by the binary "filter" may be adequate, but in general one cannot depend on fitting the multivariate joint distribution of measured chemical concentrations for a region using this simple method of selecting individual simulations. This problem may be overcome in one of two ways:

(1)   The shapes of the input parameter distributions used in the Monte Carlo simulations can be modified to try to match the observed joint distribution of the water quality variables (a trial-and-error procedure; see Chapter 7).

(2)   A subset of the acceptable simulations can be selectively chosen to match elements of the histograms that approximate the observed joint distribution of the water quality variables (a "bin filling" procedure; see Kämäri *et al.*, 1986; Posch and Kämäri, Chapter 9).

Each method may result in an adequate fit to the data, but each can be criticized as being *ad hoc*. That is, neither procedure results in an estimate of how unique or uncertain is the chosen "fit."

We undertake here an investigation of uncertainty in long-term hindcasts and forecasts of surface-water acidification in southern Norway using MAGIC (Model of Acidification of Groundwater In Catchment; Cosby *et al.*, 1985a, 1985b, 1985c). In a companion analysis (Chapter 7), we describe the conceptual model and the survey region in southern Norway; present the regionalization protocol and use the trial-and-error procedure to calibrate the model input parameter distributions for the survey region; describe a validation procedure that uses forecasts and hindcasts from the calibrated model with historical fisheries and resurvey data; and discuss the robustness of and possible biases in the calibration–validation procedures. In this chapter, we implement the "bin filling" calibration methodology; develop a resampling ("bootstrap") scheme to evaluate uncertainties in calibration, forecasts, and hindcasts; provide a brief sensitivity analysis for the input parameters; and present forecasts for the region using a different (longer-term) scenario of future deposition than was used in Chapter 7.

13.2. Analysis of Uncertainty

A quantitative analysis of the uncertainty in the simulation of the calibration data set (and uncertainty in forecasts and hindcasts) depends upon some measure of uniqueness of the final calibration values. *Ad hoc* calibration procedures, such as the trial-and-error and bin filling procedures described in this chapter and Chapter 7, make such analyses difficult. For either procedure, alternative calibrations can be found, against which the final calibration can be compared. There is no objective way, however, to decide on the likelihood of one calibration versus another. Trial-and-error procedures result in a range of values of a goodness-of-fit criterion. There is no assurance, however, that any of the achieved fits will be an exact (or very nearly exact) match to observed distributions. Bin filling procedures, on the other hand, always provide exact (or very nearly exact) matches to the observed distributions. The problem here is that many "exact" fits can be achieved, depending on which simulations are retained and which are discarded.

An obvious way to estimate the uncertainty in a bin filling calibration is to extend the selection of the Monte Carlo trials that fit the histograms defined by the measured chemical concentrations. That is, one can accumulate a large number of realizations that fall within the broad data window and then resample from these a number of times to match the observed joint histogram. The result is a number of different "fits" to the data. These can be used to infer – to bootstrap – an estimate of the uncertainty inherent in fitting the mathematical model to the regional data set.

13.3. Description of Model and Region of Application

We have developed a dynamic model of catchment response to acidic deposition called MAGIC (Cosby *et al.*, 1985a, 1985b, 1985c). MAGIC consists of:

(1)   A section in which the concentrations of major ions are assumed to be governed by simultaneous reactions involving sulfate adsorption, cation exchange, dissolution/precipitation of aluminum, and dissolution/speciation of inorganic carbon.
(2)   A mass balance section in which the flux of major ions to and from the soil is assumed to be controlled by atmospheric inputs, chemical weathering inputs, net uptake in biomass, and losses to runoff.

At the heart of MAGIC is the size of the pool of exchangeable base cations in the soil. As the fluxes to and from this pool change over time, owing to changes in atmospheric deposition, the chemical equilibria between soil and soil solution shift to give changes in surface-water chemistry. The degree and rate of change in surface-water acidity thus depend both on flux factors and the inherent characteristics of the impacted soils. Chapter 7 gives more details of the equilibrium and flux reactions incorporated in the model.

Application of MAGIC requires data for soil physical and chemical characteristics, rainfall/runoff characteristics, precipitation chemistry covering the
period from the onset of acidic deposition to the present, and estimates of base
cation weathering rates. For this regional application of the model, the distributions of these inputs for southern Norway are needed. Unfortunately, data do
not exist that adequately describe *a priori* the distributions of these inputs. The
regional application of MAGIC must be accomplished by "calibrating" some of
the input distributions for the model – particularly the weathering rates and certain soil physical–chemical properties. This procedure is described briefly below.
The derivations of distributions of average annual rainfall/runoff, soil and surface temperatures, soil and surface-water CO_2 partial pressures, and the historical sequence of precipitation chemistry for southern Norway are described in
Chapter 7. In this regional application, MAGIC is implemented using yearly
time steps and a single aggregated soil layer. Simulations are run for 130 years
(1844–1974) to reconstruct the historical acidification of lakes in the region.

Southern Norway is an area that has been severly impacted by atmospheric
acidic deposition. Surface waters are diluted and inherently sensitive to acid
deposition. In 1974–1975, water samples were obtained for 715 lakes in the
region to determine the major ion chemistry of these surface waters (Wright and
Snekvik, 1978; Cosby *et al.*, Chapter 7). Here, we consider 674 of the 715 surveyed lakes.

The region is underlain by granites and felsic gneisses. Soils are young,
podzolic, and generally thin and patchy. Information on soil depths, bulk densities, cation exchange capacities, and base saturations are derived from several
studies in southern Norway (the data and sources are summarized by Cosby *et
al.*, Chapter 7). The chemistry of precipitation in southern Norway can be depicted as a mixture of marine aerosols (Cl, Na, Mg, and to a lesser extent K, Ca,
and SO_4) and long-range transported anthropogenic pollutants (H^+, SO_4, NO_3,
and NH_4) (Wright and Dovland, 1978). These components are also deposited by
dry deposition and impaction. The concentrations of sea salt and anthropogenic
components in precipitation decrease sharply with distance from the coast. We
chose to divide the set of lakes into two classes corresponding to low sea salt
inland lakes (Cl < 100 μeq l^{-1}) and high sea salt coastal lakes (Cl > 100 μeq l^{-1}).
The regionalized model is applied to each class of lake as described below. The
results for the two classes are combined to produce the final model for southern
Norway.

13.4. Monte Carlo Methods

Fourteen parameters are varied in the Monte Carlo simulations. The first 11
parameters are related to soil hydrochemical properties. These are the annual
runoff (Q_p); the cation exchange capacity (CEC, intended here as the total area
cation exchange capacity of a system – a product of depth, bulk density, and
measured exchange capacity per unit mass); the logarithm of the solubility constant of the aluminum hydroxide solid phase (K_{Al}); the weathering rates of the
four base cations (WE_{Ca}, WE_{Mg}, WE_{Na}, and WE_K); and the logarithms of the

selectivity coefficients that specify the cation exchange affinity of soils for aluminum and the four base cations (S_{AlCa}, S_{AlMg}, S_{AlNa}, and S_{AlK}). The remaining
parameters (Cl_{atm}, SO_4^*, and NO_3^*) describe the atmospheric deposition for each
simulation. Deposition is determined by selecting a value for Cl_{atm}, using sea
salt ratios to calculate deposition of base cations and background sulfate based
on that Cl_{atm} value, and then incorporating additional sulfate (excess sulfate,
SO_4^*) and nitrate (excess nitrate, NO_3^*) to represent anthropogenic deposition of
acidic compounds.

The particular joint distribution of the 11 soil input parameters that is
"correct" for this region of southern Norway cannot be decided *a priori*. Our
approach is to define broad ranges for acceptable values of the individual soil
parameters, based on available soil data from this and other similar regions. All
soil parameter values are selected from rectangular, independent distributions
contained within these broad ranges (*Table 13.1*). The distributions of the soil
parameters are the same for each Cl class. The input distributions of the three
deposition parameters are determined from the survey data set, as described in
Chapter 7. The distributions of the deposition parameters are triangular, dependent, and differ for each Cl class.

Table 13.1. Ranges for the 11 soil hydrochemical parameters used in the Monte Carlo
simulations for southern Norway. The parameters are defined in the text. The ranges
are the same for each Cl class. The maxima and minima define independent, rectangular
distributions from which the soil parameters were sampled in the Monte Carlo simulations.

Parameter	Units	Minimum	Maximum
Q_p	cm	50.0	300.0
CEC	eq m^{-2}	10.0	500.0
WE_{Ca}	meq m^{-2} yr^{-1}	0.0	100.0
WE_{Mg}	meq m^{-2} yr^{-1}	0.0	30.0
WE_{Na}	meq m^{-2} yr^{-1}	0.0	30.0
WE_K	meq m^{-2} yr^{-1}	0.0	10.0
S_{AlCa}	logarithm (base 10)	−1.0	5.0
S_{AlMg}	logarithm (base 10)	−1.0	5.0
S_{AlNa}	logarithm (base 10)	−3.0	3.0
S_{AlK}	logarithm (base 10)	−3.0	3.0
K_{Al}	logarithm (base 10)	8.0	11.0

Given the distributions of input parameters for each Cl class, the regionalized model is calibrated by an *ad hoc* bin filling procedure. Soil and deposition
parameter values are selected at random from their specified distributions.
Simulations are run from 1844 to 1974. The simulated 1974 results are compared with the range of concentrations for each major ion observed in the
1974–1975 survey (*Table 13.2*). If the simulated concentrations fall within the
observed windows (defined by the maximum and minimum of the observed
data), the simulation is accepted; if they do not, the simulation is considered

unacceptable as a representation of a catchment in the region. This sampling–simulation procedure is repeated many times for the specified input distributions. When a large number of "acceptable" simulations are accumulated, the bootstrap stage of the Monte Carlo procedure is begun.

Table 19.2. Maximum and minimum values of lake chemistry variables observed in the lake survey of southern Norway. The survey data are divided into two classes based on Cl concentrations. These maximum and minimum values define "acceptable" simulations for each Cl class (see text). All are in $\mu eq\ l^{-1}$, except soil base saturation (BS %). The high-Cl class contains 342 lakes; the low-Cl class contains 332 lakes.

Variable	High-Cl class		Low-Cl class	
	Maximum	Minimum	Maximum	Minimum
Ca	140	5	100	5
Mg	120	20	55	5
Na	315	80	125	20
K	25	2	25	1
SO_4	210	45	180	10
Cl	350	100	100	10
NO_3	30	0	20	0
H^+	50	0	50	0
CALK[a]	80	−70	80	−70
SBC	600	115	300	45
SAA	560	165	240	25
BS %	15	1	15	1

[a]Alkalinity is calculated from ionic balance, CALK = SBC − SAA where SBC is the sum of base cation concentrations (Ca + Mg + Na + K) and SAA is the sum of acid anion concentrations (SO_4 + NO_3 + Cl).

The joint distribution of the observed sulfate, calcium, and alkalinity concentrations for each Cl class is approximated by a three-way joint histogram containing three intervals on each univariate axis. This produces a total of 27 "bins" (3 × 3 × 3) defined by the upper and lower concentrations of each of the three variables in each of the three intervals. The number of surveyed lakes in every bin is determined by stepping through the survey data set and comparing the concentrations of sulfate, calcium, and alkalinity for each lake with the limits of each bin (*Table 19.3*). The intervals on the univariate axes are selected to give approximately equal numbers of lakes in each *marginal* histogram element (*Table 19.3*). Because of strong correlations in the data, however, the numbers of lakes in the *joint* histogram elements vary widely, and some elements are devoid of lakes (*Table 19.3*).

These rather coarse 27-bin joint histograms (one for each Cl class) of the observed data are the "templates" used for calibrating the model. A higher resolution could be obtained by employing more intervals for each variable or by using more variables to define the joint histogram or by both; but, as indicated below, the coarse histogram provided adequate constraint to achieve a reasonable calibration. Despite the fact that the joint distribution of only three

Table 13.3. The number of observed lakes contained in each element of the joint histograms of observed sulfate, calcium, and alkalinity concentrations for each Cl class.

Sulfate	Calcium	High-Cl class			Low-Cl class		
		Alkalinity			Alkalinity		
Int	Int	Int 1	Int 2	Int 3	Int 1	Int 2	Int 3
1	1	32	43	15	5	38	20
	2	1	11	9	1	14	17
	3	0	0	2	0	0	10
2	1	11	3	0	37	10	1
	2	25	21	14	10	26	5
	3	4	8	21	0	4	23
3	1	0	0	0	5	0	0
	2	26	3	3	32	8	0
	3	10	19	45	17	19	29

The intervals (int) and numbers of lakes in each interval of the marginal histograms of the observed data are:

	Int	Sulfate	Calcium	Alkalinity	
High-Cl class	1	40 to 94	0 to 38	−80 to −28	μeq l^{-1}
		113	104	109	Lakes
	2	95 to 124	39 to 63	−27 to −5	μeq l^{-1}
		107	113	108	Lakes
	3	125 to 210	64 to 140	−4 to 80	μeq l^{-1}
		106	109	109	Lakes
Low-Cl class	1	0 to 37	0 to 17	−80 to −11	μeq l^{-1}
		105	116	107	Lakes
	2	38 to 57	18 to 28	−12 to 7	μeq l^{-1}
		116	113	119	Lakes
	3	58 to 180	29 to 100	8 to 80	μeq l^{-1}
		110	102	105	Lakes

variables was used (in addition to the window constraint on all variables), the simulated distributions of the remaining variables, also described below, agree reasonably well with the observed distributions. The choice of the particular three variables, however, is crucial and is discussed in Section 13.5.

Given the template for a Cl class, the accumulated ensemble of acceptable simulations for that class is randomly sampled without replacement. Each sampled simulation is assigned to an appropriate joint histogram bin, depending on the values of simulated sulfate, calcium, and alkalinity. If the number of simulations already assigned to a bin is less than the number of lakes observed in that bin, the simulation is retained; otherwise, the simulation is discarded. The random sampling continues until each bin contains exactly the same number of simulations as observed lakes. The joint distribution of input parameters associated with simulations retained in this manner form a "calibrated" joint input distribution for the model. Likewise, the joint and marginal distributions of output variables (for 1974 or any year) retained in this manner form a "calibrated" output of the model.

The random sampling and matching procedure can be repeated many times, starting with the complete acceptable ensemble of simulations, but using a new random sampling sequence. This provides a suite of bootstrapped calibrations, any one of which fits the data equally well (in the sense of the histogram template). Note that the number of simulations assigned to each bin in this procedure can be greater than the number of lakes – say, n times the number of lakes – as long as the same n is used for each bin. We chose $n = 1$ to simplify and to reduce the number of simulations needed in the acceptable ensembles to assure exact matches.

After the two Cl classes have each been calibrated, pairs of bootstrapped calibration ensembles (one from each class) are combined to produce a suite of "final calibrations" of the model for the region. The bootstrap procedure assures that ensembles from each class are appropriately weighted by the number of surveyed lakes in each Cl class (*Table 13.2*).

Given a suite of many calibrated parameter or variable distributions, the uncertainty in either parameters or variables of the regional model can be examined in a number of ways. For instance, the uncertainty over the whole range of values of a parameter or variable can be displayed by plotting all of the calibrated frequency distributions for that parameter or variable on one graph. The resulting graph of many partially overlapping lines will have a fuzzy appearance, and the width of the "fuzz" at any value will be a measure of the uncertainty in the frequency of occurrence of that value.

We present our results below by plotting the maximum and minimum frequencies of occurrence of different values of parameters or variables in the suite of final calibrations. The two lines that appear on each frequency diagram can be taken as uncertainty bounds for the distribution of the parameter or variable. The analysis protocol dictates that any value between the lines is as likely as any other – i.e., of the many calibrations achieved by this method, all are equally good, see discussion below. Thus, in comparing simulated distributions with observed distributions, the simulated distributions are plotted as a band and the observations as a single line. To the extent that the observed line lies within the simulated band, the fit of the model is considered "accurate." To the extent that the simulated band is narrow, the model is considered "precise" (not uncertain). As is usually the case, both accuracy and precision are desired.

13.5. Results

The first stage of the Monte Carlo procedure involves the generation of an acceptable ensemble of simulations for each Cl class. We ran 8,000 simulations for each class. Those simulations were retained that satisfied the window constraints (i.e., the simulated values of all output variables were within the observed ranges given in *Table 13.2*). Acceptable ensembles of 3,008 and 2,572 simulations were obtained for the low- and high-Cl classes, respectively. The joint distribution of simulated sulfate, calcium, and alkalinity for the acceptable ensemble of each Cl class was approximated by a three-way joint histogram containing three intervals on each univariate axis (*Table 13.4*). The intervals on

Table 13.4. The number of acceptable simulations contained in each element of the joint histograms defined by observed sulfate, calcium, and alkalinity concentrations for each Cl class.

Sulfate	Calcium	High-Cl class			Low-Cl class		
		Alkalinity			*Alkalinity*		
Int	*Int*	*Int 1*	*Int 2*	*Int 3*	*Int 1*	*Int 2*	*Int 3*
1	1	76	179	261	8	76	85
	2	2	32	289	0	26	131
	3	0	1	144	0	7	417
2	1	209	141	92	77	76	41
	2	40	92	198	36	76	81
	3	2	21	235	7	88	542
3	1	146	64	27	138	28	11
	2	87	61	70	107	43	24
	3	30	50	203	150	208	525

The numbers of simulations in each interval of the marginal histograms. The intervals (int) are defined in *Table 13.3*.

	Int	*Sulfate*	*Calcium*	*Alkalinity*
High-Cl class	1	988	1,199	591
	2	1,030	869	642
	3	734	684	1,519
Low-Cl class	1	745	536	527
	2	1,027	527	627
	3	1,236	1,945	1,854

each axis were the same as those used in constructing the joint histogram of the observed data for each class (*Table 13.3*).

Comparing the simulated and observed joint histograms, one can see that differences exist in both the joint and marginal fits of the acceptable simulation ensembles to the observed data. As described in Chapter 7, the Monte Carlo simulations that satisfied only the rectangular window constraints did not provide reasonable fits with the measured 1974 distributions. In particular, the marginal (univariate) distribution of calculated alkalinity was not well reproduced by the realizations that match the windows, and the simulated distribution of calcium was displaced toward higher values relative to that observed (*Tables 13.3* and *13.4*). Further adjustments were necessary, and the bin filling procedure was implemented (see Section 13.4).

Because one can argue that alkalinity is the single most important variable to consider in the assessment of effects of acidic deposition on surface waters, we chose this as a "master variable" to fit. We first attempted to remedy the deficiencies that arise in using only the binary window acceptance criterion by resampling to fit the marginal distribution of simulated alkalinity to the marginal distribution of 1974 alkalinity observations. Our reasoning was that, by fitting to this composite variable, we might also move other variables toward a better match with the observed distributions. Results proved this to be incorrect. Although the alkalinity distribution was fit nearly perfectly (10 histogram bins were used along the univariate alkalinity axis in this first attempt), the fit of the simulated distribution of calcium, for example, was worsened.

We next resampled to match the bivariate joint distribution of calculated alkalinity and calcium (using five histogram bins on each univariate axis). This procedure also yielded unacceptable results because of a strong correlation between calcium and sulfate in the simulated data. The fit to the calcium distribution was improved at the expense of significantly worsening the fit to sulfate. Thus, we arrived at our final procedure: resampling to fit the simulations to the joint distribution of alkalinity, calcium, and sulfate as measured in 1974.

Results for one application of the three-variable bin filling procedure show how well the simulated distributions match the observations (*Figure 13.1*). The simulated alkalinity distribution, although forced to fit in only three intervals in the procedure, matches the measured distribution very well on a much finer resolution. Distributions of the aggregate variables – sum of base cations and sum of strong acid anions – are also reproduced very well even though only one base cation (Ca) and one acid anion (SO_4) were used to trim the distributions (*Figure 13.1*). Even though the calibration procedure was implemented using only a coarse joint histogram of three observed variables, the correlation matrix for all simulated variables is remarkably similar to the observed correlation matrix [with the exception of NO_3; see *Table 13.5(a), (b)*].

Given the success of one application of the three-variable bin filling procedure, we next implemented the resampling scheme. Thirteen bootstrapped calibration ensembles were generated for each Cl class and combined as described above. The ranges for each parameter or variable marginal distribution (expressed as maxima and minima of the 13 resamplings) and the correlation structure induced in the parameters and variables by the calibration procedure (see Chapter 7) define the calibrated model for southern Norway.

The marginal distributions of the calibrated input parameter set can be used to infer the sensitivities of the regional model to input parameter values [(*Figure 13.2(a), (b), (c)*]. For instance, the calibrated distributions of WE_{Ca} [*Figure 13.2(b)*] and K_{Al} [*Figure 13.2(c)*] are markedly triangular even though the *a priori* distributions for these parameters are rectangular. This indicates that the model is relatively sensitive to WE_{Ca} and K_{Al} in that high (or low) values of these parameters produce simulations that are clearly not consistent with observed data. Similar triangular tendencies are apparent in the rainfall/runoff parameter [Q_p, *Figure 13.2(a)*], other weathering parameters [WE_{Mg}, WE_{Na}, and WE_K; *Figure 13.2(b)*], and the soil cation exchange parameters [with the exception of S_{AlK}; *Figure 13.2(c)*]. These findings agree with those of Cosby *et al.* (Chapter 7 in this volume), who used *a priori* triangular parameter distributions in a trial-and-error calibration of MAGIC for southern Norway. The calibrated distribution of S_{AlK} [*Figure 13.2(c)*] remained essentially rectangular, indicating that the model is insensitive to this parameter that controls the affinity of the soil for potassium. Any value of this parameter can produce simulations that agree with the observed data.

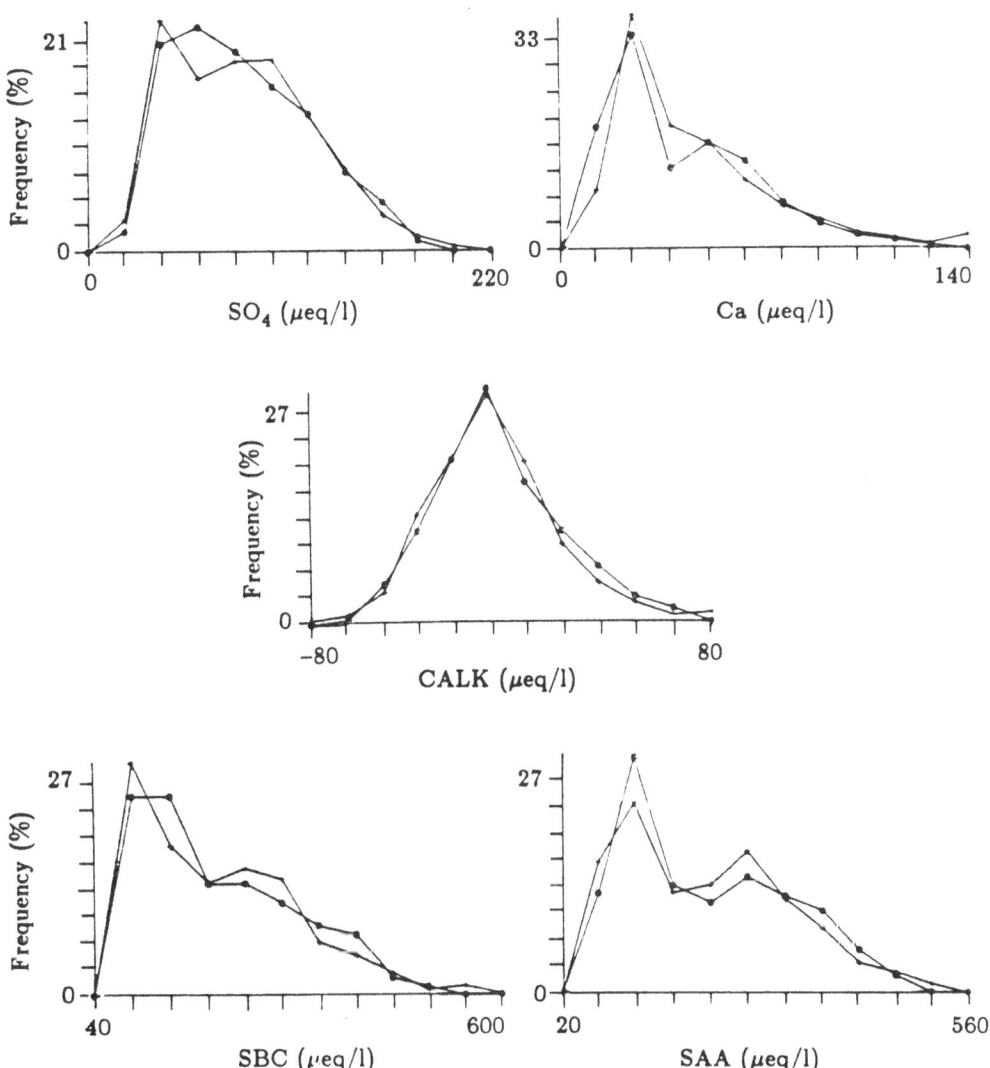

Figure 13.1. Simulated (•) and observed (+) distributions of calcium, sulfate, and alkalinity from the first calibration of the model (upper three panels). These variables were
used to define the histogram for the bin filling calibration procedure. The lower two
panels show the simulated and observed distributions of the aggregate variables: sum of
base cations and sum of acid anions.

The triangular shapes of the deposition parameter distributions $[\text{Cl}_{\text{atm}}$,
SO_4^*, and NO_3^*; *Figure 13.2(a)*] cannot be interpreted in the same manner. The *a
priori* distributions of these parameters were triangular and determined by the
survey data (see Chapter 7). The ranges of the calibrated distributions shown in
Figure 13.2(a) encompass the data-derived *a priori* distributions, leading to the
conclusion that the bin filling and bootstrap procedures did not result in biased

Table 13.5(a). Correlation matrix (r) for the chemistry variables observed in the lake survey of southern Norway.

	Ca	Mg	Na	K	SO$_4$	Cl	NO$_3$	H$^+$	CALK	SBC	SA
Ca	1.0	.71	.53	.73	.74	.53	.15	−.27	.58	.77	.62
Mg	−	1.0	.91	.71	.90	.90	.34	−.02	.15	.96	.93
Na	−	−	1.0	.61	.83	.99	.37	.12	−.06	.94	.97
K	−	−	−	1.0	.71	.62	.28	−.14	.30	.75	.68
SO$_4$	−	−	−	−	1.0	.83	.25	.17	−.01	.91	.92
Cl	−	−	−	−	−	1.0	.39	.15	−.13	.93	.98
NO$_3$	−	−	−	−	−	−	1.0	.11	−.16	.34	.39
H$^+$	−	−	−	−	−	−	−	1.0	−.67	−.02	.17
CALK	−	−	−	−	−	−	−	−	1.0	.18	−.10
SBC	−	−	−	−	−	−	−	−	−	1.0	.96
SAA	−	−	−	−	−	−	−	−	−	−	1.0

Table 13.5(b). Correlation matrix (r) for the chemistry variables simulated by the "calibrated" model.

	Ca	Mg	Na	K	SO$_4$	Cl	NO$_3$	H$^+$	CALK	SBC	SAA
Ca	1.0	.54	.56	.40	.68	.55	.00	−.25	.39	.72	.62
Mg	−	1.0	.87	.60	.80	.87	−.01	.02	.07	.91	.88
Na	−	−	1.0	.58	.84	.99	−.01	.16	−.07	.97	.98
K	−	−	−	1.0	.51	.56	.03	−.07	.17	.62	.57
SO$_4$	−	−	−	−	1.0	.83	−.03	.33	−.23	.88	.92
Cl	−	−	−	−	−	1.0	−.02	.23	−.14	.96	.98
NO$_3$	−	−	−	−	−	−	1.0	.07	−.08	−.01	.01
H$^+$	−	−	−	−	−	−	−	1.0	−.93	.05	.27
CALK	−	−	−	−	−	−	−	−	1.0	.06	−.18
SBC	−	−	−	−	−	−	−	−	−	1.0	.97
SAA	−	−	−	−	−	−	−	−	−	−	1.0

deposition inputs. That is, the calibrated deposition inputs for the model are consistent with the survey data.

The widths of the marginal distributions of the calibrated parameters provide a measure of the uncertainty in particular values of a parameter. Intuitively we expect that small uncertainties can be obtained only if the model is very sensitive to a parameter. That is, sensitive parameters are tightly constrained by the observed data. Accordingly, the *relative* sensitivities of the input parameters in this application of the model should be inversely proportional to their uncertainties, as indicated by the width of the calibrated marginal distributions in *Figures 13.2(a)*, (b), (c). By this reasoning we conclude, for example, that the regional model should be more sensitive to values K_{Al} than to values of *CEC* [*Figure 13.2(c)*], and more sensitive to values of SO_4^* than to values of Cl_{atm} [*Figure 13.2(a)*].

The ranges of the distributions of simulated water quality variables produced by the calibrated regional model closely match those measured in the lake survey for cations [*Figure 13.3(a)*], anions [*Figure 13.3(b)*], alkalinity, and pH

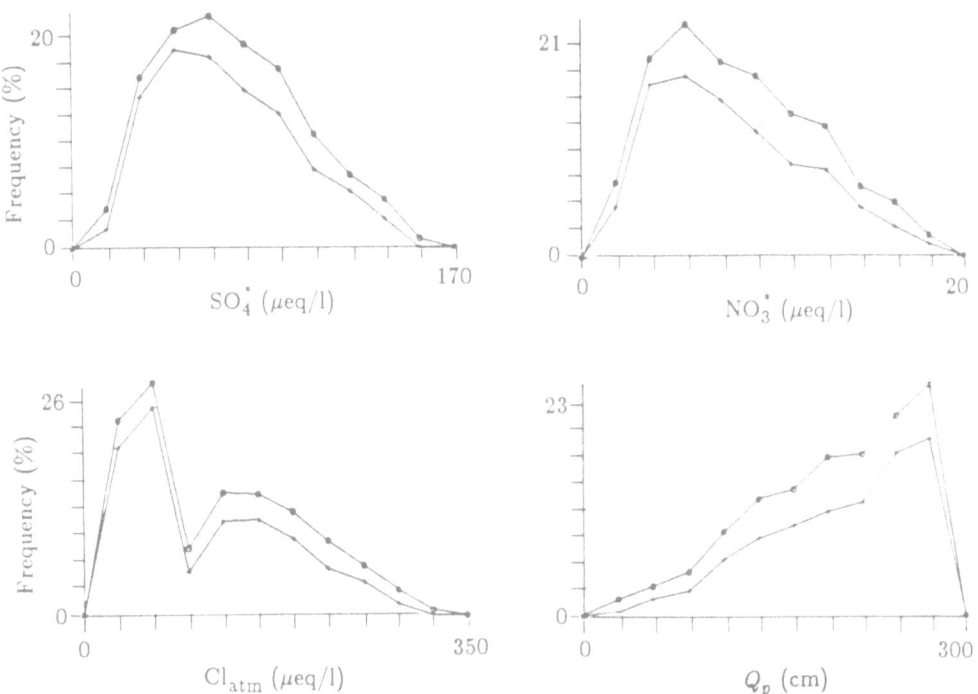

Figure 13.2(a). The ranges of the "calibrated" input distributions for the deposition and runoff parameters.

[*Figure 13.3(c)*]. The best variable fit (in the sense that the observed distribution lies almost completely within the simulated uncertainty region) is for alkalinity, the chosen master variable [*Figure 13.3(c)*]. The next best fits are for the aggregated anion and cation variables SAA and SBC. The fit of the simulated distribution of soil base saturation to the distribution derived from soil surveys in southern Norway (see site description) is remarkable in that no soil data were used in the bin filling calibration procedure.

The worst fit was for pH [*Figure 13.3(c)*]. This can be partially explained by the fact that the observed pH for nonacidic lakes in the survey is controlled to a large extent by the partial pressure of CO_2 (P_{CO_2}) in the lakes. The P_{CO_2} of natural waters tends to be greater than atmospheric, owing to CO_2 production by biomass respiration, and this results in lower pH readings for those waters. In the absence of measured P_{CO_2}, a fixed excess P_{CO_2} was assumed in the model for all lakes. The value chosen was probably too small. The fit of the pH distribution can be improved by raising the P_{CO_2} value used for the simulated lakes. In any case, the P_{CO_2} effect on pH is negligible for acidic lakes. We conclude that our estimated pH values for the acidic systems (the critical systems for evaluating the effects of reductions in deposition) are most likely correct.

Note that our characterization of good fits between simulated and observed variable distributions is entirely subjective. Given the large sample sizes of the

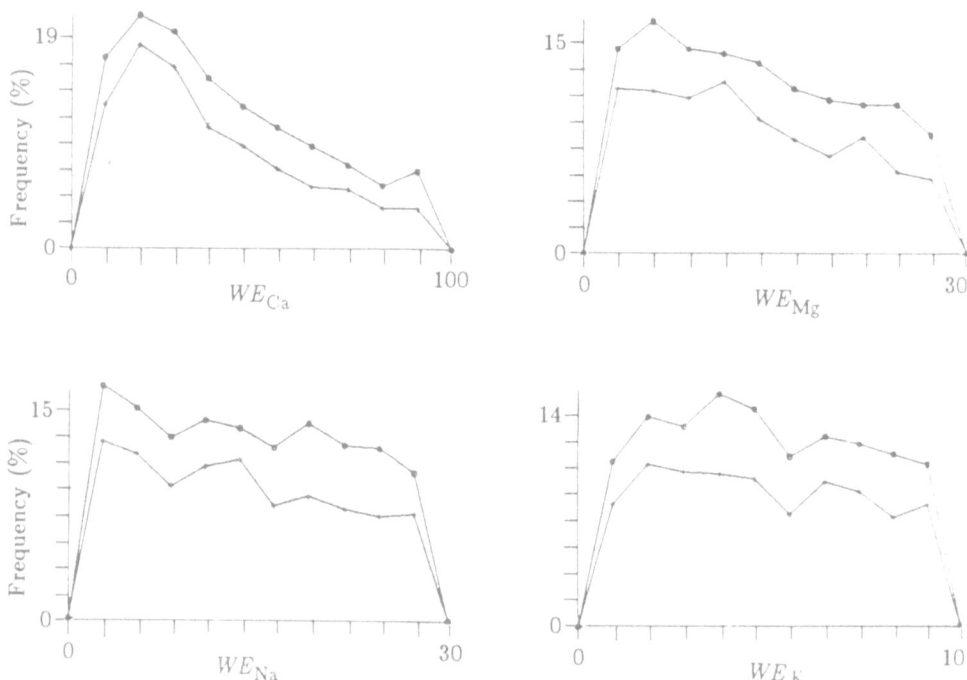

Figure 13.2(b). The ranges of the "calibrated" input distributions for the base cation weathering parameters.

survey data and model simulation ensembles, even small discrepancies in the distributions would prove to be statistically significant in an appropriate test. The purpose of the comparison in *Figure 13.3* is to demonstrate that the differences between the simulated and observed distributions are "operationally" small (even though they may be statistically significant), and the calibration, therefore, can be considered successful.

13.6. Discussion

The resampling approach to fitting the regional distributions of water chemical concentrations for southern Norway shows that the parameters in the regional application of MAGIC are tightly constrained by data defining the chemical composition of surface water at a time subsequent to significant increases in atmospheric loading, in this case 1974. This is evident in the results by the relatively narrow uncertainty bands for the distributions of parameters produced by the 13 resamplings [*Figure 13.2(a)*, *(b)*, *(c)*]. These bands would be expected to be large if many different suites of parameter values could match a given composition in 1974, whereas they would be expected to be narrow if a unique set of parameter values (with only small deviations about these values) is required to match a given composition.

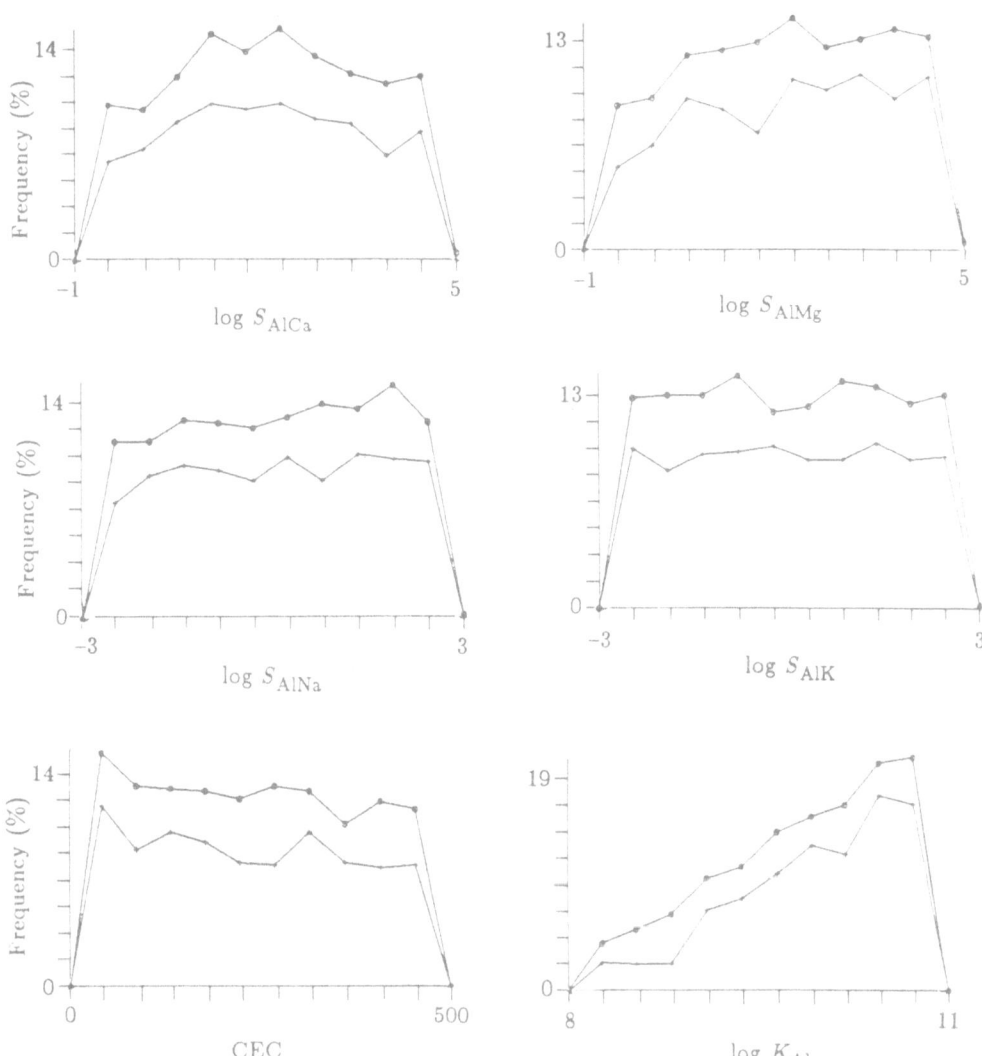

Figure 13.2(c). The ranges of the "calibrated" input distributions for the cation exchange parameters.

This observation is consistent with results obtained using automatic calibration methods for fitting MAGIC to selected catchments (Cosby *et al.*, 1987). In addition, the sensitivity analysis shows that all parameters being calibrated (with the possible exception of one soil parameter) are important in fitting the model to the survey data. This property of the model (i.e., the fact that it does not appear to be badly over-parameterized) limits the uncertainties due to parameter misspecification. Thus, the uncertainty bands shown in this chapter are all rather narrow. Of course, uncertainties that have not been taken into account here – notably, those owing to inadequacies in model structure and to

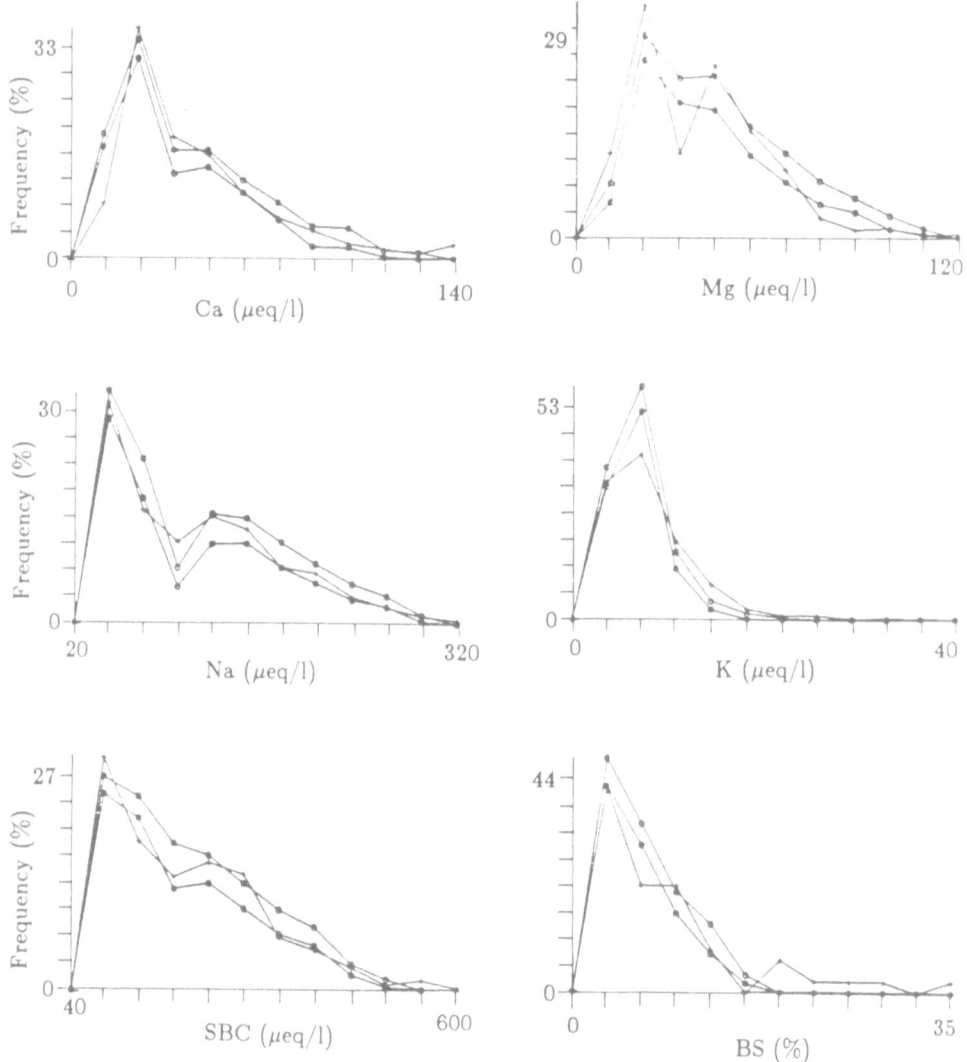

Figure 13.9(a). Simulated maximum and minimum ranges (•) and observed distributions (+) for 1974: base cation concentrations (upper four panels); sum of base cation concentrations (lower left panel); and soil base saturation (lower right panel).

lack of knowledge of historical inputs – would be expected to inflate the uncertainty estimates in this analysis.

Our resampling strategy relied on obtaining an excellent fit to alkalinity while keeping fits to other chemical constituents reasonable. The fact that fits to other variables are very good indeed suggests that the structure of MAGIC reproduces the dominant modes of behavior of these catchment systems. For example, the base saturation of the soil is not used in the resampling (and is only required to pass through a rather wide window for the original Monte Carlo simulations), yet the observed distribution of this variable is well simulated

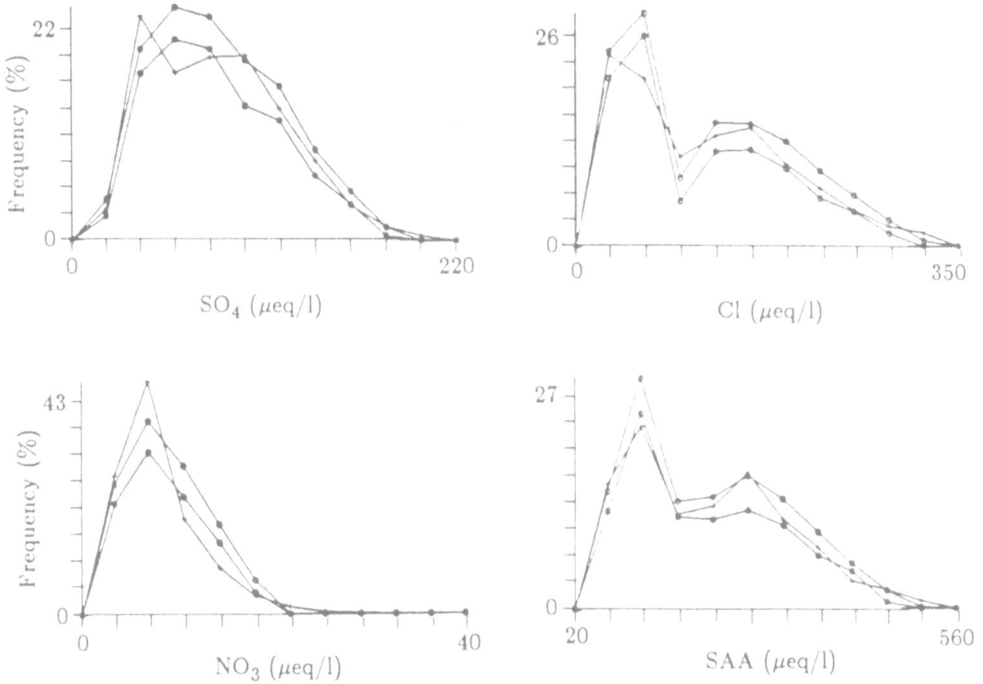

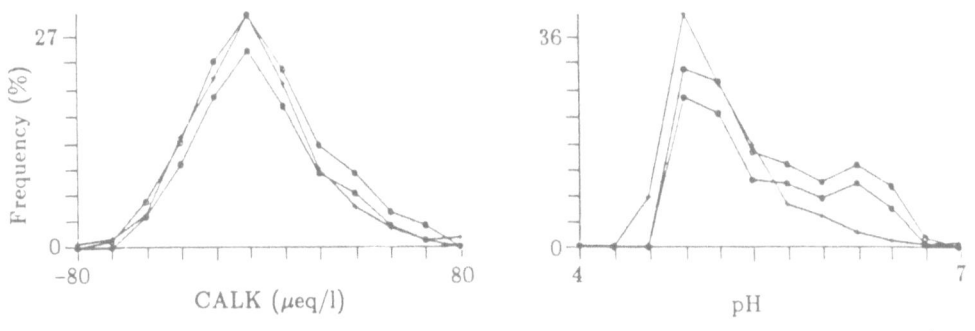

Figure 13.3(b). Simulated maximum and minimum ranges (•) and observed distributions (+) for 1974: acid anion concentrations (top and lower left panels); and sum of acid anion concentrations (lower right).

Figure 13.3(c). Simulated maximum and minimum ranges (•) and observed distributions (+) for 1974: alkalinity and pH.

[*Figure 13.3(a)*]. This is also true of the aggregated variables – the sum of base cations (SBC) and the sum of strong acid anions (SSA). Thus, even though these variables were not directly fitted in the resampling, the *calibration* uncertainty for the distributions variables is small. The *forecast and hindcast* uncertainties must now be examined.

The calibrated regional model was used to reconstruct the distribution of lake alkalinity that existed in southern Norway before the onset of acidic deposition. The historical sequence of deposition used for southern Norway is described in Chapter 7. Comparing the hindcast alkalinity distribution with the observed 1974 distribution of alkalinity in the region (*Figure 13.4*), we conclude that significant acidification of lakes in the region has occurred. That is, the observed 1974 distribution of alkalinity lies well outside the uncertainty band for the simulated 1844 alkalinity distribution. Better than half of the surveyed lakes have lost their alkalinity as a result of acidic deposition.

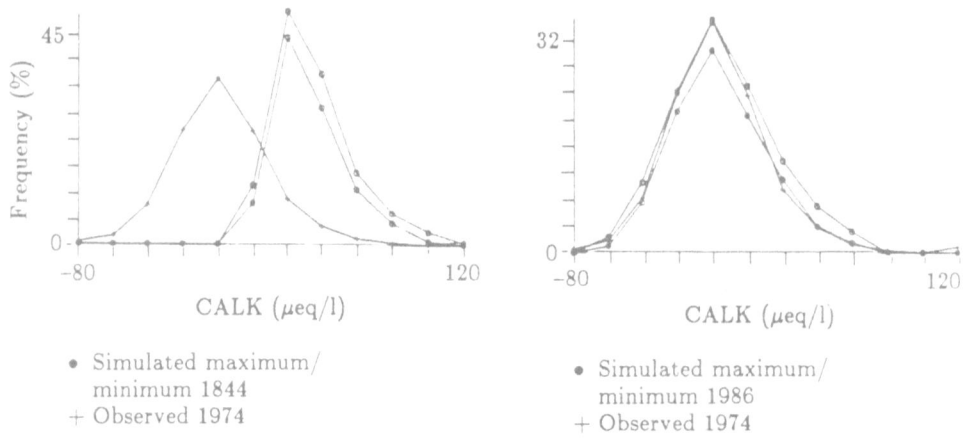

Figure 13.4. Ranges of hindcast (1844) and forecast (1986) alkalinity distributions compared with the alkalinity distribution observed in 1974.

The model was also used to forecast the changes in alkalinity distribution that occurred between 1974 and 1986. Deposition monitoring data over that period suggest that sulfur deposition in the region has declined only slightly – 7% (see Chapter 7) during the 12-year period. Using this observed decline as an input, the model was used to forecast the 1986 alkalinity distribution. Comparing the forecast distribution for 1986 with the observed distribution in 1974 (*Figure 13.4*) suggests that any changes in the alkalinity distribution that occurred as a result of the deposition decline are likely to be too small to be resolved, given the forecast uncertainty over the 12 years. That is, the observed 1974 distribution lies almost completely within the uncertainty band for the 1986 forecast, implying that no change in the distribution of alkalinity for the region by 1986 can be reliably forecast by the model (*Figure 13.4*).

Although it is not readily apparent from the plots of the simulated alkalinity distributions for 1844, 1974, and 1986 (due to changes in scale), the uncertainties in hindcast or forecast alkalinity are greater than the uncertainty in calibrated alkalinity. This is to be expected, but the fact that the forecast and hindcast uncertainties increased as little as they did underscores the utility of the regional model. It should be possible to use the model for *long-term* simulation (several decades to centuries) without losing confidence in the simulation results – subject, of course, to a reasonably correct designation of historical or future deposition scenarios.

Because the 12-year forecast involved only a slight decline in deposition over a relatively short period, we used the calibrated model to generate a second forecast for southern Norway. This forecast is based on a scenario of a 30% reduction in emissions (and, thus, deposition) for all of Europe between 1978 and 1993, followed by constant deposition at the reduced level until the year 2000. This scenario (called the "30% club" scenario) was the subject of negotiations among the European governments in the mid-1970s, although not all nations ultimately agreed. Regardless, the scenario involves a more significant decline in deposition than the observed 7% decrease used above and requires forecasts over a longer period (26 years), thus providing a more appropriate evaluation of forecast uncertainty in the model.

Given the 30% club scenario, the calibrated model forecasts a significant shift toward higher values of the alkalinity distribution for the region. The observed 1974 alkalinity distribution lies completely outside the simulated distribution for the year 2000 (*Figure 13.5*). If the forecast alkalinity uncertainty bands had encompassed the observed 1974 distribution (as was the case for the 1986 forecast above) for this scenario, we would have been forced to conclude that model imprecision would not allow a confident forecast of changes in response to the scenario. This not being the case, we conclude that forecast

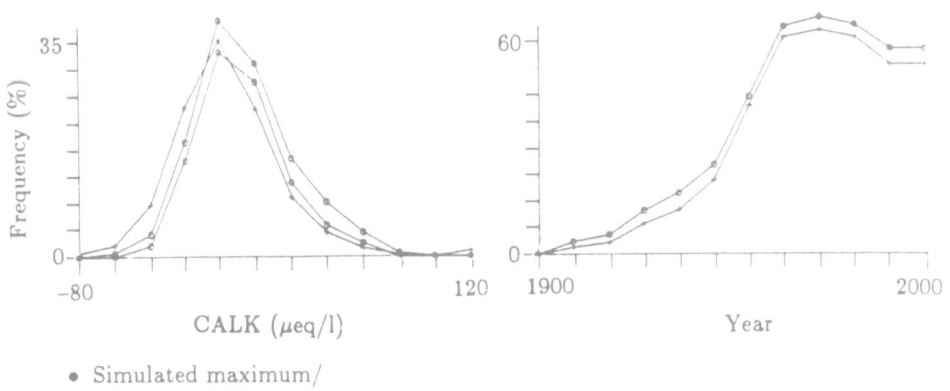

- Simulated maximum/
  minimum 2000
+ Observed 1974

Figure 13.5. The upper panel shows the range of the forecast alkalinity distribution in the year 2000 (using a 30% club scenario) compared with the observed 1974 distribution. The lower panel shows the simulated number of fishless lakes in each decade of this century. The effects of the 30% club scenario are small but noticeable.

uncertainties in the model are not too severe to prevent useful results from being obtained for reasonable (and practical) future deposition scenarios.

Cosby *et al.* (Chapter 7 in this volume) showed that simulated chemistry produced by the calibrated model can be used to reconstruct historical changes in fisheries status using an empirical relationship between chemistry and fisheries status (Wright and Henriksen, 1983). The same approach can also be used to predict future changes in fisheries in response to assumed deposition scenarios. Using the 30% club scenario and the historical deposition estimates, the number of fishless lakes expected in the region was reconstructed for the whole of the twentieth century. A cumulative plot of the percentage of fishless lakes as a function of time (*Figure 13.5*) suggests, that by 1974, approximately 60% of the surveyed lakes had lost their fish populations, with most of these losses occurring between 1940 and 1960. Wright and Snekvik (1978) report that, by 1974, some 80% of the surveyed lakes either had lost their fish populations or had sparse populations (40% in each category). The model forecasts for the 30% club scenario that approximately 10% to 15% of these lakes would recover sufficiently and that fish populations, if reintroduced, would once again be viable (*Figure 13.5*).

Our results are consistent with the intuitive notion that the largest changes in water chemical composition in southern Norway occurred over the past four or five decades and that, without significant decreases in atmospheric loading in the future, further changes are likely to be small (*Figure 13.4*). In fact, because of the asymmetry in the responses predicted by the nonlinear MAGIC, future changes are predicted to occur over a longer time span than the historical changes (Cosby *et al.*, 1985c).

In the work reported here, we have chosen calculated alkalinity as the most important variable to fit in our resampling method. Our results show that, if alkalinity is fit without considering the covariance structure in the chemical composition data, important biases in the simulations are likely to be introduced. In principle, the fit of the simulations should be to the entire multivariate distribution at the appropriate level of resolution of the histogram. The methodology outlined here can be used for this task, although the computational burden may be large even for MAGIC, which is run on a personal computer.

We consider the results presented here to be a more than adequate beginning for assessing past and future effects of acid deposition in southernmost Norway, and indicative of the general utility of regional modeling approaches. It is, however, a beginning, and much remains to be accomplished.

Acknowledgments

This research was made possible by generous grants from the Norwegian Ministry of the Environment and the Norwegian Institute for Water Research. The research was funded in part by the EPA Direct/Delayed Response Program. It has not been subjected to EPA's required peer and policy review and therefore does not necessarily reflect the views of the Agency, and no official endorsement should be inferred.

CHAPTER 14

Atmospheric Deposition Assessment Model: Applications to Regional Aquatic Acidification in Eastern North America

E.S. Rubin, M.J. Small, C.N. Bloyd, R.J. Marnicio, and M. Henrion

14.1. Introduction

US research efforts on the problem of "acid rain" are directed at improving current scientific understanding in critical areas, including sources of precursor emissions; the transport and transformation of pollutants in the atmosphere; the deposition of acidic species; and the chemical and biological effects of acid deposition on aquatic systems, materials, forests, and crops (NAPAP, 1986). The general goals of this research are to characterize the current situation and to develop analytical "models" for predicting the future response of systems to changes in key parameters.

Methods to link the various components of the problem so that the results of scientific research may be better related to the needs of policymakers also are being developed (Balson and North, 1982; Center for Energy and Environmental Studies, 1984; Alcamo *et al.*, 1987). The goal of such "assessment" methods is to provide information useful for decisions about the need for – or consequences of – policy measures to abate acid deposition. A systematic characterization of uncertainties is an important part of assessment methods development.

The analysis presented here focuses on the effect of future emissions on regional aquatic acidification in eastern North America. Specifically, we wish to estimate the degree to which further acidification of lakes, and the consequent loss of fish life, might be expected in the absence of an acid rain control program,

J. Kämäri (ed.), Impact Models to Assess Regional Acidification, 253–284.
© 1990 *International Institute for Applied Systems Analysis.*

and the extent to which such effects might be reduced or reversed as a result of policy measures to reduce emissions of sulfur dioxide (SO_2). Estimating the magnitudes and sources of uncertainties in predicted impacts is a key element of the integrated analysis.

14.2. Methodological Overview

An integrated modeling framework called the Atmospheric Deposition Assessment Model (ADAM) has been developed at Carnegie-Mellon University to analyze acid rain issues in North America (Rubin *et al.*, 1986a). A schematic of the framework is shown in *Figure 14.1*.

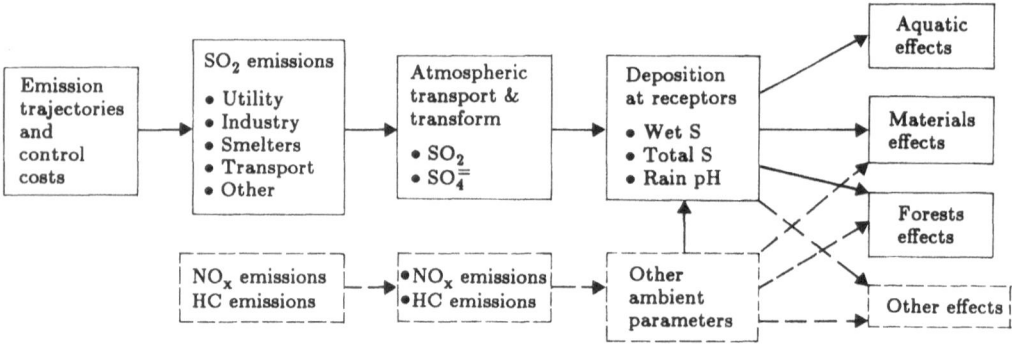

Dashed lines indicate modules and linkages
to be added in the future.

Figure 14.1. The integrated modeling framework.

The major components of the problem are linked in a computer environment designed to facilitate the testing of alternative hypotheses and the analysis of uncertainty. More detailed descriptions of the model and the rationale for its development are presented elsewhere (Center for Energy and Environmental Studies, 1984; Rubin *et al.*, 1986a; Marnicio *et al.*, 1986). The general approach is to seek simplified representations of state-of-the-art models of individual components of the acid deposition problem, and to link them systematically in an integrated framework. Simpler representations often are derived by exercising the more detailed models over a range of assumptions, then fitting the results to simpler algorithms using multivariate statistical analysis. Alternatively, results of detailed models may be employed directly in the form of a table or matrix. This process of simplification is necessary since the size, complexity, data requirements, and design of many analytical models being developed for individual components of the problem preclude their being directly linked for assessment purposes. The simplified models, however, retain the scientific credibility of the parent models on which they are based.

Figure 14.2 identifies the emission source and receptor regions currently included in the ADAM framework. The model contains up to 66 source regions, including the 48 continental states, the District of Columbia, and the 10 Canadian provinces (divided into 17 subregions). Five source sectors representing SO_2 emissions from utility combustion, industrial combustion, nonferrous smelters, transportation, and other miscellaneous sources are modeled for each region. Other pollutant species (e.g., nitrogen oxides and hydrocarbons) may be included in the emissions inventory at a future date.

The 30 receptors identified in *Figure 14.2* reflect 25 regions in the USA and five in Canada believed to be potentially sensitive to acid deposition effects on aquatic systems, materials, forests, crops, or human health. For each receptor, six deposition quantities are estimated, including ambient SO_2; ambient sulfate; wet, dry, and total sulfur deposition; and precipitation acidity (pH). The model is capable of evaluating up to three types of effects at each receptor. Effects on aquatic systems, materials, and forests currently are of prime concern. A time period of up to 50 years (1980 to 2030) may be simulated in steps of one year or more.

The model is implemented in a new software environment called DEMOS (for Decision Modeling System), developed by Henrion and co-workers (see Henrion and Wishbow, 1987), to facilitate the testing of alternative models and the analysis of uncertainty. Some ways in which the uncertainty in individual model parameters may be expressed are shown in *Figure 14.3*. These include three standard distributions (normal, uniform, and log-normal) as well as any user-specified values of discrete distribution or fractiles. These input distributions are sampled randomly using a Monte Carlo procedure to generate multiple sets of results from which probability distributions of model outputs can be obtained. Replication is performed automatically for a given sample size. This analytical capability forms the basis of the uncertainty analysis described below.

14.3. Scope of Analysis

The integrated model is used in this chapter to estimate the effects and uncertainties of acid rain control policies on lake acidity and fish viability in two potentially sensitive regions of eastern North America: the Adirondack Park area of upstate New York and the Boundary Waters region of northern Minnesota and southern Ontario.

Two scenarios are considered. The first is a "base case" situation in which no new policy initiatives are taken to abate acid deposition other than the announced Canadian plan to reduce current SO_2 emissions by approximately 40% by 1994 (*Canadian Control Program*, 1987). The second scenario examines the consequences of an acid rain control strategy for the United States typical of plans recently proposed in the US Congress (Long, 1986–1987). The methods used to model each component of the problem and to characterize key uncertainties are described below, followed by a summary of key results.

(a) Emission source centroids

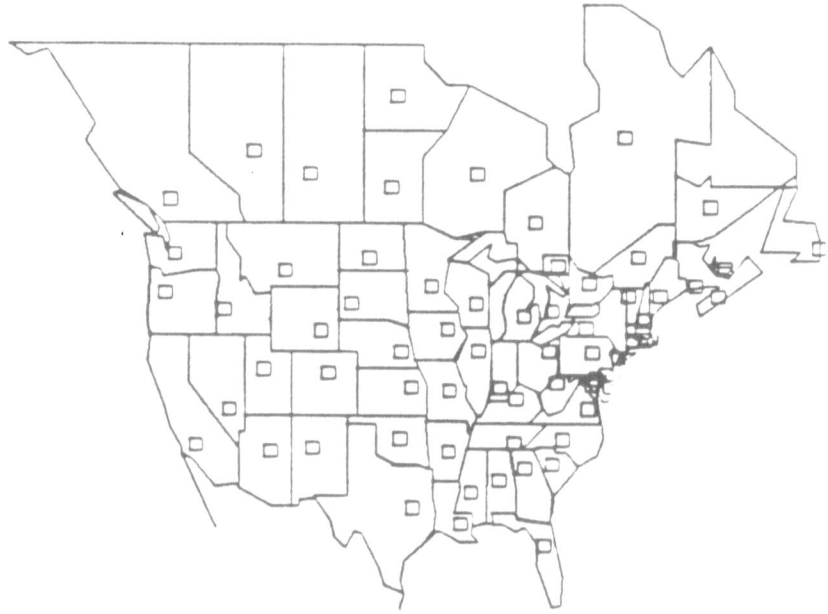

(b) Potentially sensitive receptors

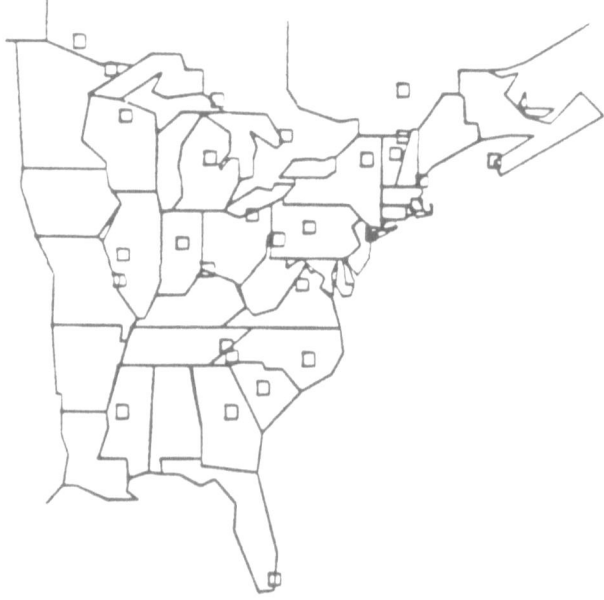

Figure 14.2. Model source (a) and receptor regions (b).

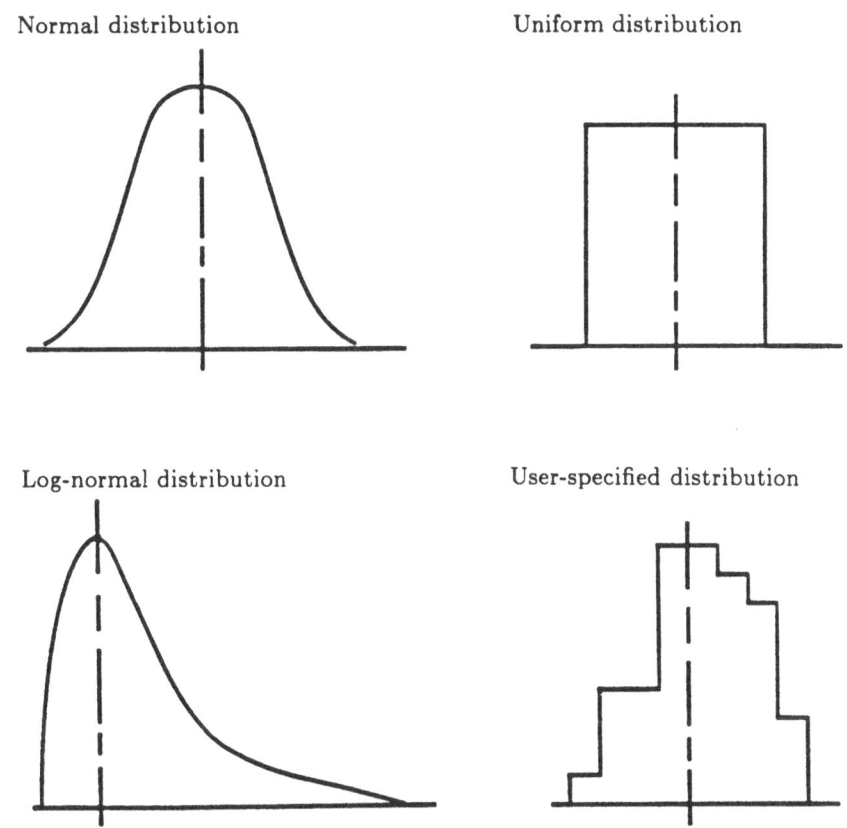

Normal distribution   Uniform distribution

Log-normal distribution   User-specified distribution

Figure 14.3. Some ways of expressing parameter uncertainty. Parameter values on x-axis versus probability of that value on y-axis. Dashed lines represent the mean (expected) value for each distribution.

14.3.1. Emissions and control costs

Acid deposition effects on aquatic systems have been linked primarily to emissions of sulfur dioxide and its reaction products.[1] To analyze aquatic impacts over the next several decades, emissions of SO_2 for the USA and Canada were estimated for the 30-year period from 1980 to 2010.[2] All emission projections were based on studies recently published by the National Acid Precipitation Assessment Program (NAPAP), the US Environmental Protection Agency (EPA), the US Department of Energy (DOE), and the Electric Power Research Institute (EPRI).

Table 14.1 shows the total base year (1980) emissions by sector (Industrial Environmental Research Laboratory, 1984). US emissions account for 84% of the North American total, with electric utility emissions the dominant source (56% of the total for North America). Industrial combustion emissions account for 17% of the total; nonferrous smelters, 11%; transportation, 3%; and other sources (industrial processes, residential, and commercial fuel combustion, etc.), 13% of the total. The geographical distribution of SO_2 emissions is weighted heavily toward the eastern USA, particularly the midwestern states containing large numbers of coal-fired power plants.

Table 14.1. The 1980 SO_2 emissions for North America (millions of short tons per year).

Sector	USA	Canada	Total
Utility	17.4	0.8	18.2 (56%)
Industrial	4.6	0.9	5.5 (17%)
Smelters	1.2	2.4	3.6 (11%)
Transport	0.9	0.2	1.1 (3%)
Other	3.1	1.0	4.1 (13%)
Total	27.2 (84%)	5.2 (16%)	32.4 (100%)

(Source: Industrial Environmental Research Laboratory, 1984.)

The uncertainty in historical and current SO_2 emissions is not well characterized, although several efforts have been initiated in the area (Chun, 1987). One measure of uncertainty comes from comparing different published inventories of 1980 state-level emissions for each sector (Industrial Environmental Research Laboratory, 1984). In general, such comparisons show good agreement for the electric utility sector (sample standard deviations on the order of 5% of the mean), with larger deviations (14 to 25% of the mean) for the non-utility sectors, which are characterized by large numbers of small sources spread over large geographical areas. While recent studies suggest that the uncertainty in emissions from individual sources can be quite high, especially for short averaging times (Chun, 1987), existing data suggest a relatively high degree of confidence in emissions from the major source category (utilities) when averaged over a period of one year and aggregated to the state level. More recent estimates of 1985 non-utility emissions (Braine and Steubi, 1987; Placet *et al.* 1986) show better agreement than earlier estimates for 1980, suggesting some improvement in inventory techniques, data quality, or both.

A much larger uncertainty is associated with projections of *future* SO_2 emissions. *Figure 14.4* shows the range of results to the year 2010 from three recent studies of the US electric utility sector (Braine and Steubi, 1987; Placet *et al.*, 1986; McGowin *et al.*, 1986). Aggregate results for the non-utility sectors also are shown, along with total Canadian emissions.

For the scenarios involving no acid rain emission controls, the projected US utility emissions in 2010 differ by approximately eight million short tons per year (Mtpy) [3], or nearly 50% of the total 1980 utility SO_2 emissions. This wide range arises principally from different assumptions made by each organization regarding the future demand for electricity, the retirement age of the coal-fired power plants (which affects the replacement of current plants with lower-emitting new plants), and the utilization of coal for power generation in the future.

The highest SO_2 emissions are projected by EPA, who anticipate extended plant lifetimes of 60 years, leading to a continually rising trend in emissions through 2010 (Braine and Steubi, 1987). DOE and EPRI scenarios, however, suggest a gradual decline of emissions in the future. The DOE National Energy Plan (Placet *et al.*, 1986) projects somewhat lower electricity growth rates and shorter plant lives than those assumed by EPA, leading to lower estimated SO_2 emissions. Still lower SO_2 emission estimates come from the EPRI "base case" scenario reflecting historical trends. This assumes an average electricity demand growth rate of 2.3%; a 20% planning reserve margin; no nuclear additions after 1994; and plant lives of 40 years for coal, oil, and gas plants and 30 years for nuclear facilities (McGowin *et al.*, 1986). Alternative EPRI scenarios (not shown in *Figure 14.4*) give different results. A scenario extending the life of fossil fuel plants to 60 years raises emissions in 2010 by 6 Mtpy, while increasing the electricity demand growth rate to 3.3% raises emissions by an additional 1 Mtyp, producing results similar to the EPA projections. Thus, the uncertainty introduced by alternative views of the future clearly dominates the "measurement" uncertainty associated with quantifying current emissions. Projections beyond the year 2010 (not considered in the present analysis) show even greater diversity.[4]

Future SO_2 emissions from non-utility sectors are related principally to the assumed use of coal in the industrial sector and the extent of emissions control from smelters. Current EPA and DOE projections for the USA (*Figure 14.4*) show relatively constant emission rates over the next two decades, although detailed studies of non-utility emission trends have not been widely undertaken.

Canadian SO_2 emissions in *Figure 14.4* are based on current projections, assuming full compliance with planned SO_2 emission reductions from smelters and other sources. The allocation of emission reductions to specific provinces was based on a detailed inventory of sources to be controlled by 1994 (*Canadian Control Program*, 1987). Since Canadian emissions are dominated primarily by smelters, rather than coal-fired power plants, future SO_2 emission trends are far less sensitive to assumed electricity demand growth than in the USA.

To characterize the magnitude and uncertainty in future SO_2 emissions, the scenario ranges for the utility and non-utility sectors in *Figure 14.4* were taken as the bounds established by current analyses. US utility emissions in any future year were assumed to be characterized by a uniform distribution ranging from a high value given by the EPA scenario to a low value given by the EPRI "base case" scenario. Thus, any value in this range was assumed to be equally probable. For the non-utility sectors, we assumed that future emissions remained constant constant at the average between the DOE and EPA estimates, and that the uncertainty was characterized by a normal distribution. A standard deviation of 14% of the mean (the sample value for these estimates) was assumed. Since future emissions were not reported on a sector-by-sector basis, all non-utility emissions were aggregated for purposes of this study.[5]

To evaluate the effects of an acid rain control policy for the USA, we modeled a scenario requiring a reduction in SO_2 emissions of 10 Mtyp below 1980 levels, to be achieved by 1995. This emission reduction plan is similar to a

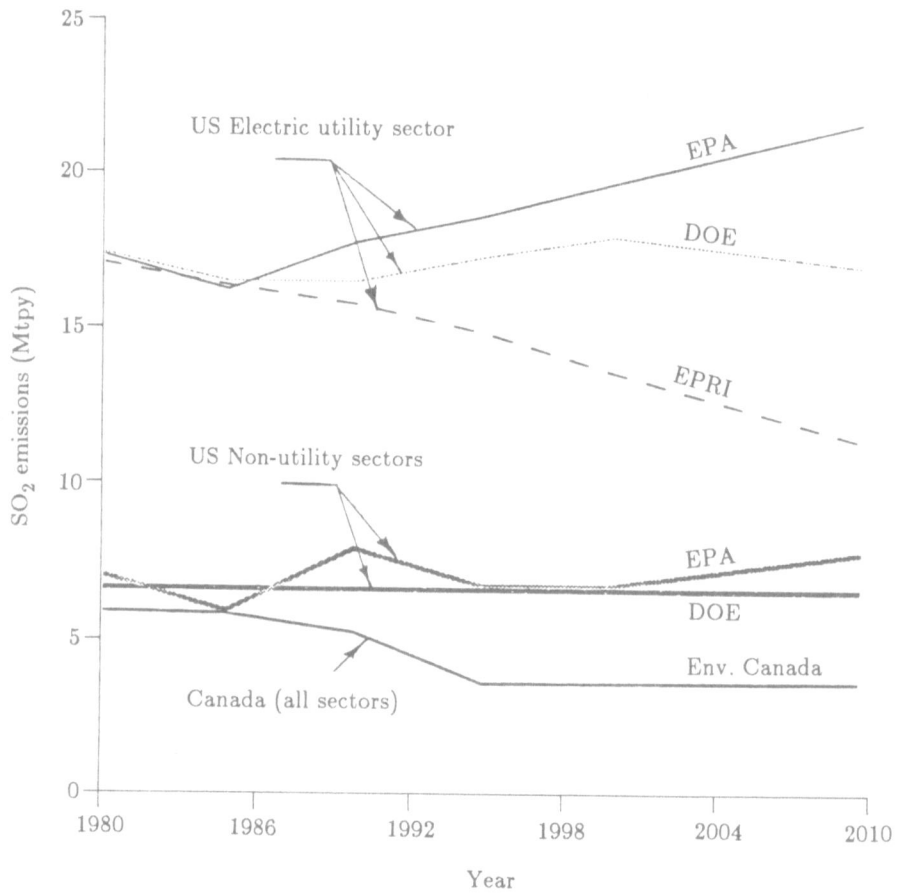

Figure 14.4. SO$_2$ emission projections for the USA and Canada (base case scenario).

number of proposals recently introduced in the US Congress (Long, 1986–1987). Emission reductions were allocated to individual states using a formula based on emissions exceeding 1.2 lbs SO$_2$/MBtu (520 ng/J) in 1980, as also proposed in several Congressional plans (Morrison and Rubin, 1985).

    For purposes of cost estimation, it was assumed that all emission reductions came from the electric utility sector. To the extent that other sectors also are controlled, emission reduction costs would be roughly compared with utility costs (Rubin, 1981), although such costs still need to be better defined. Results from a detailed utility simulation model (Rubin *et al.*, 1986b) were used to obtain a range of cost estimates reflecting different technological options for SO$_2$ emissions reduction as well as varying plant lifetimes for coal-fired units (Salmento *et al.*, 1987). Uncertainties in other technical and economic factors also affect SO$_2$ control costs and could further broaden the range of costs reported later in this chapter (Rubin *et al.*, 1986b; Cushey and Rubin, 1987).

14.3.2. Atmospheric transport and transformation

The transport and transformation of air pollutants in the atmosphere remain subjects of intense study. The acid deposition phenomenon is believed to be strongly influenced by the long-range transport of air pollutants. A number of mathematical models have been developed to predict long-range source–receptor relationships for time scales ranging from one year to episodic events lasting only days or hours. Key issues in atmospheric transport modeling include the "linearity assumption" of a constant proportionality between SO_2 emissions and sulfur deposition; the influence of year-to-year meteorological variations over large geographic areas; and the role of nitrogen oxides (NO_x), volatile organic compounds (VOC), and other species in atmospheric deposition chemistry. While NO_x and VOC are known to affect the formation of photochemical oxidants, which, in turn, may affect sulfur chemistry, the precise nature and importance of such interactions remain a subject of ongoing research (NAPAP, 1986). At present, modeling of annual average sulfur deposition is based on a linear source–receptor relationship, which appears to be reasonable for long-term regional analysis (National Research Council Committee on Atmospheric Transport and Chemical Transformation in Acid Precipitation, 1983).

Two options currently are available for representing source–receptor relationships in ADAM. One is to specify a "transfer matrix" giving the magnitude of an ambient concentration or deposition quantity at a given receptor resulting from a unit of emissions at a given source region. Total impacts then are calculated, assuming a linear relationship between SO_2 emissions and ambient concentrations or deposition on an annual average basis. The ADAM database contains default transfer matrices based on the ASTRAP model developed at Argonne National Laboratory (Shannon, 1985). This model has been widely used by EPA and DOE researchers for acid deposition analysis. Recent work with ASTRAP has focused on the uncertainties in predicted deposition due to annual meteorological variations. Niemann (1986a) used ASTRAP transfer matrices for six consecutive years (1976–1981) to calculate wet sulfur deposition rates at nine ASTRAP receptor sites in eastern North America. For constant SO_2 emission rates, variations of up to 2 to 3 kg SO_4 ha^{-1} yr^{-1} were found. Analysis of longer periods of meterological data using simplified models was recommended, and this work is in progress. Uncertainty analyses involving atmospheric chemistry parameters in ASTRAP, however, do not appear to be available.

The most straightforward way to characterize source–receptor uncertainties in ADAM using a long-range transport model like ASTRAP is to run different sets of transfer matrices reflecting different meteorological or atmospheric chemistry assumptions, or both. This requires multiple runs of the detailed model to obtain the transfer matrices for specified sources and receptors. However, if the variability and covariance structure of a transfer matrix can be specified directly, this can be used in ADAM to characterize uncertainty. This remains a difficult and cumbersome task, however, for a matrix of the required size.

An alternative method of representing source–receptor relationships in ADAM is to specify an analytical model that calculates atmospheric concentrations at a given receptor based on chemical and meteorological input parameters. This is the approach used in the present study since it allows uncertainties in both meteorological and chemical parameters to be characterized more simply and explicitly than in mechanistic long-range transport models (which are relatively difficult and expensive to use, and usually are not directly available to groups other than the model developer).

The analytical model chosen was developed by Fay *et al.* (1985), and is based on multivariate regression analysis of monitoring data for wet sulfate deposition at 109 sites in North America for the years 1980–1982. Although based on relatively simple formulations of atmospheric chemistry and meteorology, predictions from this model have been found to be comparable with those of more complex mechanistic models (Matthias and Lo, 1986). A version of this model also is used by EPA for acid deposition policy analysis (Niemann, 1986b). In this study, uncertainties were assigned to three key meteorological parameters and four chemical rate constants, affecting ambient sulfate concentration, which is used to estimate precipation acidity. Nominal parameter values and ranges, reflecting available data and reported estimates, are shown in *Table 14.2*.

Table 14.2. Source–receptor model parameter uncertainties

Parameter[a]	Base value[a]	SD[b] or Range[c]
Meterological factors		
Wind speed (horizontal)	4.0 m/s	0.74 m/s
Wind speed (vertical)	5.9 m/s	0.74 m/s
Horizontal diffusivity	$4.3 \cdot 10^6$ m^2/s	$0.43 \cdot 10^6$
Chemical factors		
Time constants for:		
Conversion ($SO_2 \rightarrow SO_4$)	$1.9 \cdot 10^5$ s	$(1.0 - 7.5) \cdot 10^5$
Wet SO_2 deposition	$11.3 \cdot 10^5$ s	$(1.5 - 14.0) \cdot 10^5$
Dry SO_4 deposition	$12.5 \cdot 10^5$ s	$(0.25 - 25.0) \cdot 10^5$
Wet SO_4 deposition	$0.6 \cdot 10^5$ s	$(0.14 - 2.5) \cdot 10^5$

[a] Based on Fay *et al.*, (1985).
[b] Standard deviations based on historical data.
[c] Ranges based on literature values.

14.3.3. Precipitation acidity

Acid deposition effects on aquatic systems are related in this study to precipitation acidity, which is determined by the annual average hydrogen concentration. An extensive study of the correlation between precipitation acidity and ambient sulfate concentrations for various regions of the USA has been undertaken by Hales (1982). The regression relationship is shown in *Table 14.3*, along with parameter values for Adirondack Park and Boundary Waters receptor regions. Mean annual rainfall values and variances for each region are based on historical data (Atmospheric Sciences and Analysis Working Group 2, 1982), while the uncertainty in regression coefficients has been estimated from the work of Hales and others.

Table 14.3. Parameter values and uncertainty estimates for precipitation acidity model.[a]

Parameter	Definition	Adirondack Park		Boundary Waters	
		Mean	SD	Mean	SD
a_i	Regression coefficient	$16.5 \cdot 10^{-6}$	$1.65 \cdot 10^{-6}$	$2.3 \cdot 10^{-6}$	$0.23 \cdot 10^{-6}$
b_i	Regression coefficient	1.618	0.162	1.573	0.157
R	Annual rainfall (mm)	1045	91	680	68

[a] Based on Hales *et al.* (1982). Precipitation acidity is given by:

$$(pH)_i = - \log \left(a_i + 3.125 \cdot 10^{-3} \frac{X_{ws}}{R} b_i \right)$$ where X_{ws} = wet sulfate deposition rate (kg S ha^{-1} yr^{-1}).

Rainfall acidity is related directly to the rate of wet sulfur deposition predicted by the atmospheric transport model. Note that because the relationships are empirically fit to current wet deposition chemistry data, they implicitly incorporate the effects of other acid and base species, such as nitrates and soil dust. In most areas of eastern North America, sulfate contributes about two-thirds of the acidity of precipitation and nitrates about one-third (Verry and Harris, 1988). Extrapolation of the pH–sulfate relationship to conditions of significant change in the level of sulfate therefore represents a source of uncertainty in the model predictions since the relationships are unlikely to be closely maintained unless there are concomitant reductions in the level of atmospheric nitrate.

Another reason for using sulfate as the driving deposition input in the integrated assessment model is that sulfate is generally quite mobile in the terrestrial environment, except in certain locations, such as the southeastern United States, where sulfate adsorption occurs in the soil. As noted by Seip (1980), nitrate is, in contrast, "taken up in the catchment, and only a small fraction . . . appear(s) in runoff." As such, nitrate in lake water is generally less than 10% of sulfate on an equivalent basis (Wright, 1988), although nitrate can play an important role during snowmelt periods (Verry and Harris, 1988; Driscoll and Schafran, 1984). Long-term lake acidification models thus commonly use sulfate as the measure of acid input (Cosby *et al.*, 1985b; Henriksen and Brakke, 1988). A more complete analysis of acid deposition and its impacts should, in the future, include delineation of the contribution of nitrogen species.

14.3.4. Deposition uncertainty

To characterize the uncertainties in sulfur deposition, the sensitivity method of Hornberger and co-workers (1985a, 1986a) was applied to the source–receptor relationships from which deposition values are calculated. The method is implemented by assuming prior distributions for uncertain model inputs, then determining which combinations of these inputs give results consistent with observed atmospheric chemistry data. This allows estimation of the posterior distributions of output results.

Ranges of values were first ascribed to the individual parameters of the Fay *et al.* (1985) model, then Monte Carlo simulation was used to calculate the precipitation acidity at six geographically dispersed locations in eastern North America (*Figure 14.5*). Combinations of the uncertain chemical rate constants that produced "acceptable" values of precipitation pH at all six receptors were chosen for use in further analyses. This method was employed to identify and thereby account for the covariance among chemical rate parameters, and the interaction between chemistry and meteorology. Given the difficulty of undertaking a more rigorous analysis of these factors (e.g., using a detailed long-range transport model), this approach was adopted to obtain a physically reasonable and computationally feasible approximation for an integrated analysis.

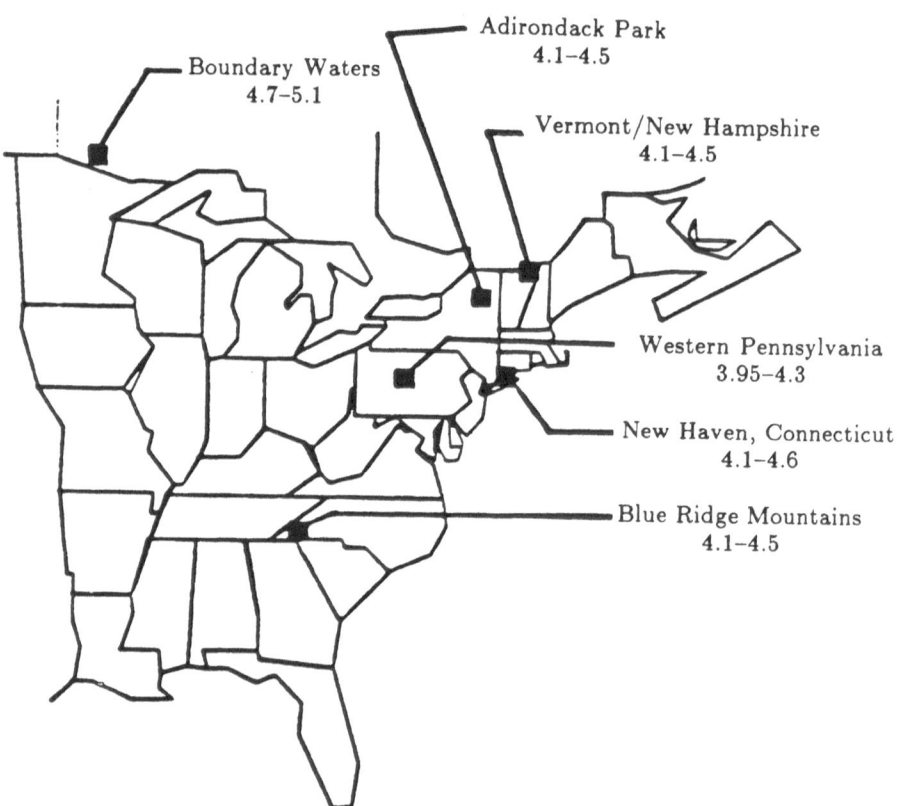

Figure 14.5. Acceptance criteria for predicted annual average precipitation acidity (pH ranges for six receptor regions).

Table 14.2 displays the range of parameter values initially used. Meteorological parameters were assumed to be independent, with "best estimate" values and variabilities characterized from historical data and published studies.

Normal distributions were assumed for the horizontal and vertical components of
annual average wind speed, horizontal diffusivity, and annual average rainfall.
Ranges of values for four chemical rate constants, governing the conversion of
SO_2 and SO_4 in wet and dry form, were obtained from a survey of the literature.
A log-uniform distribution was used to characterize each of the chemical rate
constants in the model. A series of Monte Carlo runs was conducted in which
values of the chemical rate parameters were chosen independently in conjunction
with a value for each of the meteorological parameters of the model. A subset of
these runs (100 cases) was determined to be "acceptable," based on the ability of
the model to predict precipitation pH within the ranges shown in *Figure 14.5* for
all receptor locations.[6]

The runs that produced unacceptable results illustrated the covariance
effects noted earlier. For example, *Figure 14.6* shows scatter plots of the chemical rate constant for primary conversion of SO_2 to $SO_4(\tau_c)$ versus the rate constant for wet disposition of $SO_2(\tau_{wp})$. Since high values of both parameters are
not consistent with the observed (allowable) ranges of deposition [7], such combinations are not seen in the final set of parameter values. For subsequent
analysis, the 100 sets of acceptable combinations of chemical parameter values
were used to represent the uncertainty distributions of the input parameters,
with any of the 100 sets assumed to be equally probable. Note that this implicitly incorporates the correlation structure of the acceptable parameter ranges.
Meteorological parameters continued to be taken as independent.

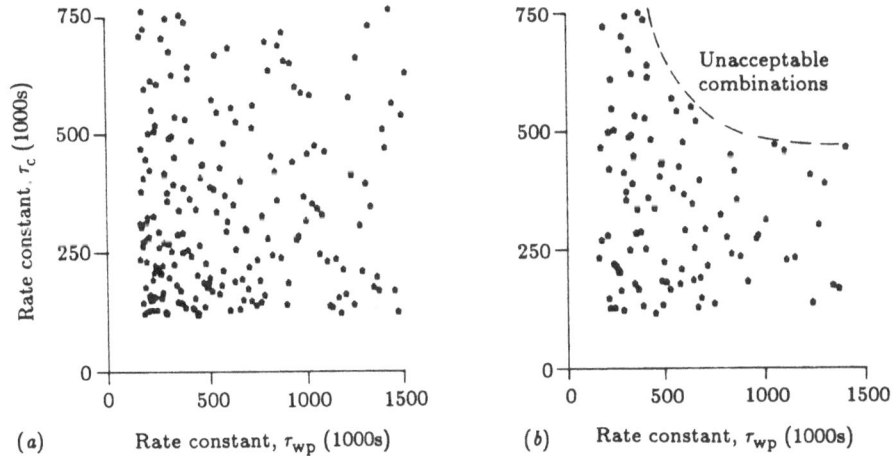

Figure 14.6. Scatter plots of initial (a) and final (b) values of two rate constants.

14.3.5. Aquatic chemistry

The response of aquatic systems to acid deposition also has been, and remains, a
subject of extensive research. Predictive models for aquatic acidification span a

range of scientific and computational complexity, in large part driven by the intended use of each model. Empirical equilibrium models provide a simple representation well suited to regional predictions; however, they lack a fundamental physical basis and do not provide dynamic predictions. Dynamic, mechanistic models predict changes over time using basic chemical equilibrium and kinetic relationships, but are difficult to apply to a region, with its many lakes. These two approaches are briefly described below, followed by a discussion of the model used in the present study – a dynamic Direct Distribution model, which combines features of the empirical equilibrium and dynamic mechanistic approaches.

Empirical equilibrium models predict the steady state alkalinity and corresponding pH in a lake by assuming that a given fraction of the acid deposition to a basin is neutralized by chemical weathering of soil and rocks in the watershed. The fraction of incoming acid that is neutralized, referred to as the "weathering factor," is empirically determined for each lake. The weathering factor can be adjusted to account for the full, complex range of chemical and biological processes in a watershed, including indirect effects of forests and soils on surface-water chemistry and in-lake processes, such as biological reduction of sulfate. In the Henriksen–Wright model (Henriksen, 1979; Wright and Henriksen, 1983), the weathering factor is assumed equal to a constant value, independent of the level of acid deposition. In the Schnoor Trickle-Down model (Schnoor and Stumm, 1984; Schnoor *et al.*, 1986a), the weathering factor varies with the level of acid deposition, based on observed mechanisms of mineral hydrolysis. Equilibrium models are relatively easy to apply to many lakes in a region, though difficulties arise when attempting to characterize the variation in weathering factors from lake to lake.

Dynamic, mechanistic models incorporate a variety of hydrologic, mass balance, and chemical equilibrium relationships to predict the evolution of lake chemistry resulting from a dynamic profile of acid deposition. Examples include the ILWAS (Integrated Lake Watershed Acidification Study) model (Chen *et al.*, 1982; Gherini *et al.*, 1985); the Birkenes model (Christophersen *et al.*, 1982); the MAGIC (Modeling Acidification of Groundwater in Catchments) model (Cosby *et al.*, 1985a, 1985b); and the RAINS (Regional Acidification INformation and Simulation) model (Kämäri and Posch, 1987). These models generally require estimates for a number of parameters for each lake, and the dynamic simulations are computationally intensive. Thus, they are difficult to apply in regional assessments. While significant progress has been made in the development of methods for regionalizing these models (Cosby and Wright, 1987; Kämäri and Posch, 1987), the computational requirements still preclude their inclusion in an integrated model, such as ADAM, where evaluations of multiple scenarios are required for policy and uncertainty analysis.

To address these shortcomings, this study uses the newly developed dynamic Direct Distribution model of Small and co-workers (Small and Sutton, 1986a, 1986b; Small *et al.*, 1987). The model uses the Henriksen–Wright equilibrium model to predict the acid-neutralizating capacity (ANC) of a watershed [8], but applies the model to the probability distribution functions of ANC and pH for a *region*. Lake-to-lake variations in the weathering factor are explicitly incorporated by the identification and use of the mean weathering factor, F, together with the variance of the weathering factor, F_{var}, for the region. A dynamic version of the model was developed by assuming that all lakes in a region approach a new equilibrium value of ANC in a manner described by an exponential equation with a characteristic time constant, t_{alk}. The resulting pH–distribution is derived from the ANC distribution at each time step, and is integrated with a pH–fish presence–absence relationship to determine the fraction of lakes able to support a fish species. The development of this relationship is described later in this chapter.

The parameters of the dynamic Direct Distribution model can be estimated from a regional mechanistic model, such as the RAINS or MAGIC model. This allows the Direct Distribution model to provide a nearly equivalent representation of a more complex mechanistic model, while maintaining the computational efficiency of an empirical equilibrium model. This procedure is illustrated in Chapter 10. Such work, however, remains for future applications. In the examples that follow, we have estimated the parameters of the Direct Distribution model from judgmental assessments based on previous studies of the regions of interest. Uncertainties in the assumed parameter values are assigned to reflect the empirical nature of this assessment.

An important question in the application of regional aquatic acidification models is whether available lake chemistry data provide a representative sample of the overall population of lakes in a region. To address this problem, the US EPA conducted the National Surface Water Survey (NSWS), in which lakes were selected and monitored from stratified samples of target sensitive regions (Environmental Protection Agency, 1986). The resulting data sets provide the most representative characterization currently available of the overall distribution of US regional lake chemistry. The parameters of these distributions defined in the Direct Distribution model include three parameters of the log-normal distribution for ANC (Θ, ξ, and ϕ^2), and two parameters of the pH–ANC relationship (p_1 and p_2). Parameter values for the NSWS subregions corresponding to the Adirondack Park and Boundary Waters (northeastern Minnesota) were evaluated by Small et al. (1988). Results are presented in Table 14.4, along with the other parameters determined for the Direct Distribution model. Comparisons of the fitted ANC distributions and pH–ANC relationships with observed values of current lake chemistry are presented in Figure 14.7.

The most critical parameter shown in Table 14.4 for determining the response of lakes to changing acid deposition is the mean weathering factor, F. Specification of the weathering factor even for a single lake is a difficult approximation (Johnson et al., 1985). F likely varies with the acidity of the deposition (Schnoor and Stumm, 1984), and, as suggested by our analysis of the RAINS and MAGIC models (Small et al., Chapter 10; Labieniec, 1988), also varies over time, even with constant deposition. It is important to note that our use of the weathering parameters considers all sources of acid neutralization in the watershed, not only base cation generation from soil weathering. Our analysis of the RAINS model applied to southernmost Finland resulted in an average value of F equal

Table 14.4. Aquatic chemistry model parameters and uncertainty.

Parameter[a]	Adirondacks Park		Boundary Water	
	Mean[a]	Range or σ[b]	Mean[a]	Range or σ[b]
ANC distribution, parameter, θ	−41	(−51/−31)	23	(13/33)
ANC distribution, parameter, ξ	496	(0.9)	5.13	(0.08)
ANC distribution, parameter, ϕ^2	1.145	(0.13)	0.941	(0.11)
Flow through ratio, FTR	0.45	(0.35/0.55)	0.35	(0.25/0.45)
Mean weathering factor, F	0.6	(0.4/0.8)	0.7	(0.5/0.9)
Weathering factor variance F_{var}	0.047	(0.027/0.067)	0.047	(0.027/0.067)
Alkalinity/weathering factor cor., ρ	0.5	(0.1/0.9)	0.5	(0.1/0.9)
Alkalinity char. time, τ_{alk}	8	(1/14)	8	(1/14)
Alkalinity/pH trans. factor, ρ_1	5.24	(5.14/5.34)	5.4	(5.2/5.6)
Alkalinity/pH trans. factor, $r\hat{h}p_2$	11.02	(1.02/21.02)	10.00	(−2.42/17.58)

[a] Based on Small and Sutton (1986b) and Small *et al.* (1987).
[b] Ranges assume a uniform distribution. A single value is the standard deviation of a normal distribution.

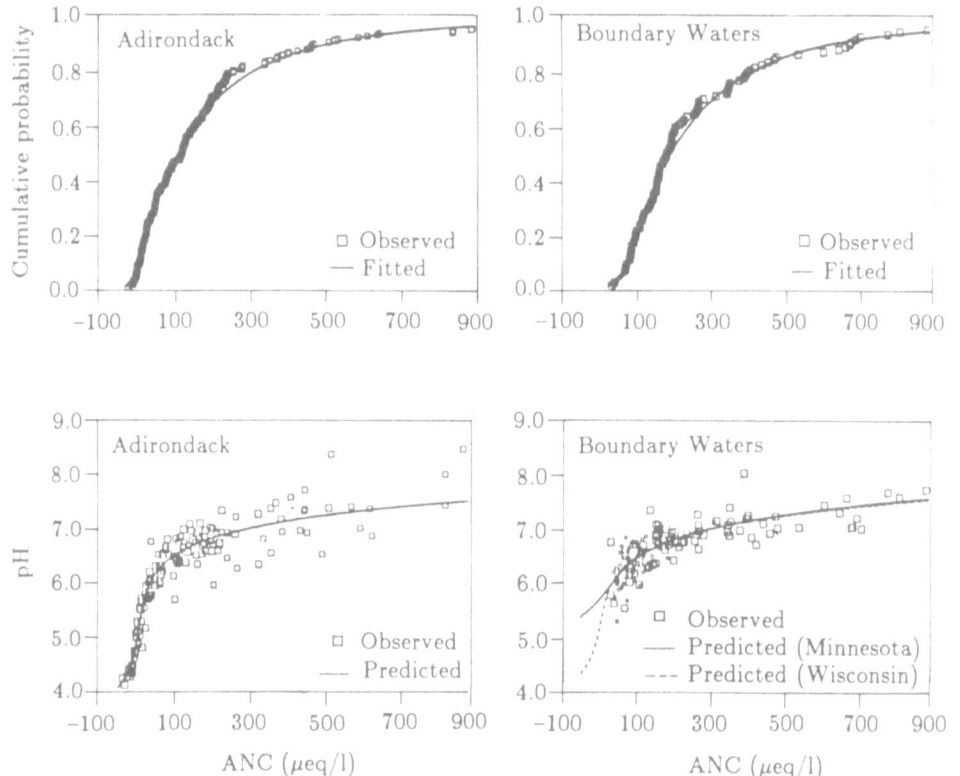

Figure 14.7. Lake *ANC* distribution and pH–*ANC* relationship for two receptor regions. (Source: Small *et al.*, 1988.)

to about 0.8 (Chapter 10). Somewhat lower values of about 0.2–0.6 were obtained from MAGIC for southernmost Norway (Labieniec, 1988). This is consistent with previous estimates for North America (Small and Sutton, 1986b; Johnson *et al.*, 1985). The assumption in *Table 14.4* of somewhat greater weathering in the Boundary Waters region relative to Adirondack Park (mean *F* of 0.7 relative to 0.6) is consistent with geological assessments of the respective regions (Eilers *et al.*, 1988; Brakke *et al.*, 1988). The estimation of *F* and its variability (F_{var}) in *Table 14.4*, however, remain highly judgmental at the present time, although, as indicated above, there is considerable opportunity to reduce the uncertainties in these parameters through additional research using mechanistic models.

The uncertainty in each parameter of the Direct Distribution model was assumed to be either normally or uniformly distributed across the range of values shown. Monte Carlo sampling from these distributions was used to characterize the overall uncertainty in the distribution of regional lake alkalinities. Because covariance effects cannot yet be rigorously assessed for lake chemistry models, we conservatively assumed that all model parameters were independent. While this generally tends to overstate the calculated estimates of uncertainty, a lack of knowledge regarding the precise nature of current alkalinity distributions for large collections of lakes mitigates in favor of this approach.

14.3.6. Fish viability

The final link in the chain from emissions to acid deposition effects is the biological effect on aquatic life associated with changes in lake acidity. A number of laboratory and field studies have been undertaken; see Haines (1981), Johnson (1982), and Muniz *et al.* (1984). However, it has been difficult to generalize these results into mechanistic models that incorporate effects at different stages of the life cycle, and to aggregate such effects to predict population dynamics. Thus, current models and measures of acid deposition effects on fish populations remain relatively rudimentary.

One measure of fish viability, developed in Canada, is the number of different species found as a function of lake acidity. Correlations of this type have been reported by Minns (in Chapter 5) for one Canadian region. However, such data on species diversity are not generally available for other locations in North America.

Studies of US fish populations have characterized fish viability in terms of the presence or absence of a given species in lakes of a given pH. A relatively simple model (*Table 14.5*) by Reckhow *et al.* (1987) used observed fish preserve data to describe the probability of finding a given fish species in lakes of a given pH. Reckhow (1987) also reported the estimation error in the two model parameters, and subsequently characterized their correlation structure. *Figure 14.8* shows the resulting distributions of fish viability for two common recreational fishing species: brook trout and lake trout. Nominal values are shown together with an 80% confidence interval calculated from the data in *Table 14.5*. While fish viability is known to depend as well on other parameters, such as calcium,

Table 14.5. Parameter values and uncertainty estimates for fish viability model.[a]

		Brook trout		Lake trout	
Parameter	Definition	Mean	Variation[b]	Mean	Variation[b]
α	Viability intercept	6.20	1.72/−0.989	5.77	2.83/−0.991
β	Viability slope	−1.05	0.28/−0.989	−0.92	0.45/−0.991

[a]Based on Reckhow *et al.* (1987). Survival probability for fish species, i, is given by :

$$P_i = \frac{1}{1 + \exp\left[\alpha_i + \beta_i\left(\text{pH}_l\right)\right]} \quad \text{where pH}_l = \text{lake pH.}$$

[b]Standard deviation of a normal distribution and the slope/intercept correlation coefficient.

magnesium, and aluminum ion concentrations, validated models of this sort are not yet available; nor are aquatic chemistry models yet able to provide valid estimates of metal ion concentrations that could serve as inputs to improve biological models. These factors contribute to the empirical uncertainty in fish response to changes in lake pH indicated in *Figure 14.8*.

14.4. Results

We turn next to a summary showing the regional acidification impacts of SO_2 emissions from the USA and Canada. The computer simulation model (ADAM) was used to link each of the previously described model components. Uncertainties were combined and propagated by sampling randomly from the specified distributions for each stochastic variable (or sets of acceptable values, in the case of atmospheric chemistry parameters), using a total of 100 iterations to obtain experimental distributions for the values of model output parameters. Examples of these results are shown in *Figure 14.9*.

These curves show the cumulative probability distributions of selected model outputs for the years 1980 to 2010 for the acid rain control scenario. Results of this type were obtained at five-year intervals for both the Adirondack Park and Boundary Waters receptors. The median (expected) value of any result corresponds to a probability of 0.50. An 80% confidence interval is bounded by cumulative probabilities of 0.10 and 0.90. This is the confidence interval illustrated in all later results. The slightly irregular shape of several distribution functions in *Figure 14.9* is a result of the finite number of simulations used to approximate the true distribution.[9]

The output quantities shown in *Figure 14.9*, together with emission control costs (not shown in that figure), represent the key results of the integrated analysis. In the discussion that follows, we summarize the principal findings for the parameters and illustrate their temporal trends over the 30-year simulation period.

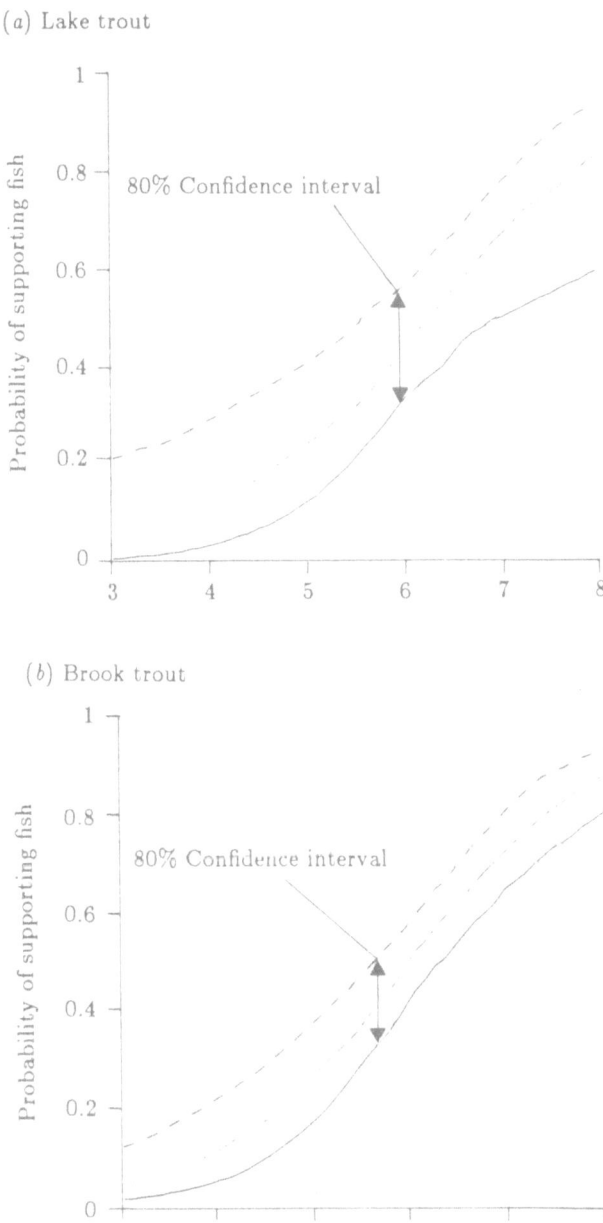

Figure 14.8. Fish viability functions for lake trout (*a*) and brook (*b*).

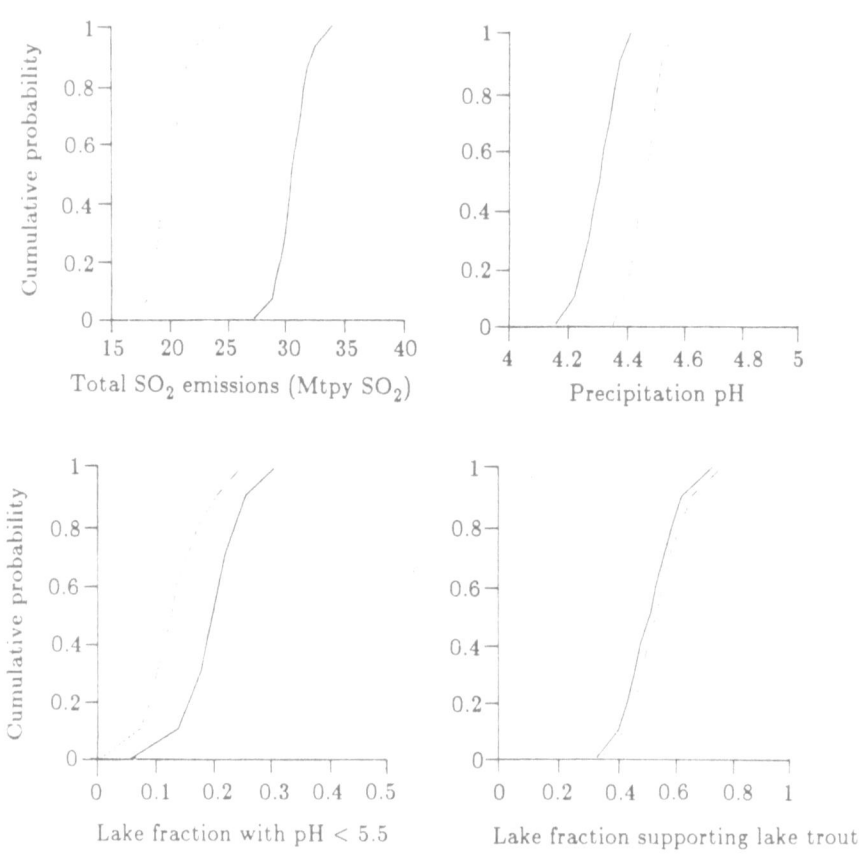

Figure 14.9. Model results for the 10 Mtpy reduction scenario, the Adirondack Park receptor (cumulative probability distributions for selected output parameters). Solid lines indicate actual, 1980 results; dashed lines indicate predicted, 2010 results.

14.4.1. Projected trends and uncertainties

The assumed trends in total SO_2 emissions from US and Canadian sources are again shown in *Figure 14.10* for the base case scenario (no change in current policy) and the acid rain control scenario (reducing US emissions to 10 Mtpy below 1980 levels). In contrast to the scenario ranges shown in *Figure 14.4*, an 80% confidence interval about the mean is shown here based on the frequency distributions illustrated in *Figure 14.9*.

For the emission reduction scenario, *Figure 14.11* shows the corresponding range of SO_2 control costs that would be incurred if all US emission reductions were achieved at coal-fired power plants.[10] The cost ranges cover four cases. The first assumes that only coal switching, coal cleaning, and current flue gas desulfurization (FGD) processes are available for retrofit applications. A second case assumes that two dry SO_2 removal processes now under development for

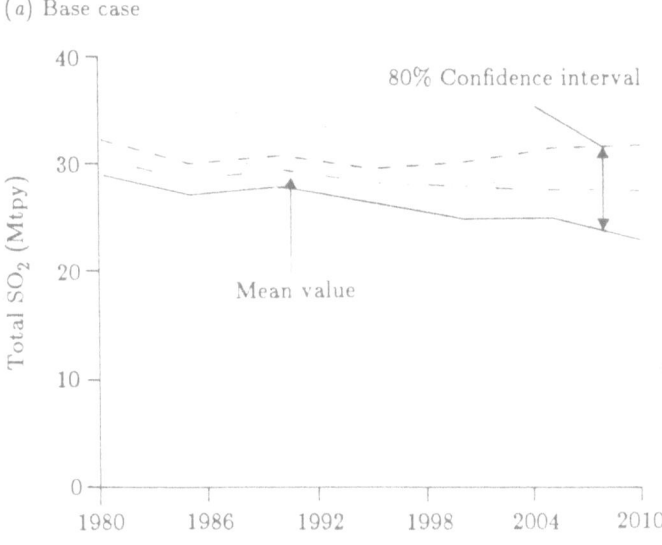

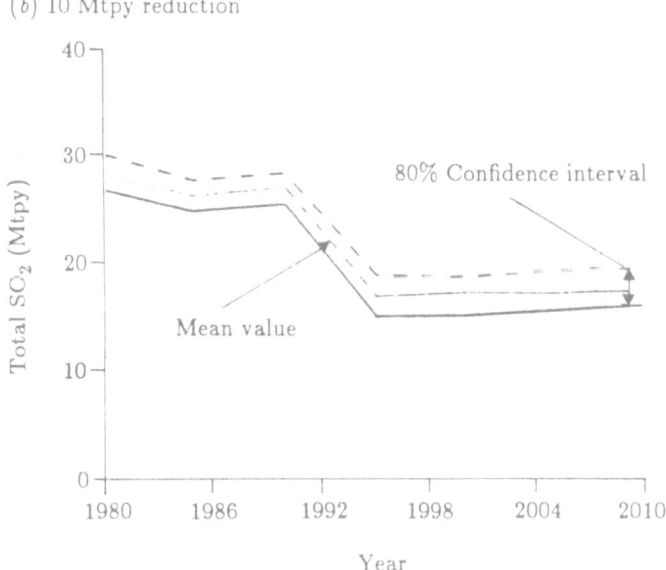

Figure 14.10. Total SO_2 emissions for the base case and acid rain control scenarios (mean values with 80% confidence intervals).

(a) Capital costs for FGD retrofits

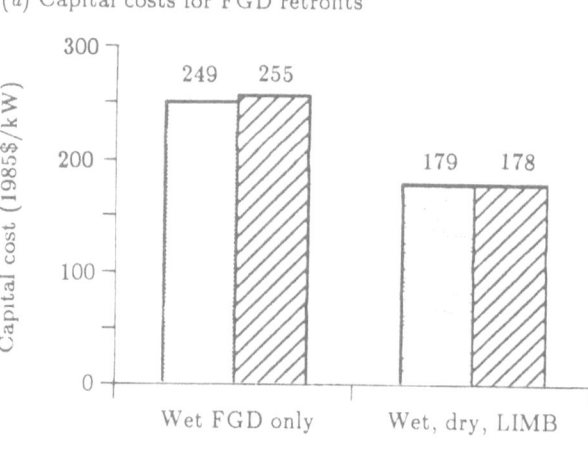

(b) Total leveled cost for emission reduction
(All coal-fired power plants)

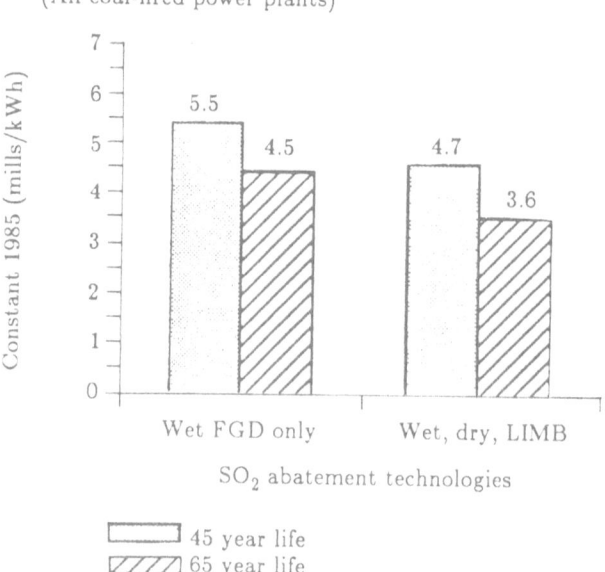

☐ 45 year life
▨ 65 year life

Figure 14.11. Cost for the 10 Mtpy acid rain control scenario.

medium- and high-sulfur coal applications (lime spray dryers and furnace lime-stone injection, or LIMB) also will be commercially available by the mid-1990s. For both cases, two power plant lifetimes are analyzed: a nominal (historical) case of 45 years, and an extended life of 65 years for all but the smallest units. The average increase in levelized electricity cost for these four cases varies from 3.6 to 5.5 mills/kWh (in constant 1985 dollars). This is roughly 10% to 14% of current generation costs for coal-fired plants in the eastern USA. As noted

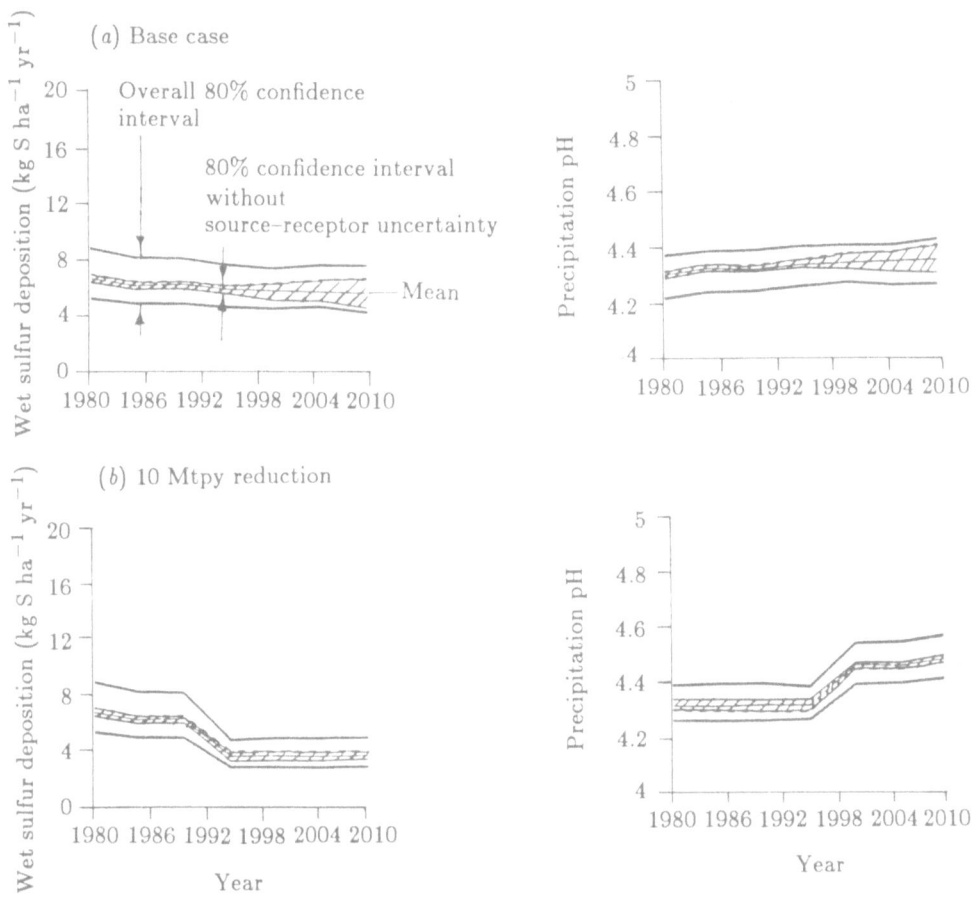

Figure 14.12. Atmospheric deposition quantities for Adirondack Park (expected values with 80% confidence intervals).

earlier, other technical and economic factors could introduce greater uncertainty in the cost of emission reductions (Rubin *et al.*, 1986b).

Figure 14.12 next shows the predicted values of wet sulfur deposition and precipitation pH computed at the Adirondack Park receptor. An 80% confidence interval is show first for an uncertainty only in SO_2 emissions, then for the combined uncertainty in both SO_2 emissions and source–receptor relationships. For the base case, the uncertainty in future emissions increasingly dominates the total uncertainty as time goes on; while for the emission reduction scenario, the atmospheric transport uncertainties are generally more significant. In the year 2010, the overall 80% confidence interval encompasses wet sulfur deposition rates 4.1 to 7.5 kg S ha^{-1} yr^{-1} in the base case and 2.8 and 4.9 kg S ha^{-1} yr^{-1} for the 10 Mtpy reduction scenario.

For both scenarios, the generally decreasing trend in sulfur deposition corresponds to the downward trend in average emissions across the range of scenarios considered. The 80% confidence interval for the base case encompasses a range of wet sulfur deposition rates similar to that found at Adirondack Park with the ASTRAP model when both annual emissions and meteorology were varied over a recent six-year period (Niemann, 1986a). A 90% to 95% confidence interval would encompass a larger range of deposition rates (see *Figure 14.9*). An analysis of the calculated source contributions to wet sulfur deposition at the Adirondack Park and Boundary Waters receptors showed that both regions were affected principally by emissions from midwestern states along the Ohio River Valley in the USA, and by sources in southern Ontario in Canada.

The changes in wet deposition lead to changes in precipitation pH (*Figure 14.12*). The 80% confidence interval for Adirondack Park encompasses a range of approximately ±0.14 pH units. A similar pattern of change was found at the Boundary Waters receptor (not shown), where the pH profile remained about 0.5 pH units above that predicted for Adirondack Park throughout the simulation period.

The effect of changing deposition rates on regional aquatic systems is reflected by a shift in the distribution of lake *ANC*. This is illustrated in *Figure 14.13* for the Adirondack Park receptor. The shift toward higher values of *ANC* implies a reduction in the fraction of lakes with a pH below some given value. Trends in the fraction of lakes with a pH of 5.5 or less are shown in *Figure 14.14*. This pH value was chosen to represent lakes where effects on fish life are likely to be pronounced. For the base case, the expected value of lakes with pH less than

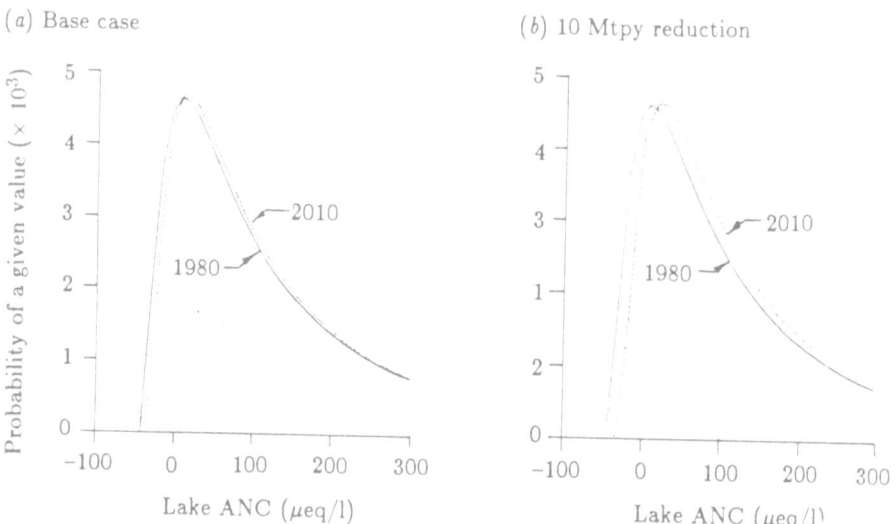

Figure 14.13. Current and projected distributions of lake alkalinity for Adirondack Park.

5.5 falls from about 20% in 1980 to 17% in 2010. For the SO_2 reduction scenario, this fraction falls to approximately 13% in 2010.

The confidence intervals in *Figure 14.14* first show the cumulative effect of uncertainties in all model components prior to the aquatic chemistry model (i.e.,

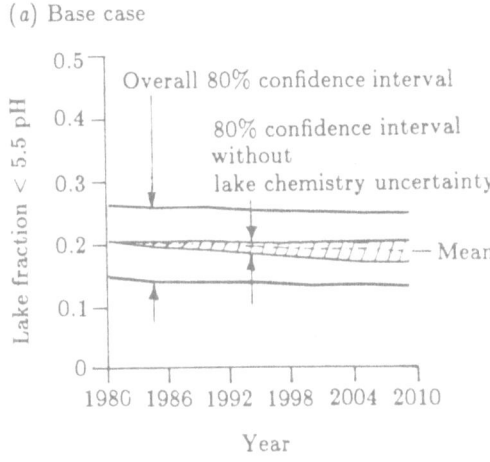
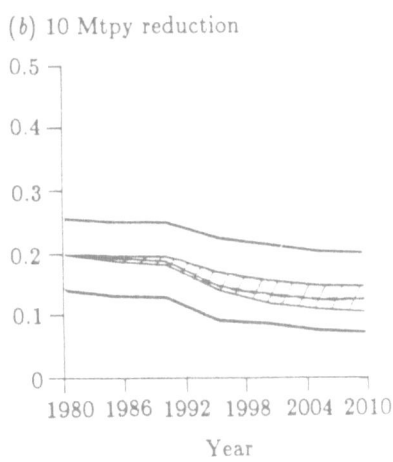

Figure 14.14. Fraction of Adirondack Park lakes with pH below 5.5 (80% confidence bands).

the emissions and atmospheric transport models), then the combined effect of the additional uncertainty in aquatic chemistry. This shows that the incremental effect of aquatic uncertainties dominates overall result. Because the frequency distribution of the fraction of lakes with pH less than 5.5 is not symmetric with respect to the median value (*Figure 14.9*), the 80% confidence interval also is not symmetric.

For both scenarios, the overall uncertainty interval about the mean remains relatively constant with time, reflecting primarily the uncertainties in atmospheric transport and lake chemistry. The effect of the 1995 emission reduction scenario, however, is clearly seen in the trend lines for each confidence level, which show a declining fraction of acidic lakes. This suggests that, although there is considerable uncertainty as to the actual number of acidic lakes (even today, reflecting uncertainty in the current resource inventory), the likelihood of recovery as a result of emission reductions is more robust. For example, over the 25 years from 1985 to 2010, the fraction of lakes with pH less than 5.5 decreases by 5% to 6% for all confidence levels. For the base case scenario, the expected change is a decrease of 1% to 2% for the assumed range of future SO_2 emissions.[11]

Results for the Boundary Waters receptor indicated much smaller effects. A negligible fraction of the lakes there currently has a pH as low as 5.5, and only a third has a pH below 6.7 (Environmental Protection Agency, 1986). By 2010, this fraction is projected to fall by only 0.5% in the base case and by 1% in the emission reduction case. The 80% confidence bands encompass values from approximately 20% to 50% of the lakes with pH below 6.7.

The effect of changing lake acidity on the expected fraction of the lakes in the two study regions able to support fish life is shown in *Figure 14.15*, based on the presence/absence relationships described earlier. For brook trout and lake trout at Adirondack Park, there is a small increase of 0.6% from 1985 to 2010 for the base case scenario, with a larger increase of 2.4% for the emission reduction scenario. These changes are similar for the 10% and 90% confidence levels. Thus, the results indicate that the number of Adirondacks lakes capable of supporting trout will most probably remain stable and even improve slightly in the *absence* of acid rain controls (given the assumed range of projected emissions), but would improve more significantly with an emissions reduction policy. A further *decline* in fishable lakes would be expected, however, if the increasing emission levels reflected in the EPA assumptions were judged to be the most likely case. *Figure 14.15* shows the overall 80% confidence intervals extending to approximately ±7% of the median values for lakes able to support brook trout and ±11% for lakes able to support lake trout at Adirondack Park. For Boundary Waters, the comparable uncertainty intervals are ±8% for brook trout and ±14% for lake trout.

Perhaps the most appropriate way of characterizing the impact of an acid rain control program on future lake acidification is to compare the expected

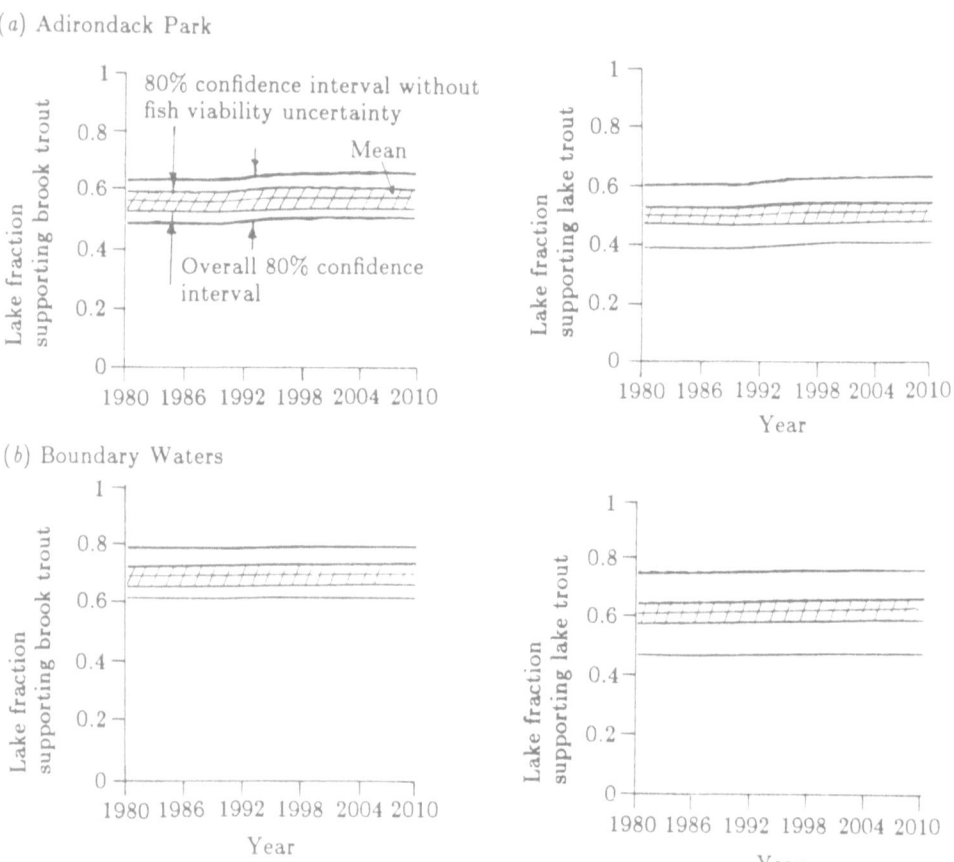

Figure 14.15. Fraction of lakes able to support brook trout and lake trout (80% confidence intervals for 10 Mtpy reduction scenario).

effects in some future year relative with the base case. For 2010, this yields an expected improvement of 1.7% in the number of Adirondack Park lakes able to support either brook or lake trout. For Boundary Waters, the analogous figure is 0.3%.

To put these numbers into perspective, the stratified samples upon which the NAPAP lake survey data were based represented approximately 1,500 lakes in both the Adirondack Park and Boundary Waters region. The total lake population in these two regions, however, is nearly twice as large, although the distribution of *ANC* among the total population may not necessarily reflect that of the subset chosen for the survey. Based on the total population, however, the 1.7% difference in the Adirondack Park lakes supporting brook and lake trout in 2010 corresponds to approximately an additional 50 lakes in that region that would be expected to recover as a result of the 10 Mtpy emission reduction. For Boundary Waters, an increase of six lakes supporting trout would be expected.

The implication of these figures is that the effects of an acid rain control program on fish viability could well be masked by the significant uncertainty that exists as a result of inadequate data, incomplete scientific understanding, and the natural variability of the atmosphere and ecosystems. In view of these results, it is useful to explore further the factors that contribute most to the overall uncertainty. In doing this, we introduce another measure for characterizing model results – namely, the probability distribution of *changes* in key parameters from 1980 to 2010.

14.4.2. Major sources of uncertainty

The procedure used in this study to model the effects of uncertainty is a stochastic simulation in which a 30-year trajectory linking emissions to aquatic effects is repeated 100 times, choosing different values of key model parameters from specified probability distributions. While the preceding discussions have emphasized the distribution of results predicted for any given year, one can also look at the overall *change* between the first and last years predicted by each of these 100 iterations. This is done in *Figure 14.16*, which shows the probability distribution function of the change from 1980 to 2010 in the percentage of the Adirondack Park lakes supporting lake trout. Both the base case and 10 Mtpy SO_2 reduction scenarios are shown.

Figure 14.16 also illustrates how the overall change is influenced by uncertainties in each model component. The cumulative distribution function is decomposed to show first the effect of uncertainty only in the projected SO_2 emissions, with no uncertainty in other model components. Then, uncertainty in atmospheric transport–transformation is added, followed by uncertainty in precipitation pH and so on.

For the base case scenario, the expected change is less than 1% in lakes able to support lake trout, as discussed earlier. The computer uncertainty distribution, however, indicates a 10% chance the changes may be as low as zero or as high as 2%. The negative range corresponds to a decrease in the number of fishable lakes (which has a lower probability of occurrence).

(a) Base case

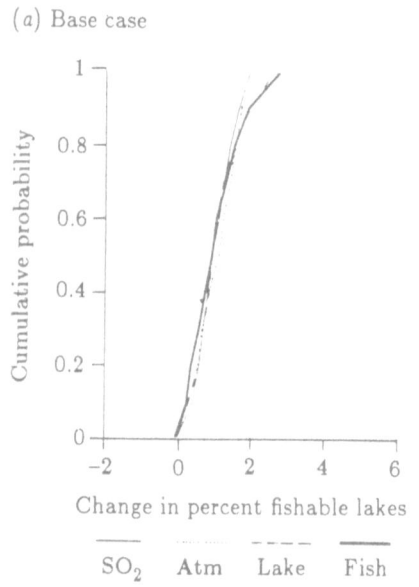

(b) 10 Mtpy reduction

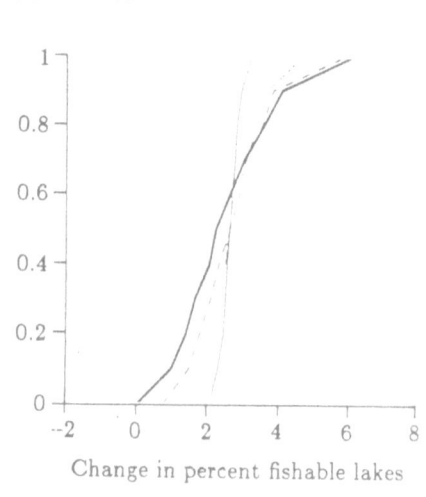

SO₂   Atm   Lake   Fish

Figure 14.16. Change from 1980 to 2010 in fraction of the Adirondack Park lakes able to support lake trout: cumulative contribution of component uncertainties.

The overall distribution of values is much wider for the acid rain control scenario. An 80% confidence interval includes changes of 1% to 4%, compared with the expected value of 2.3%. This range corresponds to an increase of roughly 30 to 120 additional lakes able to support trout fishing. The range would be still larger for higher degrees of confidence.

The decomposition of uncertainty by component shows that the most significant factors in predicting the change in lakes able to support fish life are the uncertainties in lake chemistry and fish viability functions. By comparison, uncertainties in SO_2 emissions and atmospheric transport have notably smaller impacts, particularly for the emission reduction scenario (where full compliance with the reduction requirement is assumed).

If only the degree of lake acidification is of interest, rather than the biological effects on fish, the dominant uncertainties lie in the modeling of lake chemistry, as seen in *Figure 14.17*. Thus, our results suggest that a better understanding of fish biology together with improved models of regional lake response are most essential to reduce the uncertainties in predicted impacts of emission reduce on aquatic resources.

14.5. Caveats and Conclusions

This chapter has attempted to illustrate how an integrated model of the acid deposition process can be used to estimate the impacts and uncertainties of

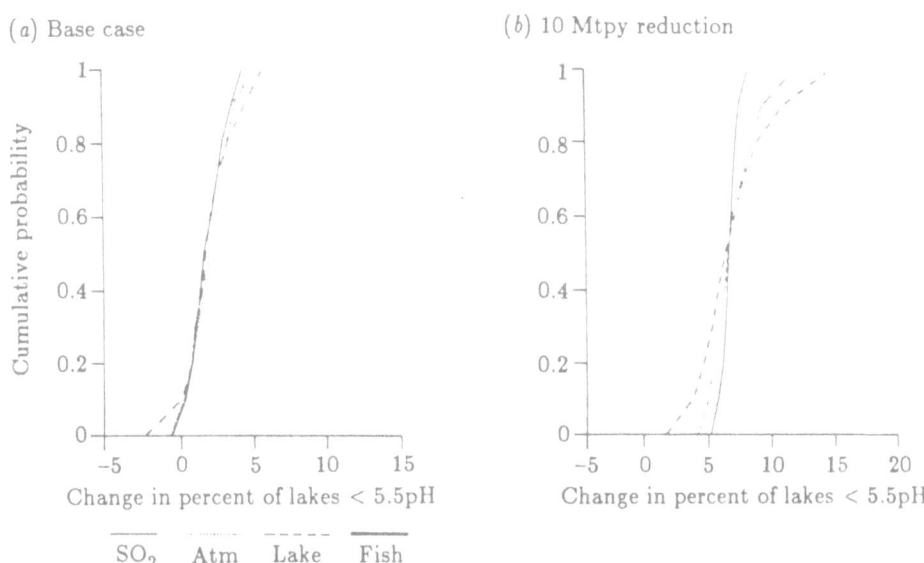

Figure 14.17. Change from 1980 to 2010 in fraction of the Adirondack Park lakes with pH below 5.5: cumulative contribution of component uncertainties.

policy measures currently under consideration to abate acid deposition in North America. We have focused on the problem of regional lake acidification and the consequent loss of fish life, which is the problem that initially brought "acid rain" to the forefront of public attention.

While many significant advances in our understanding of the acid deposition phenomenon have been derived from research over the past decade, we have seen that considerable uncertainties remain. Some of these dominate the ability to predict or discern the benefits of an acid rain control program, or the consequences of taking no action, over the next several decades. For aquatic effects, the present analysis suggests that the greatest uncertainties lie in our understanding of the biological response of different fish species to acidic waters and the processes governing the response of aquatic systems to changes in acid deposition at the regional level. These two factors – particularly the former – contribute most to the overall uncertainty in the estimated impact of emission reductions on the number of lakes potentially able to support fish life in two regions of eastern North America. The additional uncertainties estimated for atmospheric transport–transformation processes and the level of future SO_2 emissions were of less significance to the overall uncertainty for the range of cases examined.

Important caveats, of course, include the assumptions of annual average atmospheric chemistry as a linear process, the adoption of single values to characterize regional deposition levels, and the use of empirically based models as reasonable representations for this level of analysis. Such assumptions can only be refined through additional research and data acquisition, and the development of improved models in the future. Indeed, the extent to which current models may give biased or inaccurate predictions cannot be assessed until verifiably better models are available. Thus, many uncertainties simply cannot be quantified at this time.

It must also be recognized that uncertainty estimates inevitably reflect various technical or professional judgments that are an inherent part of the analysis. For example, the range of scenarios characterizing future SO_2 emission levels must be accepted as reflecting a reasonable characterization of uncertainty based on current estimates or analyses by major parties at interest. Although research *can* help improve analytic *methods*, additional research cannot significantly reduce the uncertainty in estimating SO_2 emissions 30 or 50 years from now. Rather, the power of the modeling framework described in this chapter is to allow the consequences of alternative assumptions about the future to be examined easily and, by this means, to help inform the judgments and opinions important for policy analysis and research.

In the present case, our analysis suggests that despite significant uncertainties, the process of regional lake acidification in the Boundary Waters region of the upper Midwest and the Adirondack Park region of New York appears likely to proceed slowly over the next two to three decades in the absence of any new policy initiatives. The *direction* of change is most likely to follow future SO_2 emission trends, which are highly uncertain. An acid rain control program lowering SO_2 emissions in the eastern USA to 10 Mtpy below 1980 levels would accelerate the recovery of acidic lakes in Adirondack Park, although the magnitude of the recovery (an expected increase over the "no control" case of less than 2% or about 50 lakes capable of supporting trout) could well be masked by the uncertainties and variability in atmospheric processes, lake chemistry, and biological processes. Changes at Boundary Waters would be far less discernible, according to the present analysis.

Our analysis does demonstrate a potential for more dramatic changes in regional aquatic acidification, though with lower probabilities of occurrence. Thus, the need for policy actions inevitably must be weighed against acceptable levels of risk. Further research is expected to reduce current uncertainties and to expand the geographical extent over which regional acidification effects may be quantitatively estimated. Additional research also will allow other potential effects of atmospheric deposition (on materials, forests, etc.) to be evaluated in future analyses.

Acknowledgments

This work was sponsored by a grant to Carnegie-Mellon University from the Claude Worthington Benedum Foundation. The development of the models described here was sponsored in part by a subcontract from Oak Ridge National Laboratory, with funds provided by the US Environmental Protection Agency as part of the National Acid

Precipitation Assessment Program. All results, opinions and conclusions expressed in this chapter, however, are those of the authors alone. Support in model development was provided by Mark Cushey, Cliff Davidson, Lester Lave, Gregory McRae, Dan Stoops, Michael Sutton, and Sarosh Talukdar of CMU. The cooperation and assistance of individuals at ANL, DOE, EPA, EPRI, and ORNL during the course of this work are gratefully acknowledged.

Notes

[1]   Although other species also are involved in acid deposition chemistry, their effects on aquatic systems appear to be indirect via terrestrial systems or, in the case of nitrogen oxide species, related more to episodic events rather than the long-term average lake acidity modeled in this chapter. The impacts of episodic and seasonal events are implicitly reflected in the longer-term acidity and fish viability models.

[2]   Some studies project emissions and effects over a time period of 50 years or more. Because of the very large uncertainty in emission estimates over such periods, however, we chose a somewhat shorter time period, believed to be more relevant for current policy decisions regarding acid rain controls.

[3]   The English system of units is used in this chapter, reflecting prevalent practice in the USA. To convert short tons to metric tons, multiply by 0.9.

[4]   A number of studies project emissions to the year 2030: Braine and Steubi (1987), Placet et al. (1986), and McGowin et al. (1986). In that time period, the effects of power plant lifetime disappear as all older plants are eventually retired. Emission estimates are then dominated by the assumed demand for electricity and the commercial use of "clean coal technology," offering lower emission levels than current plant designs. Current EPA, DOE, and EPRI projections beyond 2010 continue to assume electricity demand growth rates of 2% to 3% per year. Historically, however, rates several times higher than that also have been sustained (e.g., 7% per year during the 1950s and 1960) and cannot necessarily be ruled out in the future.

[5]   This does not affect any subsequent part of the analysis since all emissions are aggregated in the source–receptor model.

[6]   These values were judged to represent the widest range of pH that could occur at these locations with current emissions, based on current and historical deposition data.

[7]   High values of τ_c imply low transformation rates of SO_2 to SO_4 (which deposits more rapidly than SO_2 in wet form), together with high values of τ_{wp} (low wet deposition rates of SO_2) and would produce very low sulfate deposition values and associated high pH values outside the acceptable range.

[8]   Note that the term ANC is used interchangeably with alkalinity in this chapter.

[9]   Computer experiments indicated that 100 iterations were sufficient to determine consistent estimates of confidence intervals of 90% or less. A larger number of iterations could be expected to yield some differences in the extreme tails of the distributions.

[10] Control costs are reported to illustrate the capabilities of the integrated assessment model. The reader certainly should not infer that these costs would be incurred solely for the purpose of mitigating aquatic effects in the two regions analyzed in this chapter.

[11] Of course, for either of the bounding scenarios – i.e., the EPA projections on the high side or the EPRI scenario on the low – the expected changes would be different from those shown in *Figure 14.9*. Because future emissions in this analysis are assumed to lie anywhere within a range (with uniform probability),

the bounding scenario values themselves are reflected primarily in the tails of the overall distributions for lake acidity shown earlier in *Figure 14.9*.

Conclusions

J. Kämäri

Regionalization Methods

Scientists more and more recognize systems analysis as contributing significantly to the progress of science as well as to the management decisions being formulated on the basis of scientific evidence. No better mechanism than that provided by systems analysis seems to be available for evaluating the variety of environmental consequences provoked by human activities. Systems analysis is here understood broadly to include both the organization of data and information into models and the testing and exploration necessary for assessing the model's performance.

This volume deals with a very well-defined and rather narrowly focused problem area. The authors have employed a carefully designed research strategy and followed the objective of providing useful information for the decision-making process. To accomplish that objective, new kinds of mathematical constructs have been developed that are no longer calibrated just for individual sites, field plots, stands, or catchments, but can make use of regional data. How results from synoptic surveys, or similar regional studies, have been utilized by the models varies from application to application. These approaches chosen for extrapolating from one single site to a larger regional scale – a process referred to as "regionalization" – are discussed in the following.

Dose–Response relationships. Examining regional patterns in surface-water acidification has made it possible to construct empirical dose–response relationships for the phenomenon (Chapters 5 and 6). The inherent assumption in these simple models is that all regions are in steady state with respect to loading, and that all regions behave similarly under the same atmospheric loading. In other words, the shift in a regional distribution under changing atmospheric deposition can be calculated on the basis of the empirical relationships.

J. Kämäri (ed.), Impact Models to Assess Regional Acidification, 285–289.

Identification of a simplified model from mechanistic models. In this approach, a simple distribution-based dose–response model, which has a time dependency incorporated in its structure, is calibrated on the basis of results of a more complex mechanistic model (Chapter 10). The parameters of the simplified model are derived from the observed behavior of the more complex model. There are obvious advantages of this kind of simplification procedure – faster computing, allowing interactive use, being not the least important. The disadvantages – mismatches in the model results between the simplified version and the mechanistic version – are still to be investigated.

Scaling. The most straightforward and perhaps the most widely used methodology for regionalization involves modeling the behavior of a number of individual sites and then scaling the results to a larger population of sites. A prerequisite for this approach is that the representativeness of the sample site population be known. Most often, the modeled sites treated this way typify the region under consideration, and can thus with good reason be assumed to represent most of the sites in the area (Chapters 2 and 3). In practice, however, the representativeness is very seldom precisely known, and so-called typical sites cannot be named. If that is the case, the number of sites to be modeled, to guarantee representativeness, should be large. In such cases this methodology, owing to the need for extensive data collection, becomes an expensive approach.

Grid-based application. In the grid-based application, the area under consideration is subdivided into a number of equal-sized grid cells. All parameter values, initial conditions for all variables, as well as the time pattern for all driving variables have to be determined separately for each individual grid. Either direct measurements or indirect information, from which model inputs are derived, may be utilized. When using indirect information, so-called transfer functions may be applied to interpret the required parameter value from available large-scale information.

The grid-based application is not, however, suitable for all models. Environmental models that include parameters or variables that are difficult or impossible to estimate for a large number of grids should avoid this type of application. A soil acidification model is a good prospect for a grid-based application (Chapter 1). Large-scale soil and geologic maps may be used to derive the initial conditions and the rate constants for the model.

Monte Carlo applications. Monte Carlo parameter estimation is a trial-and-error procedure for the solution of the inverse problem, i.e., for estimating poorly known input and parameter values derived from the comparison of model outputs with available measurements. In the Monte Carlo simulation, constraints on the output are formulated, describing the expected satisfactory behavior of the model. Next, probability distributions are defined for all poorly known input and parameter values. The Monte Carlo program then randomly samples the parameter vectors from these distributions, runs the simulation model through a selected period of time, and finally tests for violations of the constraints. This process is repeated for a large number of trials.

When the Monte Carlo simulation framework is used for regional applications, not only the site-specific parameters, but also those exhibiting spatial variation are varied. Two fairly similar ways to utilize the Monte Carlo program in the regional applications have been introduced in this volume. In both cases, all inputs and parameters, even those that are spatially varied, are assigned ranges broad enough so that any reasonable value for an input can be selected. Monte Carlo simulations are then carried out by randomly selecting a set of input values from these designated ranges and integrating the equations from a suitable starting time. The accepted Monte Carlo runs in a way produce a set of theoretical lakes, which are assumed to be representative of real catchments for that region. These accepted ensembles can then be used for predictions of the response of systems to different patterns of future acidic deposition.

In the first type of Monte Carlo application, those randomly selected parameter ensembles are accepted for predictions, which produce values for the state variables that simultaneously fall within the prespecified ranges observed in the region (Chapter 7). In the second type of regional application, the Monte Carlo simulation is used to determine the combinations of parameters that produce an acceptable distribution of one or two state variables observed in the study region (Chapter 9). The difference between these two approaches is that the latter type attempts to fit a prespecified *distribution* as closely as possible, and, thus, the desired distribution is obtained by force. The former type accepts all trials that simultaneously hit several prespecified *ranges*. In this case, the inherent model structure is expected to produce (and, in fact, produces) the observed regional distributions for state variables.

Areas for Improvement

The authors of the preceeding chapters have already outlined some important directions for future research as well as some major sources of uncertainty bearing on the reliability of the model results. These specific questions will not be addressed here; instead, I will identify a few areas, put forward in the Task Force Meeting in Warsaw, which have not received much attention so far. These recommendations should be considered feedback from the substantial body of experience in applying systems analysis to acidification.

Model reliability. A very basic question addressed to the acidification models – "How well does the model perform?" – seems to be very difficult to answer. There is, however, a clear need to explore this question in quantitative terms. The whole issue of uncertainty is closely linked to the possible use of the models in a decision-making context. The person who uses the model should be offered information on how to interpret the results.

Many issues of uncertainty analysis have been resolved over the past five to ten years. Modelers are, however, facing new uncertainty issues as they move toward more generalized models and a larger spatial scale. An additional source of error is introduced in models that attempt to forecast broad-scale effects as a function of site-specific estimates, i.e., models that depend upon extrapolation of site-specific results to the region. The extrapolation error may be critical in predicting acidification on a large spatial scale, and it is therefore important for researchers to address this concern.

For example, determining the boundaries of the study regions quite possibly also influences the outcome. The question here is what variable should one use as a basis for establishing the environmental regions to minimize the possible prediction errors. There is no information available on how the predictions are affected, or how they could be improved, by developing regional definitions on the basis of homogeneity of processes.

Data derivation. It has been very fortunate that many of the acidification models developed so far have been process-oriented. That has allowed the models to be used as research tools, indicating gaps in knowledge and directing further research. Applications of process-oriented regional models for acidification have demonstrated that there are but a few important measurable parameters that largely determine the long-term behavior of the systems. The national and international survey programs certainly welcome this information and concentrate on the most important variables first.

Surveys for determining the state of acidification (e.g., soil base saturation, lake alkalinity) are naturally important programs for regional models, but research should continue on a process level as well. Specific soil research is needed to derive parameter values describing the rates of some major soil processes (e.g., mineralization rate constants, weathering rate constants). This type of research should be directed to dominant soil types within the region under consideration.

For modelers, it has become regrettably apparent that survey programs proceed slowly. Therefore, we recognize the need to utilize existing regional data as much as possible. More research can be conducted on deriving reliable transfer functions between soil properties needed in the models (e.g., CEC) and soil characteristics available on large-scale soil maps (e.g., soil types). The lack of regional data, moreover, implies that data have to be extrapolated from one region to another. This extrapolation introduces some uncertainty to predictions that should be identified and, where possible, quantified.

The quantity and quality of regional data will always vary between regions. This variability in data evidently implies that simple models with less extensive input requirements may be more useful than more detailed models. For large-scale regional applications, models should be simplified, within the limits of scientific knowledge, to such degree that data requirements are in balance with data availability. An important task for future research would be to determine the optimal level of model sophistication (process descriptions versus empirical relationships) that would maximize the use of relevant information, but minimize the errors associated with regionalized predictions.

Model use. When looking back some 15 years and evaluating the progress in modeling, particularly in acidification modeling, it is easy to state that some models today have reached a level of acceptance that justifies their use for management purposes. When comparing modeling with other kinds of scientific work, it is obvious that the expected end product of the modeling research is at present much more than a scientific paper. There seems to be a clear demand to construct our models in such a way that they are usable by governmental officers, managers, politicians, and technical advisors – those who are usually referred to as decision makers.

Much too often, we hear about models that have not been used in the real world. In many cases, the reason has been the lack of documentation. It should be self-evident that poorly documented programs are like complex tools without instructions on how to use them. It would be overly optimistic to propose that a software support team should always be maintained to answer all specific questions by the potential model users. It would be even more unrealistic to expect the development team actually to perform the applications. Nevertheless, we have surely learned by now that there is hardly any use to be seen for a set of statements in a computer language after they are written. At a minimum, such statements must be translated into a more common language. Better yet they ought to be thoroughly explained. Above all, clear instructions should be provided on how to proceed at each step of model application: data derivation, calibration, verification, and prediction.

Scientists have now taken the important initiative of developing models that attempt to establish the link between large-scale data and large-scale consequences. Nobody can, however, force the use of these models. Scientists can only advise the decision makers, whose task is to take into account facts (model results) and human values (public opinion), weigh those two, and derive laws, standards, and policies. The more credible and usable our models, the more heavily will the results weigh in the decision-making process.

References

Aastrup, M. and Johnson, J., 1986, in C. Bernes, ed., *Monitor 1986: Sura och försurade vatten*, SNV, Solna, Sweden [in Swedish].

Abrahamsen, G., Bjor, K., and Teigen, O., 1976, *Field Experiments with Simulated Acid Rain in Forest Ecosystems*, Research Report, FR 4/76, SNSF-Project, Box 61, 1432 Ås-NLH, Norway.

Adler, G.H. and Wilson, M.L., 1985, Small Mammals on Massachusetts Islands: The Use of Probability Functions in Clarifying Biogeographic Relationships, *Oecologia* **66**: 178–186.

Ågren, G.I. and Kauppi, P., 1983, *Nitrogen Dynamics in European Forest Ecosystems: Considerations Regarding Anthropogenic Nitrogen Depositions*, CP-83-28, International Institute for Applied Systems Analysis, Laxenburg, Austria.

Ahl, T., 1986, in C. Bernes, ed., *Monitor 1986: Sura och försurade vatten*, SNV, Solna, Sweden [in Swedish].

Alcamo, J., Hordijk, L., Kämäri, J., Kauppi, P., Posch, M., and Runca, E., 1985, Integrated Analysis of Acidification in Europe, *Journal of Environmental Management* **21**: 47–61.

Alcamo, J., Amann, M., Hettelingh, J.-P., Holmberg, M., Hordijk, L., Kämäri, J., Kauppi, L., Kauppi, P., Kornai, G., and Mäkelä, M., 1987, Acidification in Europe: A Simulation Model for Evaluating Control Strategies, *Ambio* **16**: 232–245.

Almer, B., Dickson, W., Ekström, C., and Hörnström, E., 1978, Sulphur Pollution and the Aquatic Ecosystem, in J.O. Nriagu, ed., *Sulphur in the Environment*, John Wiley, New York, NY, USA.

Atmospheric Sciences and Analysis Working Group 2, 1982, *US–Canadian Memorandum of Intent on Transboundary Air Pollution*, US Environmental Protection Agency, Washington, DC, USA.

Aust, H., 1983, The Groundwater Resources of the Federal Republic of Germany, in *Groundwater in Water Resources Planning*, UNESCO **1**: 15–33.

Baker, L.A., Brezonik, P.L., and Pollman, C.D., 1986, Model of Internal Alkalinity Generation: Sulphate Retention Component, *Water, Air, and Soil Pollution* **31**: 89–94.

Balson, W.E. and North, D.W., 1982, *Acid Deposition: Decision Framework*, (2 Vols.) EA-2540, Prepared by Decision Focus Incorporated, Los Altos, CA, for Electric Power Research Institute, Palo Alto, CA, USA.

Barbour, C.D. and Brown, J.H., 1974, Fish Species Diversity in Lakes, *American Naturalist* **108**: 473–489.

Bard, Y., 1974, *Nonlinear Parameter Estimation*, Academic Press, New York, NY, USA.

Bartell, S.M., Gardner, R.H., O'Neill, R.V., and Giddings, J.M., 1983, Error Analysis of Predicted Fate of Anthracene in a Simulated Pond, *Environmental Toxicology and Chemistry*, **2**: 19–28.

Bartell, S.M., Gardner, R.H., and O'Neill, R.V., 1986, The Influence of Bias and Variance in Predicting the Effects of Phenolic Compounds in Ponds, in B.T. Fairchild, ed., *Proceedings for the 1986 Eastern Simulation Conference*, Vol. 17, No. 2, Supplementary Proceedings.

Bathurst, J.C., 1986, Physically Based Distributed Modeling of an Upland Catchment Using the Système Hydrologique Européen, *Journal of Hydrology* **87**: 79–102.

Battarbee, R.W. and Flower, R.J., 1986, *Paleoecological Evidence for the Timing and Causes of Lake Acidification in Galloway, South West Scotland*, Working Paper 8, University College, London, UK.

Beggs, G.L., Gunn, J.M., and Olver, C.H., 1985, *The Sensitivity of Ontario Lake Trout (Salvelinus namaycush) and Lake Trout Lakes to Acidification*, Ontario Fisheries Technical Report Series No. 17, Ontario, Canada.

Belmans, C., Wesseling, J.G., and Feddes, R.A., 1981, *Simulation Model of the Water-balance of a Cropped Soil Providing Different Types of Boundary Conditions (SWATRE)*, ICW-note, 1257.

Benjamin, J.R. and Cornell, C.A., 1970, *Probability, Statistics, and Decision for Civil Engineers*, McGraw-Hill, New York, NY, USA.

Bernes, C., ed., 1986, *Monitor 1986: Sura och försurade vatten*, SNV, Solna, Sweden [in Swedish].

Bettelheim, J. and Littler, A., 1979, *Historical Trends of Sulphur Oxide Emissions in Europe Since 1865*, Central Electricity Generating Board, London, Report No. CEGB, PL-GS/E/1/79, London, UK.

Bhat, K.K., Flowers, T.N., and O'Callaghan, J.R., 1980, A Model for the Simulation of the Fate of Nitrogen in Farm Wastes on Land Application, *Journal of Agricultural Science* 94(1): 183–193.

Bloom, P.R. and Grigal, D.F., 1985, Modeling Soil Response to Acidic Deposition in Nonsulfate Adsorbing Soils, *Journal of Environmental Quality* 14: 489–495.

Bolt, G.H., Bruggenwert, M.G.M., and Kamphorst, A., 1976, Adsorption of Cations by Soil, in G.H. Bolt and M.G.M. Bruggenwert, eds., *Soil Chemistry A: Basic Elements*, Elsevier, Amsterdam, Netherlands.

Bouma, J., Van Lanen, H.A.J., Breeuwsma, A., Wösten, J.H.M., and Kooistra, M.J., 1986, Soil Survey Data Needs When Studying Modern Land Use Problems, *Soil Use and Management* 2(2): 125–130.

Braine, B. and Stuebi, R., 1987, Memorandum to Paul Schwengels of EPA on Interim EPA Base Case Forecasts, US Environmental Protection Agency, Washington, DC, USA.

Brakke, D.F., Landers, D.H., and Eilers, J.M., 1988, Chemical and Physical Characteristics of Lakes in the Northeastern United States, *Environmental Science and Technology* 22: 155–163.

Breemen, N. van, de Visser, P.H.B., and van Grinsven, J.J.M., 1986, Nutrient and Proton Budgets in Four Soil–Vegetation Systems Underlain by Pleistocene Alluvial Deposits, *Journal of the Geological Society* 143: 659–666.

Breeuwsma, A. and Schoumans, O.F., 1987, Forecasting Phosphate Leaching on a Regional Scale, in W. van Duijvenbooden and H.G. van Waegeningh, eds., *Vulnerability of Soil and Groundwater to Pollutants: International Conference*, Noordwijk aan Zee, Netherlands, Proceedings and Information No. 38, TNO-CHO/RIVM, The Hague, Netherlands.

Breeuwsma, A., Wösten, J.H.M., Vleeshouwer, J.J., Van Slobbe, A.M., and Bouma, J., 1986, Derivation of Land Qualities to Assess Environmental Problems from Soil Surveys, *Soil Science Society of America Journal* 50(1): 186–190.

Briggs, J.C., 1986, Introduction to the Zoogeography of North American Fishes, in C.H. Hocutt and E.O. Wiley, eds., *The Zoogeography of North American Freshwater Fishes*, Wiley, New York, NY, USA.

Brodersen, K., Mortensen, P.B., and Petersen, T., 1986, *ECCES: A Model for Calculation of Environmental Consequences from Energy Systems, Predicting Ion Concentrations and Acidification Effects in Terrestrial Ecosystems*, Risø-M-2615, Denmark.

Burns, J.C., Coy, J.S., Tervet, D.J., Harriman, R., Morrison, B.R.S., and Quine, C.P., 1984, The Loch Dee Project: A Study of the Ecological Effects of Acid Precipitation and Forest Management on an Upland Catchment in South West Scotland, *Fisheries Management* 15: 145–167.

Canadian Control Program, The, 1987, Environment Canada, Ottawa, Ontario, Canada.

Carter, A.D., Palmer, R.C., and Monkhouse, R.A., 1987, Mapping the Vulnerability of Groundwater to Pollution from Agricultural Practice, Particularly with Respect to Nitrate, in W. van Duijvenbooden and H.G. van Waegeningen, eds., *Vulnerability of Soil and Groundwater to Pollutants*, The Hague, Netherlands.

Center for Energy and Environmental Studies, 1984, *A Conceptual Framework for Integrated Assessments of Acid Deposition*, Final Report from CEES, Carnegie Mellon University, Pittsburgh, PA, to US Environmental Protection Agency, Washington, DC, USA.

Chen, C.W., Dean, J.D., Gherini, S.A., and Goldstein, R.A., 1982, Acid Rain Model: Hydrological Module, *Journal of Environmental Engineering* 108: 455–472.

Chen, C.W., Gherini, S.A., Hudson, R.J.M., and Dean, J.D., 1983, *The Integrated Lake–Watershed Acidification Study*, Vol. 1: *Model Principles and Application Procedures*, Electric Power Research Institute, Final Report, EPRI EA-3221, Research Project 1109–5, Palo Alto, CA, USA.

Christensen, B., Mortensen, P.B., and Petersen, T., 1985, *Illustration of the Present Capabilities of the ECCES Program System*, M-2501, Report of the Risø National Laboratory, Denmark.

Christophersen, N., Seip, H.M., and Wright, R.F., 1982, A Model for Streamwater Chemistry at Birkenes, Norway, *Water Resources Research* 18: 977–996.

Christophersen, N., Dymbe, L.H., Johannessen, M., and Seip, H.M., 1984, A Model for Sulphate in Streamwater at Storgama, Southern Norway, *Ecological Modelling* 21: 35–61.

Chun, K.C., 1987, *Uncertainty Data Base for Emissions-Estimation Parameters: Interim Report*, ANL/EES-TM-328, Argonne National Laboratory, Argonne, IL, USA.

Clark, J.S. and Hill, R.G., 1964, The pH–Percent Base Saturation Relationships of Soils, *Soil Science Society America Proceedings* 28: 490–492.

Cook, R.B., Jones, M.L., Marmorek, D.R., Elwood, J.W., Malanchuk, J.L., Turner, R.S., and Smol, J.P., 1988, *The Effects of Acidic Deposition on Aquatic Resources in Canada: An Analysis of Past, Present, and Future Effects*, ORNL/TM-10405, Oak Ridge National Laboratory, Oak Ridge, TN, USA.

Cosby, B.J. and Wright, R.F., 1987, A Regional Model of Surface Water Acidification in Southern Norway, in M.B. Beck, ed., *Systems Analysis in Water Quality Management*, Pergamon Press, New York, NY, USA.

Cosby, B.J., Wright, R.F., Hornberger, G.M., and Galloway, J.N., 1985a, Modeling the Effects of Acid Deposition: Assessment of a Lumped Parameter Model of Soil Water and Streamwater Chemistry, *Water Resources Research* 21: 51–63.

Cosby, B.J., Wright, R.F., Hornberger, G.M., and Galloway, J.N., 1985b, Modeling the Effects of Acid Deposition: Estimation of Long-Term Water Quality Responses in a Small Forested Catchment, *Water Resources Research* 21: 1591–1601.

Cosby, B.J., Hornberger, G.M., Galloway, J.N., and Wright, R.F., 1985c, Time Scales of Catchment Acidification, *Environmental Science and Technology* 19: 1141–1149.

Cosby, B.J., Hornberger, G.M., Wright, R.F., and Galloway, J.N., 1986a, Modeling the Effects of Acid Deposition: Control of Long-Term Sulfate Dynamics by Soil Sulfate Adsorption, *Water Resources Research* 22: 1283–1291.

Cosby, B.J., Whitehead, P.G., and Neal, R., 1986b, A Preliminary Model of Long-Term Changes in Stream Acidity in Southwestern Scotland, *Journal of Hydrology* 84: 381–401.

Cosby, B.J., Hornberger, G.M., Wolock, D.M., and Ryan, P.F., 1987, Calibration and Coupling of Conceptual Rainfall-Runoff/Chemical Flux Models for Long-Term Simulation of Catchment Response to Acidic Deposition, in M.B. Beck, ed., *Systems Analysis in Water Quality Management*, Pergamon Press, New York, NY, USA.

Cushey, M.A. and Rubin, E.S., 1987, Simplified Models of Acid Rain Control Costs for the Electric Utility Sector, in *Proceedings of the 80th Annual Meeting, Air Pollution Control Association*, Pittsburgh, PA, USA.

Dale, V.H., Jager, H.I., Gardner, R.H., and Rosen, A.E., 1989, Using Sensitivity and Uncertainty Analysis to Improve Predictions of Broad-Scale Forest Development, *Ecological Modelling*, 42: 169–178.

Davidson, J.M., Graetz, D.A., Rao, P.S.C., and Selim, H.M., 1978, *Simulation of Nitrogen Movement and Uptake in the Plant Rootzone*, Ecological Research Series 600/3–78–029, Environmental Protection Agency, Athens, GA, USA.

de Grosbois, E., Dillon, P.J., Seip, H.M., and Seip, R., 1986, Modeling Hydrology and Sulphate Concentration in Small Catchments in Central Ontario, *Water, Air, Soil Pollution* 31: 45–57.

Devore, J.L., 1982, *Probability and Statistics for Engineering and the Sciences*, Brooks/Cole Publishing Company, Monterey, CA, USA.

de Vries, W., 1987, The Role of Soil Data in Assessing the Large-Scale Impact of Atmospheric Pollutions on the Quality of Soil Water, in W. van Duijvenbooden and H.G. van Waegening, eds., *Vulnerability of Soil and Groundwater to Pollutants*, International Conference, Noordwijk aan Zee, March 30–April 3, Proceedings and Information No. 38, TNO–CHO/RIVM, The Hague, Netherlands.

de Vries, W. and Breeuwsma, A., 1987, The Relation Between Soil Acidification and Element Cycling, *Water, Air, and Soil Pollution* 35: 293–310.

de Vries, W., Heijnnen, M.M.T., Balkema, W., and Sjardijn, R.C., forthcoming, Kinetics and Mechanisms of Buffer Processes in Acid Sandy Soils.

Dickson, W., 1985, unpublished article, SNV, Solna, Sweden.

Driscoll, C.T. and Schafran, 1984, Short-Term Changes in the Base Neutralizing Capacity of an Acid Adirondack Lake, New York, *Nature* 310: 308–310.

Eadie, J.McA. and Keast, A., 1984, Resource Heterogeneity and Fish Species Diversity in Lakes, *Canadian Journal of Zoology* 62: 1689–1695.

Eadie, J.McA., Hurly, T.A., Montgomerie, R.D., and Teather, K.L., 1986, Lakes and Rivers as Islands: Species–Area Relationships in the Fish Faunas of Ontario, *Environmental Biology of Fishes* 15: 81–89.

Eager, C., 1984, Review of the Biology and Ecology of the Balsam Wooly Aphid in Southern Appalachian Spruce–Fir Forests, in P.S. White, ed., *The Southern Appalachian Spruce-Fir Ecosystem: Its Biology and Threats*, National Park Service Research/ Resources Management Report SER-71, 36–50.

Edmunds, W.M. and Kinniburgh, D.G., 1986, The Susceptibility of UK Groundwaters to Acid Deposition, *Journal of the Geological Society of London* 143: 707–720.

Edmunds, W.M. and Kinniburgh, D.G., 1986a, *Regional Hydrogeochemical Survey of Groundwater Acidity in Parts of Scotland*, Interim Report, British Geological Survey, Wallingford, UK.

Eilers, J.M., Glass, G.E., and Webster, K.E., 1983, Relationships Between Susceptibility of Lakes to Acidification and Factors Controlling Lake Water Quality: Hydrology as a Key Factor, *Canadian Journal of Fisheries and Aquatic Sciences* 40: 1896–1904.

Eilers, J.M., Lien, G.L., and Berg, R.G., 1984, *Aquatic Organisms in Acidic Environments: A Literature Review*, Wisconsin Department of Natural Resources, Technical Bulletin 150.

Eilers, J.M., Brakke, D.F., and Landers, D.H., 1988, Chemical and Physical Characteristics of Lakes in the Upper Midwest, United States, *Environmental Science and Technology* 22: 164–172.

Eliassen, A. and Saltbones, J., 1983, Modelling of Long-Range Transport of Sulphur Over Europe: A Two-Year Model Run and Some Model Experiments, *Atmospheric Environment* 17: 1457–1473.

Environmental Protection Agency, 1986, *Characteristics of Lakes in the Eastern United States*, EPA/600/4-86/007a,b,c, US Environmental Protection Agency, Washington, DC, USA.

Falkengren-Grerup, V., Linnermark, N., and Tyler, G., 1987, Acidity and Exchangeable Cation Pools of South Swedish Soils from 1949 to 1985, *Chemosphere* 16:2239–2248.

FAO–UNESCO, 1974, *Soil Map of the World*, Vols. I and V, Paris, France.

Fay, J., Golomb, D., and Kumar, S., 1985, Source Apportionment of Wet Sulfate Deposition in Eastern North America, *Atmospheric Environment* 19: 1773–1782.

Fedra, K., 1983, *Environmental Modeling under Uncertainty: Monte Carlo Simulation*, RR-83-28, International Institute for Applied Systems Analysis, Laxenburg, Austria.

Flower, R.J., Rippey, B., and Tervet, D.J., 1985, *Thirty-four Galloway Lochs, Bathymetry, Water Quality, and Surface Sediment Diatom Assemblages*, Working Paper No. 14, University College, London, UK.

Fölster, H., 1985, Proton Consumption Rates in Holocene and Present-Day Weathering of Acid Forest Soils, in J.I. Drever, ed., *The Chemistry of Weathering*, D. Reidel, Dordrecht, Netherlands.

Forsberg, C., Morling, G., and Wetzel, R.G., 1985, Indications of the Capacity for Rapid Reversibility of Lake Acidification, *Ambio* 14: 164–166.

Forsius, M., 1987, *Suomen järvien alueellinen happamuustilanne*, [Summary: Regional Extent of Lake Acidification Situation in Finland], National Board of Waters and Environment, Report 9, Helsinki, Finland.

Frank, J., 1980, *Soil Survey at Birkenes, a Small Catchment in Aust-Agder County, Southern Norway*, Technical Note, SNSF-Project, TN 60/80, SNSF-Project, Box 61, 1432 Ås-NLH, Norway.

Fukunaga, K. and Koontz, W.L.G., 1970, Application of the Karhunen–Loeve Expansion to Feature Selection and Ordering, *IEEE Transactions on Computers* C-19: 311–318.

Galloway, J.N., Norton, S.A., and Church, M.R., 1983, Fresh Water Acidification from Atmospheric Deposition of Sulfuric Acid: A Conceptual Model, *Environmental Science and Technology* 17: 541A–545A.

Gardner, R.H. and Trabalka, J.R., 1985, *Methods of Uncertainty Analysis for a Global Carbon Dioxide Model*, US Department of Energy Technical Report, DOE/OR/21400–4, Washington, DC, USA.

Gardner, R.H., O'Neill, R.V., Mankin, J.B., and Kumar, D., 1980, Comparative Error Analysis of Six Predator–Prey Models, *Ecology* 61: 323–332.

Gherini, S.A., Mok, L., Hudson, R.J.M., Davis, G.F., Chen, C.W., and Goldstein, R.A., 1985, The ILWAS model: Formulation and Application, *Water, Air, and Soil Pollution* 26: 425–460.

Granat, L., 1986, in C. Bernes, ed., *Monitor 1986. Sura och försurade vatten*, SNV, Solna, Sweden [in Swedish].

Greig, D.C., 1971, *British Regional Geology: The South of Scotland*, HMSO, Edinburgh, Scotland.

Hagin, J. and Amberger, A., 1974, *Contribution of Fertilizers and Manures to the N and P Load of Waters: A Computer Simulation Model*, Final Report submitted to the Deutsche Forschungsgemeinschaft.

Haines, T.A., 1981, Acid Precipitation and Its Consequences for Aquatic Ecosystems, *Transactions of the American Fisheries Society* 110: 669–707.

Hales, J.M., 1982, The MAP3S/RAINE Precipitation Chemistry Network: Statistical Overview for the Period 1976–1980, *Atmospheric Environment* 16: 1603–1631.

Hallbäcken, L. and Tamm, C.O., 1986, Changes in Soil Acidity from 1927 to 1982–84 in a Forest Area of Southwest Sweden, *Scandinavian Journal of Forest Research* 1.

Hari, P., Kaipiainen, L., Korpilahti, E., Mäkelä, A., Nilson, T., Oker-Blom, P., Ross, J., and Salminen, R., 1985, *Structure, Radiation, and Photosynthetic Production in Coniferous Stands*, University of Helsinki, Department of Silviculture, Research Notes 54, Helsinki, Finland.

Harriman, R., Morrison, B., Caines, C.A., Collen, P., and Watt, A.W., 1987, Long-term Changes in Fish Populations of Acid Streams and Lochs in Galloway, Southwest Scotland, *Water, Air, and Soil Pollution* 32:89–112.

Harvey, H.H., 1975, Fish Populations in a Large Group of Acid-Stressed Lakes, *Int. Ver. Theor. Ang. Limnol. Verh.* 19: 2406–2417.

Helgeson, H.C., Murphy, W.M., and Aagaard, P., 1984, Thermodynamic and Kinetic Constraints on Reaction Rates among Minerals and Aqueous Solutions, II: Rate Constants, Effective Surface Area and the Hydrolysis of Feldspar, *Geochimica et Cosmochimica Acta* 48: 2405–2432.

Henriksen, A., 1979, A Simple Approach for Identifying and Measuring Acidification in Freshwater, *Nature* 278: 542–545.

Henriksen, A., 1980, Acidification of Freshwaters – A Large Scale Titration, in D. Drabløs and A. Tollan, eds., *Proceedings International Conference Ecological Impact of Acid Precipitation*, SNSF–Project, Sandefjord, Norway.

Henriksen, A., 1982a, *Changes in Base Cation Concentrations Due to Freshwater Acidification*, NIVA Rep. OF–81623, Norwegian Institute of Water Research.

Henriksen, A., 1982b, Susceptibility of Surface Waters to Acidification, in R.E. Johnson, ed., *Acid Rain/Fisheries, Proceedings International Symposium on Acidic Precipitation and Fishery Impacts in Northeastern North America*, American Fisheries Society, Bethesda, MD, USA.

Henriksen, A., 1983, Forsuringsmodeler – kan de brukes? in *19. Nordiska Symposiet om Vattenforskning, Vääksy*, Nordforsk, Finland [in Norwegian].

Henriksen, A., 1984, Changes in Base Cation Concentrations Due to Freshwater Acidification, *Verh. Internat. Verein. Limnol.* **22**: 692–698.

Henriksen, A. and Brakke, D.F., 1988, Sulfate Deposition to Surface Waters, *Environmental Science and Technology* **22**:8–14.

Henriksen, A. and Seip, H.M., 1980, Strong and Weak Acids in Surface Waters of Southern Norway and Southwestern Scotland, *Water Research* **14**: 809–813.

Henrion, M. and Wishbow, N., 1987, *DEMOS User's Manual: 3*, Department of Engineering and Public Policy, Carnegie-Mellon University, Pittsburgh, PA, USA.

Hettelingh, J.-P. and Hordijk, L., 1986, Environmental Conflicts: The Case of Acid Rain in Europe, *Annals of Regional Science* **20**: 38–52.

Hoeks, J., 1986, Acidification of Groundwater in the Netherlands, in *Proceedings International Conference on Water Quality Modeling in the Inland Natural Environment*, 10–13 June, Bournemouth, UK.

Hoekstra, C. and Poelman, J.N.B., 1982, *Dichtheid van gronden gemeten aan de meest voorkomende bodemeenheden in Nederland*, STIBOKA Report 1582.

Holling, C.S., ed., 1978, *Adaptive Environmental Assessment and Management*, Wiley, Chichester, UK.

Holmberg, M. and Hari, P., 1987, On the Influence of Evapotranspiration and Precipitation on Long-Term Changes in Exchangeable Base Cations in Forest Soil [English summary], *Aquilo* **25** I: 13–19.

Holmberg, M., Mäkelä, A., and Hari, P., 1985, Simulation Model of Ion Dynamics in forest Soil, in C. Troyanowsky, ed., *Air Pollution and Plants*, VCH, Weinheim.

Holmberg, M., Johnston, J., and Maxe, L., 1987, Assessing Aquifer Sensitivity to Acid Deposition, in W. van Duijvenbooden and H.G. van Waegeningen, eds., *Vulnerability of Soil and Groundwater to Pollutants*, The Hague, Netherlands.

Holtan, H., ed., 1986, *Norwegian Water Encyclopedia*, Norwegian Hydrologic Committee, Oslo, Norway.

Hornberger, G.M. and Cosby, B.J., 1985a, Selection of Parameter Values in Environmental Models Using Sparse Data: A Case Study, *Applied Mathematics and Computation* **17**: 335–355.

Hornberger, G.M. and Cosby, B.J., 1985b, Evaluation of a Model of Long-Term Response of Catchments to Atmospheric Deposition of Sulfate, in H.A. Barker and P.C. Young, eds., *Proceedings IFAC Symposium on Identification and System Parameter Estimation*, York, England, 3–7 July, Pergamon Press, New York, NY, USA.

Hornberger, G.M. and Spear, R.C., 1981, An Approach to the Preliminary Analysis of Environmental Systems, *Journal of Environmental Management* **12**: 7–18.

Hornberger, G.M. Cosby, B.J., and Galloway, J.N., 1986a, Modeling the Effects of Acid Deposition: Uncertainty and Spatial Variability in Estimation of Long-Term Sulfate Dynamics in a Region, *Water Resources Research* **22**(8): 1293–1302.

Hornberger, G.M., Cosby, B.J., and Rastetter, E.B., 1986b, Regionalization of Prediction of Effects of Atmospheric Deposition on Surface Waters, in *Proceedings International Conference on Water Quality Modelling in the Inland Natural Environment*, Bournemouth, England, 10–13 June, BHRA, Fluid Engineering Centre, Bedford, UK.

Hornberger, G.M., Cosby, B.J., and Wright, R.F., 1987, Analysis of Historical Surface Water Acidification in Southern Norway Using a Regionalized Conceptual Model (MAGIC), in M.B. Beck, ed., *Systems Analysis in Water Quality Management*, Pergamon Press, New York, NY, USA.

Hovmand, M.J. and Petersen, L., 1985, Forsuringsprojektet: Jordforsuring, Miljostyrelsen [part of a review of acidification problems in Denmark, issued by the Danish Environmental Protection Agency].

Howells, G.D., 1984, Fishery Decline: Mechanisms and Predictions, *Phil. Trans. R. Soc. London B* 305: 529–547.

Hunsaker, C.T., Christensen, S.W., Beauchamp, J.J., Olson, R.J., Turner, R.S., and Malanchuk, J.L., 1987, *Empirical Relationships between Watershed Attributes and Headwater Lake Chemistry in the Adirondack Region*, ORNL/TM-9838, Oak Ridge National Laboratory, Oak Ridge, TN, USA.

Iman, R.L. and Conover, W.J., 1982, A Distribution-Free Approach to Inducing Rank Correlations Among Input Variables, *Communication in Statistics: Simulation and Computation* 11: 311–334.

Industrial Environmental Research Laboratory, 1984, *Status Report on the Development of the NAPAP Emission Inventory for the 1980 Base Year and Summary of Preliminary Data*, EPA-600/7-84-091, US Environmental Protection Agency, Research Triangle Park, NC, USA.

Jacks, G. and Knutsson, G., 1982, *Känsligheten för grundvattenförsurning i olika delar av landet (Huvudrapport)*, Projekt Kol-Hälsa-Miljö, Teknisk Rapport 49, Sweden.

Jacks, G., Knutsson, G., Maxe, L., and Fylkner, A., 1984, Effect of Acid Rain on Soil and Groundwater in Sweden, in B. Yaron, *et al.*, eds., *Pollutants in Porous Media*, Springer-Verlag, Berlin, Heidelberg, Federal Republic of Germany, and New York, NY, USA.

Järvinen, O., 1986, *Laskeuman laatu Suomessa 1971-1982*, [Deposition in Finland in 1971–1982; in Finnish], Mimeograph Report 408, National Board of Waters, Helsinki, Finland.

Järvinen, O. and Haapala, K., 1980, *Sadeveden Laatu Suomessa 1971-1977*, Report National Board of Waters, 198, Helsinki, Finland.

Jefferies, D.S., Wales, D.L., Kelso, J.R.M., and Linthurst, R.A., 1986, Regional Chemical Characteristics of Lakes in North America, Part I: Eastern Canada, *Water, Air, and Soil Pollution* 31: 551–567.

Johnson, R.E., ed., 1982, *Acid Rain/Fisheries*, American Fisheries Society, Bethesda, MD, USA.

Johnson, M.G., Leach, J.H., Minns, C.K., and Olver, C.H., 1977, Limnological Characteristics of Ontario Lake in Relation to Associations of Walleye *Stizostedion vitreum vitreum*, Northern Pile *Esox lucius*, Lake Trout *Salvelinus namaycush*, and Smallmouth Bass *Micropterus dolomieni*, *Journal of the Fisheries Research Board of Canada* 34: 1592–1601.

Johnson, D.W., Nilsson, I.S., Reuss, J.O., Seip, H.M., and Turner, R.S., 1985, *Predicting Soil and Water Acidification: Proceedings of a Workshop*, ORNL/TM-9258, Oak Ridge National Laboratory, Oak Ridge, TN, USA.

Jones, M.L., Marmorek, D.R., Rattie, L.P., Wedeler, C.H.R., and Slocombe, D.S., 1987, *Derivation and Analysis of a Model to Predict the Effect of Acidic Deposition on Surface Waters*, Report by ESSA Ltd. to Departments of Fisheries and Oceans, and Environment.

Jones, M.L., Minns, C.K., Marmorek, D.R., and Elder, F.C., 1989, Assessing the Potential Extent of Damage to Inland Fisheries in Eastern Canada, 2: Application of the Regional Model, *Canadian Journal of Fisheries and Aquatic Sciences* 46.

Kaitala, V., Hari, P., Vapaavuoti, E., and Salminen, R., 1982, Dynamic Model for Photosynthesis, *Annals of Botany* 50: 385–396.

Kämäri, J., 1986, Linkage between Atmospheric Inputs and Soil and Water Acidification, in J. Alcamo and J. Bartnicki, eds., *Atmospheric Computations to Assess Acidification in Europe: Work in Progress*, RR-86-5, International Institute for Applied Systems Analysis, Laxenburg, Austria.

Kämäri, J., 1987, Prediction Models for Acidification, in *Proceedings International Symposium on Acidification and Water Pathways*, 4–8 May, Bolkesjø, Norway.

Kämäri, J. and Posch, M., 1987, Regional Application of a Simple Lake Acidification Model in Northern Europe, in M.B. Beck, ed., *Systems Analysis in Water Quality Management*, Pergamon Press, New York, NY, USA.

Kämäri, J., Posch, M., and Kauppi, L., 1985, *A Model for Analyzing Lake Water Acidification on a Large Regional Scale*, Part I: *Model Structure*, CP-85-48, International Institute for Applied Systems Analysis, Laxenburg, Austria.

Kämäri, J., Posch, M., Gardner, R.H., and Hettelingh, J.-P., 1986, *A Model for Analyzing Lake Water Acidification on a Large Regional Scale: Part 2, Regional Application*, WP-86-66, International Institute for Applied Systems Analysis, Laxenburg, Austria.

Kämäri, J., Hettelingh, J.-P., Posch, M., and Holmberg, M., 1989, Regional Freshwater Acidification: Sensitivity and Long-Term Dynamics, in J. Alcamo, *et al.*, eds., *The RAINS Model of Acidification: Science and Strategies in Europe*, Kluwer, Dordrecht, Netherlands.

Kaniciruk, P., Eilers, J.M., McCord, R.A., Landers, D.H., Brakke, D.F., and Linthurst, R.A., 1986, *Characteristics of Lakes in the Eastern United States*. Vol. III: *Data Compendium of Site Characteristics and Chemical Variables*, EPA/600/4-86/007c, US Environmental Protection Agency, Washington, DC, USA.

Kauppi, P., Kämäri, J., Posch, M., Kauppi, L., and Matzner, E., 1986, Acidification of Forest Soils: Model Development and Application for Analyzing Impacts of Acidic Deposition in Europe, *Ecological Modelling* 33: 231–253.

Kelly, C.A., Rudd, J.W.M., Hesslein, R.H., Schindler, D.W., Dillon, P.J., Driscoll, C.T., Gherini, S.A., and Hecky, R.E., 1987, Prediction of Biological Acid Neutralization in Acid-Sensitive Lakes, *Biogeochemistry* 3: 129–140.

Kelso, J.R.M., 1985, Standing Stock and Production of Fish in a Cascading Lake System on the Canadian Shield, *Canadian Journal of Fisheries and Aquatic Sciences* 42: 1315–1320.

Kent, C. and Wong, J., 1982, An Index of Littoral Zone Complexity and Its Measurement, *Canadian Journal of Fisheries and Aquatic Sciences* 39: 847–853.

Kleijn, C.E. and de Vries, W., 1987, Characterizing Soil Moisture Composition in Forest Soils, in W. van Duijvenbooden and H.G. van Waegening, eds., *Proceedings International Conference on Vulnerability of Soil and Groundwater to Pollutants*, Noordwijk aan Zee, March 30–April 3, Proceedings and Information No. 38, TNO–CHO/ RIVM, The Hague, Netherlands.

Kolenbrander, G.J., 1974, Efficiency of Organic Manure in Increasing Soil Organic Matter Content, *Proceedings 10th Congress International Soil Sciences*, Vol. 2, Moscow, USSR.

Konsten, C.J.M., Tiktak, A., and Bouten, W., 1987, Monitoring Soil Chemical and Physical Parameters under Douglas Fir in the Netherlands, in *Proceedings International Symposium on Acidification and Water Pathways*, II, 4–8 May, Bolkesjø, Norway.

Kukkola, M. and Saramäki, J., 1983, Growth Response in Repeatedly Fertilized Pine and Spruce Stands on Mineral Soils, *Comm. Inst. For. Fenn.* **114**: 1–55.

Labieniec, P.A., 1988, Identification of the Direct Distribution Model from Regionalized Mechanistic Models of Aquatic Acidification, Master's Thesis, Department of Civil Engineering and Public Policy, Carnegie Mellon University, Pittsburgh, PA, USA.

Leach, J.H., Dickie, L.M., Shuter, B.J., Borgmann, U., Hyman, J., and Lysack, W., 1987, A Review of Methods for Prediction of Potential Fish Production with Application to the Great Lakes and Lake Winnipeg, *Canadian Journal of Fisheries and Aquatic Sciences* **44** (Supplement 2): 471–485.

Lee, Y.H., 1980, The Linear Plot: A New Way of Interpreting Titration Data, Used in Determination of Weak Acids in Lake Water, *Water, Air, and Soil Pollution* **14**: 287–298.

Legendre, P. and Legendre, V., 1983, Postglacial Dispersal of Freshwater Fishes in the Quebec Peninsula, *Canadian Journal of Fisheries and Aquatic Sciences* **41**: 1781–1802.

Lehmhaus, J., Saltbones, J., and Eliassen, A., 1986, *A Modified Sulphur Budget for Europe for 1980*, EMEP/MSC-W Report 1/86.

Lepistö, A., Whitehead, P.G., Neal, C., and Cosby, B.J., 1988, Modeling the Effects of Acid Deposition: Estimation of Long-Term Water Quality Responses in Forested Catchments in Finland *Nordic Hydrology*, **19**: 99–120.

Likens, G.E., Bormann, F.H., Pierce, R.S., Eaton, J.S., and Johnson, N.M., 1977, *Biogeochemistry of a Forested Ecosystem*, Springer-Verlag, New York, NY, USA.

Linthurst, R.A, Landers, D.H., Eilers, J.M., Brakke, D.F., Overton, W.S., Meir, E.P., and Crowe, R.E., 1986, *Characteristics of Lakes in Eastern United States*, Vol. I: *Population Descriptions and Physicochemical Relationships*, EPA-600/4-86-007A, US Environmental Protection Agency, Washington, DC, USA.

Long, C., 1986–1987, Washington Report, *Journal of the Air Pollution Control Association* **36–37**: Oct.–Apr. [Specific proposals include Senate Bill S2203 (Mitchell–Stafford) and HR 4567 (Waxman)].

Lotse, E. and Otabbong, E., 1985, *Physicochemical Properties of Soils at Risdaldheia and Sogndal: RAIN Project*, Report No. 0-82073, Norwegian Institute for Water Research, Blindern, Oslo 3, Norway.

Ludwig, D., Jones, D.D., and Holling, C.S., 1978, Qualitative Analysis of Insect Outbreak Systems: The Spruce Budworm and Forest, *Journal of Animal Ecology* **47**: 315–332.

MacArthur, R.H. and Wilson, E.O., 1967, *The Theory of Island Biogeography*, Princeton University Press, Princeton, NJ, USA.

Mahon, R. and Balon, E.K., 1977, Fish Community Structure in Lakeshore Lagoons on Long Point, Lake Erie, Canada, *Environmental Biology of Fishes* **2**: 71–82.

Mankin, J.B., O'Neill, R.V., Shugart, H.H., and Rust, B.W., 1977, The Importance of Validation in Ecosystem Analysis, in G.S. Innis, ed., *New Directions in the Analysis of Ecological Systems*, Simulation Councils, La Jolla, CA, USA.

Marmorek, D.R., Jones, M.L., Minns, C.K., and Elder, F.C., 1989, Assessing the Potential Extent of Damage to Inland Fisheries in Eastern Canada, 1: Development and Evaluation of a Simple Site Model, *Canadian Journal of Fisheries and Aquatic Sciences*, **46**.

Marnicio, R.J., Rubin, E.S., Henrion, M., Small, M.J., McRae, G.J., and Lave, L.B., 1985, A Comprehensive Modeling Framework for Integrated Assessments of Acid Deposition, *Proceedings 78th Annual Meeting of Air Pollution Control Association*, Paper 85-1B.2, Pittsburgh, PA, USA.

Marnicio, R.J., Rubin, E.S., Small, M.J., and Henrion, M., 1986, The Acid Deposition Assessment Model: An Integrated Framework for Benefit and Cost Analysis, in *Proceedings of the 7th World Clean Air Congress*, IUAPPA, Sydney, Australia.

Matthias, C.S. and Lo, A.K., 1986, Application and Evaluation of the Fay and Rosenzweig Long-Range Transport Model, *Atmosphric Environment* **20**: 1913–1921.

Mayer, R. and Ulrich, B., 1974, Conclusions on Filtering Action of Forest from Ecosystem Analysis, *Oecologica Plantarum* **9**: 157–168.

McGowin, C.R. *et al.*, 1986, *Sensitivity Analysis of Electric Utility SO_2 Emissions in the US*, Electric Power Research Institute, Palo Alto, CA, USA.

Meentemeijer, V., 1979, Climatic Regulation of Decomposition Rates in Terrestrial Ecosystems, in D.C. Adrano and I. Lehr Brisbin, eds., *Proceedings Symposium on Environmental Chemistry and Cycling Processes*, Conf. 760429, US Department of Energy, Washington, DC, USA.

Mills, K.H. and Schindler, D.W., 1986, Biological Indicators of Lake Acidification, *Water, Air, and Soil Pollution* **30**: 779–789.

Minns, C.K., 1984, Analysis of Lake and Drainage Area Counts and Measures for Selected Watershed across Ontario's Shield Region, *Canadian MS Report of Fisheries and Aquatic Sciences* **1748**.

Minns, C.K., 1986a, Acid Rain: A Further Estimate of the Risk to Ontario's Inland Fisheries, *Canadian MS Report of Fisheries and Aquatic Sciences* **1874**.

Minns, C.K., 1986b, A Simple Whole-Lake Phosphorus Model and a Trial Application to the Bay of Quinte, in C.K. Minns *et al.*, eds., Project Quinte: Point-Source Phosphorus Control and Ecosystem Response in the Bay of Quinte, Lake Ontario, *Canadian Special Publication on Fisheries and Aquatic Sciences* **86**: 84–90.

Minns, C.K. and Kelso J.R.M., 1986, Estimates of Existing and Potential Impact of Acidification on the Freshwater Fishery Resources and Their Uses in Eastern Canada, *Water, Air, and Soil Pollution* **31**: 1079–1090.

Minns, C.K., Kelso, J.R.M., and Johnson, M.G., 1986, Large-Scale Risk Assessment of Acid Rain Impacts on Fisheries: Models and Lessons, *Canadian Journal of Fisheries and Aquatic Sciences* **431**: 900–921.

Minns, C.K., Millard, E.S., Cooley, J.M., Johnson, M.G., Hurley, D.A., Nicholls, K.H., Robinson, G.W., Owen, G.E., and Crowder, A., 1987, Production and Biomass Size-Spectra in the Bay of Quinte, and Eutrophic Ecosystem, *Canadian Journal of Fisheries and Aquatic Sciences*, **44** (Supplement 2): 148–155.

Mohren, G.M.J., 1987, *Simulation of Forest Growth, Applied to Douglas Fir Stands in the Netherlands*, Agricultural University of Wageningen, Netherlands.

Morrison,, M.B. and Rubin, E.S., 1985, A Linear Programming Model for Acid Rain Policy Analysis, *Air Pollution Control Association Journal* **35**: 1137–1148.

Mulder, J.R., van Dobben, H.F., de Visser, P.H.B., Booltink, H.G.W., and van Breemen, N., 1987, Effect of Vegetation Cover (Pine Forest Vs. No Vegetation) on
Atmospheric Deposition and Soil Acidification, in *Proceedings International Symposium on Acidification and Water Pathways*, Vol. I, 4–8 May, Bolkesjø, Norway.

Müller, M.J., 1982, *Selected Climatic Data for a Global Set of Standard Stations for
Vegetation Science*, Dr. W. Junk, The Hague, Netherlands.

Muniz, I.P., 1984, The Effects of Acidification on Scandinavian Freshwater Fish Fauna,
Phil. Trans. R. Soc. London B. 305: 517–528.

Muniz, I.P., Seip, H.M., and Sevaldrud, I.H., 1984, Relationship Between Fish Populations and pH for Lakes in Southernmost Norway, *Water, Air, and Soil Pollution*
23: 97–113.

Musgrove, T.J. and Whitehead, P.G., 1988, Validation of MAGIC by Reference to
Paleoecological Trends and Regional Patterns of Acidification, *DOE Conference
Proceedings*, Lancaster, UK.

NAPAP, 1986, *The National Acid Precipitation Assessment Program, Annual Report,
1986*, National Acid Precipitation Assessment Program, Washington, DC, USA.

National Research Council Committee on Atmospheric Transport and Chemical
Transformation in Acid Precipitation, 1983, *Acid Deposition: Atmospheric
Processes in Eastern North America – A Review of Current Scientific Understand-
ing*, National Academy Press, Washington, DC, USA.

Neal, C., Whitehead, P.G., and Cosby, B.J., 1986, Modeling the Effects of Acidic Deposition and Conifer Afforestation on Steam Acidity in the British Uplands, *Journal
of Hydrology*, 86: 15–26.

Niemann, B.L., 1986a, Further Evaluation and Application of the MOI Long-Range
Transport Models Using Improved Data Bases, in *Proceedings of the 2nd Interna-
tional Speciality Conference on Methodology of Acidic Deposition*, Air Pollution
Control Association, Pittsburgh, PA, USA.

Niemann, B.L., 1986b, Regional Acid Deposition Calculations with the IBM PC LOTUS
1-2-3 System, *Environmental Software* 1: 175–181.

Nilsson, J., ed., 1986, *Critical Loads for Sulphur and Nitrogen*, Nordisk ministerråd,
Miljörapport 1986: 11.

Norton, S.A. and Henriksen, A., 1983, The Importance of CO_2 in Evaluation of Effects
of Acidic Deposition, *Vatten* 39: 346–354.

Norton, S.A., Hanson, D.W., and Campana, R.J., 1980, *The Impact of Acidic Precipitation and Heavy Metals on Soils in Relation to Forest Ecosystems*, International
Symposium on Effects of Air Pollutants, June 22–27, Riverside, CA, USA.

Nuotio, T., Hyyppä, J., and Kämäri, J., 1985, Buffering Properties of Soils in 53
Forested Catchments in Southern Finland, *Aqua Fennica* 15: 35–40.

Olson, M.P., Voldner, E.C., and Oikawa, K.K., 1983, Transfer Matrices from the
AES–LRT Model, *Atmosphere–Ocean* 21(3): 344–361.

O'Neill, R.V. and Gardner, R.H., 1979, Sources of Uncertainty in Ecological Models, in
B.P. Zeigler, *et al.*, eds., *Methodology in Systems Modeling and Simulation*, North-
Holland Publishing, Amsterdam, Netherlands.

Overrein, L.N., Seip, H.M., and Tollan, A., 1981, *Acid Precipitation: Effects on Forest
and Fish*, Final Report of the SNSF-Project 1972–1980, SNSF-Project, Box 61,
1432 Ås-NLH, Norway.

Overton, W.S., Kaniciruk, P., Hook, L.A., Eilers, J.M., Landers, D.H., Brakke, D.F., Blick, D.J., Linthurst, R.A., DeHaan, M.D., and Omernik, J.M., 1986, *Characteristics of Lakes in the Eastern United States*, Vol. II: *Lakes Sampled and Descriptive Statistics for Physical and Chemical Variables*, EPA/600/4-86/007b, US Environmental Protection Agency, Washington, DC, USA.

Pauwels, S.J. and Haines, T.A., 1986, Fish Species Distribution in Relation to Water Chemistry in Selected Maine Lakes, *Water, Air, and Soil Pollution* 30: 477–488.

Perdue, E.M., 1985, Acidic Functional Groups of Humic Substances, in G.R. Aiken *et al.*, eds., *Humic Substances in Soil, Sediment, and Water*, Wiley, New York, NY, USA.

Petersen, T., 1984, *Development of the Program System ECCES for Calculating Environmental Consequences from Energy Systems Status*, M–2447, Risø National Laboratory, Denmark.

Placet, M., Streets, D.G., and Williams, E.R., 1986, *Environmental Trends Associated with the Fifth National Energy Policy Plan*, ANL/EES-TM-323, Argonne National Laboratory, Springfield, VA, USA.

Rago, P.J. and Wiener, J.G., 1986, Does pH Affect Fish Species Richness When Lake Area is Considered? *American Fisheries Society Transactions* 115: 438–447.

Rapp, G., Allert, J.D., Liukkonen, B.W., Ilse, J.A., Looucks, O.L., and Glass, G.E., 1985, Acid Deposition and Watershed Characteristics in Relation to Lake Chemistry in Northeastern Minnesota, *Environment International* 11: 425–440.

Reckhow, K.H., 1987, Private Communication from Dr. Reckhow, School of Forestry, Duke University, Durham, NC, USA.

Reckhow, K.H., Black, R.W., Stockton, T.B., Vogt, J.D., and Wood, J.G., 1985, *Empirical Models of Fish Response to Lake Acidification*, unpublished manuscript of report to US–EPA on NAPAP Project E3–02.

Reckhow, K.H., Black, R.W., Stockton, T.B., Jr., Vogt, J.D., and Wood, J.G., 1987, Empirical Models of Fish Response to Lake Acidification, *Canadian Journal of Fisheries and Aquatic Sciences* 45: 1432–1442.

Reuss, J.O., 1983, Implications of the Ca-Al Exchange System for the Effect of Acid Precipitation on Soils, *Journal of Environmental Quality* 12: 591–595.

Reuss, J.O. and Johnson, D.W., 1985, Effect of Soil Processes on the Acidification of Water by Acid Deposition, *Journal of Environmental Quality* 14: 26–31.

Reuss, J.O. and Johnson, D.W., 1986, *Acid Deposition and the Acidification of Soils and Waters*, Ecological Studies 59, Springer-Verlag, Berlin, Heidelberg, Federal Republic of Germany, and New York, NY, USA.

Reuss, J.O., Christophersen, N., and Seip, H.M., 1986, A Critique of Models for Freshwater and Soil Acidification, *Water, Air, and Soil Pollution* 30: 909–930.

Ricklefs, R.E., 1987, Community Diversity: Relative Role of Local and Regional Processes, *Science* 235: 167–171.

Rodhe, H., 1982, Emission, Transport, Deposition, in *Acidification Today and Tomorrow*, Ministry of Agriculture, Environment '82, Stockholm, Sweden.

Rosen, A.E., Olson, R.J., Gruendling, G.K., Bogucki, D.J., Malanchuk, J.L., Durfee, R.C., Turner, R.S., Adams, K.B., Wilson, D.L., Coleman, P.R., Brandt, C.C., and Hunsaker, C.T., 1988, *An Adirondack Watershed Data Base: Attribute and Mapping Information for Regional Acidic Deposition Studies*, ORNL/TM-10144, Oak Ridge National Laboratory, Oak Ridge, TN, USA.

Rubin, E.S., 1981, Air Pollution Constraints on Industrial Coal Use: An International Perspective, *Air Pollution Control Association Journal* 31: 349–360.

Rubin, E.S., Marnicio, R.J., Henrion, M., Small, M.J., and McRae, G.J., 1986a, *Development of an Operational Level II Acid Deposition Assessment Model*, Center for Energy and Environmental Studies, Carnegie-Mellon University, Pittsburgh, PA, USA.

Rubin, E.S. *et al.*, 1986b, Controlling Acid Deposition: The Role of FGD, *Environmental Science and Technology* **20**: 960–969.

Salmento, J.S., Rubin, E.S., and Dowlatabadi, H., 1987, Power Plant Life Extension Impacts on Emission Reduction Strategies, in *Proceedings of the 80th Annual Meeting*, Air Pollution Control Association, Pittsburgh, PA, USA.

Scavia, D., Powers, W.F., Canale, R.P., and Moody, J.L., 1981, Comparison of First-Order Error Analysis and Monte Carlo Simulation in Time-Dependent Lake Eutrophication Models, *Water Resources Research* **17**: 1051–1059.

Schnoor, J.L. and Stumm, W., 1984, Acidification of Aquatic and Terrestrial Systems, in W. Stumm, ed., *Chemical Processes in Lakes*, Wiley, New York, NY, USA.

Schnoor, J.L. and Stumm, W., 1986, The Role of Chemical Weathering in the Neutralization of Acidic Deposition, *Swiss Journal of Hydrology* **48**(2): 171–195.

Schnoor, J.L., Palmer, W.D., and Glass, G.E., 1984, Modeling Impacts of Acid Precipitation for Northeastern Minnesota, in J.L. Schnoor, ed., *Modeling of Total Acid Precipitation Impacts*, Butterworth Publishers, Boston, MA, USA.

Schnoor, J.L., Nair, D.R., Kuhns, M.A., Lee, S.J., and Nikolaidis, N.P., 1985, *Lake Resources at Risk due to Acid Deposition in the Northeastern United States*, Department of Civil and Environmental Engineering, University of Iowa, Iowa City, IA, USA.

Schnoor, J.L., Nikolaidis, N.P., and Glass, G.E., 1986a, Lake Resources at Risk to Acidic Deposition in the Upper Midwest, *Water Pollution Control Federation Journal* **58**(2): 139–148.

Seip, H.M., 1980, Acidification of Freshwater: Sources and Mechanisms, in D. Drabløs and A. Tollan, eds., *Proceedings of an International Conference on the Ecological Impacts of Acid Precipitation*, SNSF, Ås-NLH, Norway.

Seip, H.M, Abrahamsen, G., Gjessing, E.T., and Stuanes A., 1979a, *Studies of Soil-, Precipitation- and Runoff Chemistry in Six Small Natural Plots ("Mini-Catchments")*, Internal Report, SNSF-Project, IR 46/69, SNSF-Project, Box 61, 1432 Ås-NLH, Norway.

Seip, H.M., Gjessing, E.T., and Kamben, H., 1979b, *Importance of the Composition of the Precipitation for the pH in Runoff – Experiments with Artificial Precipitation on Partly Soil-Covered "Mini-Catchments,"* Internal Report, SNSF-Project, IR 47/79, SNSF-Project, Box 61, 1432 Ås-NLH, Norway.

Shannon, J.D., 1985, *User's Guide for the Advanced Statistical Trajectory Regional Air Pollution (ASTRAP) Model*, EPA/600/8-95/016, US Environmental Protection Agency, Research Triangle Park, NC, USA.

Shaw, R.W., 1986, A Proposed Strategy for Reducing Sulphate Deposition in North America, II: Methodology for Minimizing Costs, *Atmospheric Environment* **20**: 201–206.

Shugart, H.H. and West, D.C., 1981, Long-term Dynamics of Forest Ecosystems, *American Scientist* **69**: 647–652.

Small, M.J. and Sutton, M.C., 1986a, A Regional pH–Alkalinity Relationship, *Water Resources Research* **20**: 335–343.

Small, M.J. and Sutton, M.C., 1986b, A Direct Distribution Model for Regional Aquatic Acidification, *Water Resources Research* **22**: 1749–1758.

Small, M.J., Sutton, M.C., and Labieniec, P.A., 1987, Modeling Distributions of Aquatic Chemistry in Regions Impacted by Acid Deposition, in M.B. Beck, ed., *Systems Analysis in Water Quality*, Pergamon Press, New York, NY, USA.

Small, M.J., Sutton, M.C., and Milke, M.W., 1988, Parametric Distributions of Regional Lake Chemistry: Fitted and Derived, *Environmental Science and Technology* **22**.

Somers, K.M. and Harvey, H.H., 1984, Alteration of Fish Communities in Lakes Stressed by Acid Deposition and Heavy Metals Near Wawa, Ontario, *Canadian Journal of Fisheries and Aquatic Sciences* **41**: 20–29.

Soveri, J., 1985, *Influence of Meltwater on the Amount and Composition of Groundwater in Quarternary Deposits in Finland*, Water Research Institute, 63, Helsinki, Finland.

Spear, R.C. and Hornberger, G.M., 1983, Control of DO in a River Under Uncertainty, *Water Resources Research* **19**: 1226–1270.

Spear, R.C. and Hornberger, G.M., 1983, Eutrophication in Peel Inlet 2: Identification of Critical Uncertainties Via Generalized Sensitivity Analysis, *Water Resources Research* **14**:43–49.

Steur, G.G.L., de Vries, F., and Van Wallenburg, C., 1986, A New Small-Scale Soil Map of the Netherlands, in *Proceedings 13th Congress International Society of Soil Science*, Vol. 3, 13–20 August, Hamburg, Germany, F.R.

Strong, D.R., Jr., Simberloff, D., Abele, L.G., and Thistle, A.B., eds., 1984, *Ecological Communities: Conceptual Issues and the Evidence*, Princeton University Press, Princeton, NJ, USA.

Stuanes, A. and Sveistrup, T.E., 1979, *Field Experiments with Simulated Acid Rain in Forest Ecosystem, 2: Description and Classification of the Soils Used in Field, Lysimeter and Laboratory Experiments*, Research Report, SNSF-Project, FR 15/79, SNSF-Project, Box 61, 1432 Ås-NLH, Norway.

Stumm, W. and Morgan, J.J., 1981, *Aquatic Chemistry: An Introduction Emphasizing Chemical Equilibria in Natural Waters*, 2nd ed., Wiley, New York, NY, USA.

Swedish Meteorological and Hydrological Institute, 1983, *Svenskt sjöregister*, SMHI, Norrköping, Sweden [in Swedish].

Swedish Standards Institution, 1981, *Swedish Standard Methods, SS0281 Series*, Stockholm, Sweden [in Swedish].

Thompson, M.E., 1982, The Cation Denudation Rate as a Quantitative Index of Sensitivity of Eastern Canadian Rivers to Acidic Atmospheric Precipitation, *Water, Air, and Soil Pollution* **18**: 215–226.

Thornley, J.H.M., 1974, Light Fluctuations and Photosynthesis, *Annals of Botany* **38**: 363–373.

Tonn, W.M. and Magnuson, J.J., 1982, Patterns in the Species Composition and Richness of Fish Assemblages in Northern Wisconsin Lakes, *Ecology* **63**: 1149–1166.

UAH-UNESCO, 1970–1985, *International Hydrogeological Map of Europe* (unpublished maps, Bundesanstalt für Geowissenschaften und Rohstoffe, Hannover, Federal Republic of Germany).

Ulrich, B., 1983a, A Concept of Forest Ecosystem Stability and of Acid Deposition as Driving Forces for Destabilization, in B. Ulrich and J. Pankrath, eds., *Effects of Accumulation of Air Pollutants in Forest Ecosystems*, D. Reidel, Dordrecht, Netherlands.

Ulrich, B., 1983b, Soil Acidity and Its Relations to Acid Deposition, in B. Ulrich and S. Pankrath, eds., *Effects of Accumulation of Air Pollutants in Forest Ecosystems*, D. Reidel, Dordrecht, Netherlands.

Ulrich, B. and Matzner, E., 1983, *Abiotische Folgewirkungen der weiträumigen Ausbreitung von Luftverunreinigung*, Umweltforschungsplan des Bundesministers des Innern, Forschungsbericht 1040 26 15, Federal Republic of Germany.

UNESCO, 1972, *International Geological Map of Europe and the Mediterranean Region*, Bundesanstalt für Bodenforschung, Hannover, UNESCO, Paris, France.

Urvas, L. and Erviö, R., 1974, Influence of the Soil Type and the Chemical Properties of Soil on the Determining of the Forest Type [English summary], *Journal of the Agricultural Society of Finland* 3: 307–319.

Veen, J.A. van, 1977, The Behavior of Nitrogen in Soil: A Computer Simulation Model, Thesis, Free University, Amsterdam, Netherlands.

Veen, J.A. van, Breteler, H., Olie, J.J., and Frissel, M.J., 1981, Nitrogen and Energy Balance of a Short-Rotation Poplar Forest System, *Netherlands Journal of Agricultural Science* 29: 163–172.

Verry, E.S. and Harris, A.R., 1988, A Description of Low- and High-Acid Precipitation, *Water Resources Research* 24: 481–492.

Von Brömssen, U., 1986, Acidification and Drinking Water – Groundwater, in T. Schneider, ed., *Acidification and Its Policy Implications*, International Conference, Amsterdam, Netherlands.

Wales, D.L. and Beggs, G.L., 1986, Fish Species Distribution in Relation to Lake Acidity in Ontario, *Water, Air, and Soil Pollution* 30: 601–609.

Ward, J.H., 1963, Hierarchical Grouping to Optimize an Objective Function, *Journal of the American Statistical Association* 58: 236–244.

Warren Springs Laboratory, 1983, *Acid Deposition in the United Kingdom*, Warren Springs Laboratory Report, Stevenage, UK.

Welsh, W.T., Tervet, D.J., and Hutchinson, P., 1986, *Acidification Investigations in Southwest Scotland*, Solway River Purification Board, Dumfries, UK.

Weltforstatlas [World Forestry Atlas], 1975, Verlag Paul Parey, Hamburg and Berlin, Federal Republic of Germany.

Whitehead, P.G. and Young, P.C., 1979, Water Quality in River Systems: Monte Carlo Analysis, *Water Resources Research* 15: 451–459.

Whitehead, P.G., Neale, R., and Paricos, P., 1987, Modeling Long-Term Stream Acidification Trends in Upland Wales, *Proceedings of the International Symposium on Acidification and Water Pathways*, 4–8 May, Bolkesjø, Norway, Vol. 1.

Wieting, J., 1986, Water Acidification by Air Pollutants in the Federal Republic of Germany, *Water, Air, and Soil Pollution* 31: 247–256.

Wösten, J.H.M., Bannink, M.H., De Gruijter, J.J., and Bouma, J., 1986, A Procedure to Identify Different Groups of Hydraulic–Conductivity and Moisture–Retention Curves for Soil Horizons, *Journal of Hydrology* 86: 133–145.

Wright, R.F., 1983, *Predicting Acidification of North American Lakes*, Acid Rain Res. Rep. 4/1983, NIVA, Oslo, Norway.

Wright, R.F., 1988, Acidification of Lakes in the Eastern United States and Southern Norway: A Comparison, *Environmental Science and Technology* 22: 178–182.

Wright, R.F. and Dovland, H., 1978, Regional Surveys of the Chemistry of the Snowpack in Norway Late Winter, 1973, 1974, 1975, and 1976, *Atmospheric Environment* 12: 1755–1768.

Wright, R.F. and Henriksen, A., 1979, *Regional Survey of Lakes and Streams in southwest Scotland*, Internal Report IR 72/80, SNSF-Project, Ås-NLH, Norway.

Wright, R.F. and Henriksen, A., 1983, Restoration of Norwegian Lakes by Reduction in Sulphur Deposition, *Nature* 305: 422–424.

Wright, R.F. and Snekvik, E., 1978, Acid Precipitation: Chemistry and Fish Populations in 700 Lakes in Southernmost Norway, *Verh. Internat. Verein. Limnol.* **20**: 765–775.

Wright, R.F., Torstein, D., Henriksen, A., Hendrey, G.R., Gjessing, E.T., Johannessen, M., Lysholm, C., and Storen, E., 1977, *Regional Surveys of Small Norwegian Lakes*, IR33/77, SNSF-Project, Box 61, Ås-NHL, Norway.

Wright, R.F., Harriman, R., Henriksen, A., Morrison, B., and Caines, L.A., 1980a, Acid Lakes and Streams in the Galloway area, South-western Scotland, in D. Drabløs and A. Tollan, eds., *Ecological Impact of Acid Precipitation*, SNSF-Project, Ås-NLH, Norway.

Wright, R.F., Conroy, N., Dickson, W.T., Harriman, R., Henriksen, A., and Schofield, C.L., 1980b, Acidified Lake Districts of the World: A Comparison of Water Chemistry of Lakes in Southern Norway, Southern Sweden, Southwestern Scotland, the Adirondack Mountains of New York and Southeastern Ontario, in D. Drabløs and A. Tollan, eds., *Ecological Impact of Acid Precipitation*, SNSF-Project, Ås-NLH, Norway.

Wright, R.F., Cosby, B.J., Hornberger, G.M., and Galloway, J.N., 1986, Comparison of Paleolimnological with MAGIC Model Reconstruction of Water Acidification, *Water, Air, and Soil Pollution* **30**: 307–380.

Yearbook of Forest Statistics, 1979, Official Statistics of Finland XVII A: 11, Folia Forestalia 430, Helsinki, Finland.

Authors

S.M. Bartell
Oak Ridge National Laboratory
Oak Ridge, TN
USA

C. Bernes
National Environmental
 Protection Board
Solna
Sweden

C.N. Bloyd
Argonne National Laboratory
Argonne, IL
USA

K. Brodersen
Risø National Laboratory
Risø
Denmark

B.J. Cosby
Department of Environmental Sciences
University of Virginia
Charlottesville, VA
USA

H. Christiansen
Risø National Laboratory
Risø
Denmark

W. de Vries
Soil Survey Institute
Wageningen
The Netherlands

R.H. Gardner
Oak Ridge National Laboratory
Oak Ridge, TN
USA

P. Hari
Department of Silviculture
University of Helsinki
Helsinki
Finland

M. Henrion
Carnegie-Mellon University
Pittsburgh, PA
USA

J.-P. Hettelingh
IIASA
Laxenburg
Austria

M. Holmberg
Department of Silviculture
University of Helsinki
Helsinki
Finland

G.M. Hornberger
Department of Environmental Sciences
University of Virginia
Charlottesville, VA
USA

J. Johnston
Department of Civil Engineering
University of California
Davis, CA
USA

J. Kämäri
National Board of Waters
 and the Environment
Helsinki
Finland

P.A. Labieniec
Department of Civil Engineering
Carnegie-Mellon University
Pittsburgh, PA
USA

B. Larsen
Risø National Laboratory
Risø
Denmark

R.J. Marnicio
Ebasco Services, Inc.
Dublin, OH
USA

L. Maxe
Royal Institute of Technology
Stockholm
Sweden

C.K. Minns
Great Lakes Laboratory for
 Fisheries and Aquatic Sciences
Burlington, Ontario
Canada

T.J. Musgrove
Institute of Hydrology
Wallingford, Oxon
UK

T. Petersen
Risø National Laboratory
Risø
Denmark

M. Posch
IIASA
Laxenburg
Austria

E.S. Rubin
Carnegie-Mellon University
Pittsburgh, PA
USA

M.J. Small
Department of Civil Engineering
Carnegie-Mellon University
Pittsburgh, PA
USA

M.C. Sutton
Department of Civil Engineering
Carnegie-Mellon University
Pittsburgh, PA
USA

E. Thörnelöf
National Environmental
 Protection Board
Solna
Sweden

P.G. Whitehead
Institute of Hydrology
Wallingford, Oxon
UK

R.F. Wright
Norwegian Institute of Water Research
Oslo
Norway